Lipids in Nanotechnology

Lipids in Nanotechnology

Editor
Moghis U. Ahmad

AOCS
PRESS

AOCS Mission Statement

To be a global forum to promote the exchange of ideas, information, and experience, to enhance personal excellence, and to provide high standards of quality among those with a professional interest in the science and technology of fats, oils, surfactants, and related materials.

Library of Congress Cataloging-in-Publication Data

Lipids in nanotechnology / editor, Moghis Ahmad.
p. cm.
Includes bibliographical references and index.
1. Lipids. 2. Nanotechnology. I. Ahmad, Moghis.

TP453.L56L58 2012
660'.294514--dc23

2011022800

CONTENTS

PREFACE

Nanotechnology is a rapidly expanding field due to multidisciplinary support from researchers in the academic, industry, and federal sectors. The broad areas that are covered under nanotechnology include fundamental nanoscale phenomena and processes, nanomaterials, nanoscale devices and systems, nanomanufacturing, and benefits and risks of nanotechnology. Research in chemistry, physics, biology, and engineering drives the development and exploration of nanotechnology field. Applications of nanotechnology to the agriculture and food sector are relatively recent compared with their use in pharmaceuticals and drug delivery, cosmetics, microelectronics, and aerospace. Food nanotechnology is one industrial sector where nanotechnology will play an important role in the future.

Drug delivery dominates the use of nanotechnology in pharmaceuticals. Applications include organic nanoplatforms such as polymers and lipids (e.g., liposomes, nanoemulsions, and solid-lipid nanoparticles). The term lipids is used generally (in a very broad sense) and includes triglycerides, partial glycerides, PEGyalated lipids, fatty acids, steroids and waxes. Phospholipids (lecithin) play an important role in the field of nanotechnology. Numerous published articles discuss the role of lipids in drug delivery, diagnostics, nutraceuticals, and cosmetics. The application of new classes of lipids (e.g., marine phospholipids, milk phospholipids) is emerging that illustrates the principal advantage of choosing marine phospholipids and milk phospholipids as liposomal carriers. Liposomes are generally prepared with phospholipids isolated from soy or egg; but in recent years, there has been a growing interest in the health benefits and functional properties of milk phospholipids that may offer some advantages in the manufacture of nanoliposomes for the entrapment of bioactive compounds in food systems.

There are plenty of data available in literature on the role of lipids in nanotechnology; this is the time to summarize what is known to date. To the best of my knowledge this book will serve as a valuable reference and resource for those interested in the field of nanotechnology; from basic research to engineering aspects of nanoparticles. The purpose in writing this book was to gather together various aspects of nanotechnology from many of the leaders in the field. The contributing authors have published articles in the field and are eminently qualified to summarize their own work and related work in their field of expertise. The authors were encouraged to provide comprehensive reviews of their areas of interest. Each chapter is presented in a way to stand on its own without the need for extensive references to other chapters or sources. When this book is read in its entirely, it covers from thermodynamics to engineering aspects of nanoparticles or nanoemulsions; synthesis and applications of surface active lipids to food and cosmetics; and from pharmaceutical applications to nanomedicine. To understand the techniques more completely, it is beneficial to see all aspects, as discussed in the variety of applications demonstrated herein.

I hope that the readers will find it valuable to read this work from those authors who are already recognized for their leadership in the field of lipids and nanotechnology. Also described in this book are new developments that will be applied to the nanotechnology field in the next decade and hopefully beyond. I hope that this book will find its deserved place and good use in the hands of scholars, scientists, and technologists who are interested in the field of lipid nanotechnology.

Moghis U. Ahmad

1

NANOTECHNOLOGY: EMERGING INTEREST, OPPORTUNITIES, AND CHALLENGES

Moghis U. Ahmad
Jina Pharmaceuticals Inc., 28100 N. Ashley Circle, Suite 103,
Libertyville, IL, 60048, USA

INTRODUCTION

The field of nanotechnology was first predicted by Richard Feynman (Nobel Laureate in Physics, 1965) in his pioneering lecture entitled "There's plenty of room at the bottom" at the 1959 meeting of the American Physical Society. Nanoparticles can range in size from 1 to 100 nanometers (nm). One nanometer is equal to one billionth of a meter (1nm = 10^{-9} m). The National Science Foundation and the National Nanotechnology Initiative define nanotechnology as understanding and control of matter at dimensions of 1–100 nm, where unique phenomena enable novel applications (NNI, 2006).

A nanometer-sized particle measures one billionth of a meter and one can imagine how small it is when a human hair measures 80,000 nm, a DNA strand is 2.5 nm wide, and a protein chain is 5 nm in diameter. Applications with structural features on the nanoscale level have physical, chemical, and biological properties that are substantially different from their macroscopic counterparts; nanotechnology can be beneficial on various levels. Nanotechnology has become a rapidly growing field with potential applications from electronics to food, cosmetics, and pharmaceuticals. In this chapter I will focus on present and future applications of nanotechnology in food and nutraceuticals systems and in cosmetics.

Research in chemistry, physics, biology, and engineering drives the development and exploration of the nanotechnology field. The applications to the agriculture and food sector are relatively recent compared with the use of nanotechnology in cosmetics, drug delivery and pharmaceuticals, microelectronics, and aerospace. Certain industries such as microelectronics, aerospace, and pharmaceuticals have already begun manufacturing commercial products of nanoscale size. Even though the food industry is just beginning to explore its applications, nanotechnology exhibits a great potential (Tarver, 2006).

Food nanotechnology is an area of emerging interest and opens up a whole universe of new possibilities for the food industry. Food undergoes a variety of post-harvest and processing-induced modifications that affect its biological and biochemical makeup, so nanotechnology developments in the fields of biology and biochemistry eventually influence the food industry. Ideally, the systems with structural features in the nanometer-length range could affect aspects from food safety to molecular synthesis (Chen et al., 2006).

There are commonly two distinguished forms of nanofood applications: food additives (nano inside) and food packaging (nano outside). Nanoscale food additive, for example, may be used to influence product shelf life, texture, flavor, nutrient composition, or even detect food pathogens and provide functions as food quality indicators. Nanotechnologies in the area of food packaging are mainly considered to increase product shelf life, indicate spoilt ingredients, or generally increase product quality (say, for example, by preventing gas flow across product packaging) (Nickols-Richardson & Piehowski, 2008).

It is a subject of concern that if changing the size of materials can lead to radical, useful properties; can we be sure how size will affect other properties and, mainly, the toxicity of such materials (Weiss et al., 2006). It is likely that the products of nanotechnology intended for food consumption are to be classified as novel products and will be cleared after vigorous testing; there are concerns, though, particularly in the area of food contact materials, that there could be inadvertent release and ingestion of nanoparticles of undetermined toxicity (Tiede et al., 2008). Such concerns need to be addressed because the ultimate success of food nanotechnology products depends on consumer acceptance.

Similarly, there are a number of classes of nanoparticles used or proposed for use in cosmetic applications. Applications of nanotechnology can be found in many cosmetic products, such as moisturizers, hair care products, make up, and sunscreen. The application of nanomaterials in cosmetic products has been the subject of discussion in the media, within research groups, and among policy makers during the past few years. There is a lack of agreement among researchers on whether nanomaterials are safe for dermal use, and toxicity is the main issue. Currently, there are two main uses of nanotechnology in cosmetics. The first is the use of nanoparticles as UV-filters. The second use is nanotechnology for delivery; liposomes and niosomes are used in the cosmetic industry as delivery vehicles.

This chapter will review some of the nanotechnologies used in food and cosmetic industries and focus on the opportunities and challenges in these areas.

NANOTECHNOLOGY IN FOOD APPLICATIONS

POTENTIAL FOOD APPLICATIONS

The potential for food nanotechnology appears unlimited. All aspects of the food industry, from ingredients to packaging to food analysis, are researching viable nanotechnology applications. All aspects of the food industry, from food ingredient to food packaging to food analysis, are resulting in numerous promising applications for improved food production, processing, packaging, and storage (Graveland-Bikkerand & de Kruif, 2006; Vernikov et al., 2009; Sozer & Kokini, 2009). Identification of bacteria and quality monitoring using biosensors, improved food packaging systems, and nanoencapsulation of bioactive food components are a few examples of emerging applications of nanotechnology for the food industry. Carbon nanotubes can be used in food packaging to improve its mechanical properties. Carbon nanotubes exhibited antimicrobial effects and *Escherichia coli* bacteria died on immediate direct contact with aggregates of carbon nanotubes; in fact, the long, thin nanotubes puncture *E.coli* cells, causing cellular damage (Kang et al., 2007). For the detection of carcinogenic pathogens new tools and techniques are being developed using nanotechnology, and biosensors are being developed for improved and contamination-free food (Shrivastava & Dash, 2009).

Some achievements have applications in many sectors of the food industry, such as harnessing the casein micelle, a natural nanovehicle of nutrients, for delivering hydrophobic bioactives; a unique nanotube based on enzymatic hydrolysis of α-lactalbumin; a novel encapsulation technique based on cold-set gelation for delivering heat-sensitive bioactives; developments and use of Maillard reaction-based conjugates of milk proteins and polysaccharides for encapsulating bioactives; introduction of β-lactoglobulin-pectin nanocomplexes for delivery of hydrophobic nutraceuticals in clear acid beverages; development of core-shell nanoparticles made of heat-aggregated β-lactoglobulin, nanocoated by beet-pectin for bioactive delivery; application of milk proteins for drug targeting, including lactoferrin or bovine serum albumin conjugated nanoparticles for effective in vivo drug delivery across the blood–brain barrier; beta casein nanoparticles for targeting gastric cancer; fatty acid-coated bovine serum albumin nanoparticles for intestinal delivery; and Maillard conjugates of casein and resistant starch for targeting of colon (Liveny, 2010). Nanocharcoal® adsorbent is used for discoloration of food products (Augustin & Hemar, 2009).

In the food industry, several novel applications of nanotechnology have become apparent, including the use of nanoparticles such as liposomes, micelles, nanoemulsions, and biopolymeric nanoparticles.

NANODISPERSIONS AND NANOCAPSULES. Nanodispersions and nanocapsules are ideal for delivery of functional ingredients in food. Functional ingredients such as vitamins, antimicrobials, antioxidants, preservatives, and food flavor come in various molecular and physical forms, and functional ingredients are part of the delivery system. A delivery system has numerous functions: the first one is to transport a functional ingredient to its desired site, other functions include protecting an ingredient from chemical or biological degradation, such as oxidation, and controlling the rate of release of functional ingredients under specific environmental conditions.

NANOEMULSION. A nanoemulsion is an emulsion in which the diameters of the dispersed droplets measure <500 nm, encapsulate functional ingredients within their droplets, and minimize the chemical degradation (McClements & Decker, 2000). Different type of nanoemulsions, such as nanostructures multiple emulsions or nanostructures multilayer emulsions, offer multiple encapsulating abilities from a single delivery system carrying multiple functional components. In such structures, a functional component could be released in response to a specific environmental effect. Nanoemulsions have recently received a lot of attention from the food industry due to their high clarity. These enable the addition of nanoemulsified bioactives and flavors to a beverage without a change in product appearance. Solid-lipid nanoparticles are formed by controlled crystallization of food nanoemulsions and have been reported for delivery of bioactives (Weiss et al., 2006; Weiss et al., 2008). The major advantages of solid-lipid nanoparticles include large-scale production without using organic solvents, high concentration of functional compounds in the system, long term stability, and the ability to be spray-dried into powder form.

NANOENCAPSULATION. Nanoencapsulation is defined as a technology to pack substances in miniature, make use of techniques, such as nanocomposite, nanoemulsification, and nanoestructuration, and provide final product functionality that includes controlled release. Nanoencapsulation techniques help in the protection of bioactive compounds, such as vitamins, antioxidants, proteins, lipids, and carbohydrates, for the production of functional foods with better functionality and stability. Encapsulation and controlled release technologies, as well as the principle of novel food delivery systems, have been summarized (Huang et al., 2009). Lipid-based nanoencapsulation enhances the performance of antioxidants by improving their solubility and bioavailability, in vivo and in vitro stability, and preventing their unwanted interactions with other food components.

Lipid-based nanoencapsulation systems that can be used for the protection and delivery of food and nutraceuticals are nanoliposomes, nanocochleates, and archaesomes. Nanoliposome technology facilitates areas such as encapsulation and controlled release of food materials, and enhanced bioavailability, stability, and shelf-life of sensitive ingredients. In food systems nanoliposomes can be used as carrier vehicles of nutrients, nutaceuticals, enzymes, food additives, and food antimicrobials (Mozafari et al., 2008).

Nanoencapsulation technologies have the potential to meet food industry challenges related to the effective delivery of health functional ingredients and controlled release of flavor compounds. Soy lecithin is the main structural ingredient in the formation of aqueous nanodispersions that carry high loads of water-insoluble actives, including water-insoluble nutraceuticals, fat-soluble vitamins, and flavors. Encapsulated actives disperse easily into water-based products, showing improved stability and increased bioavailability.

Nanocochleates are nanocoiled particles that wrap around micronutrients and stabilize and protect an extended range of micronutrients, and they have the potential to increase the nutritional value of processed foods. Nanocochleates consist of purified soy phospholipids that contain about 75% by weight of lipids. The phospholipids may be phosphotidylserine, dioleoylphosphatidylserine, phosphatidic acid, phosphatidylinositol, phosphatidyl glycerol, or a mixture of one or more of these lipids with other lipids. Alternatively, the lipid can include phosphatidylcholine, phosphatidylethanolamine, diphosphotidylglycerol, dioleoyl phosphatidic acid, disteroyl phosphatidylserine, dimyristoyl phosphatidylserine, and dipalmitoyl phosphatidylglycerol.

BIOPOLYMERIC NANOPARTICLES. Food-grade biopolymers such as proteins or polysaccharides are used to produce nanometer-sized particles (Gupta & Gupta, 2005; Ritzoulis et al., 2005). A single biopolymer separates into smaller nanoparticles by using an aggregative (attraction) or segregative (repulsion) interaction. These nanoparticles are used to encapsulate functional ingredients and release them in response to distinct environmental triggers. The most common component of many biodegradable biopolymeric nanoparticles is polylactic acid (PLA), which needs associative compounds like polyethylene glycol (PEG) for better delivery of active components in different parts of the body (Riley et al., 1999).

EDIBLE NANOCOATINGS

Edible nanocoating is as thin as 5 nm and is commercially viable for the food industry. Edible nanocoatings could be used on a wide variety of foods, such as meats, cheese, fruit and vegetables, confectionery, bakery, and fast foods; could act as a vehicle to deliver colors, flavors, antioxidants, enzymes, and antibrowning agent; and could also increase the shelf life of manufactured foods (Morillon et al., 2002; Cagri et al., 2004; Azeredo et al., 2009). Nanocoatings are manufactured from polysaccharides, proteins, and lipids. Polysaccharide- and protein -based nanocoatings are good barriers against oxygen and carbon dioxide but are poor performers against moisture. Lipid-based nanocoating is good at protecting food from moisture but gives limited resistance to gases and has poor mechanical strength. In order to get the desired properties in edible coatings, research is underway to identify additives such as polyols that can improve quality.

Coating of foods with thin films involves either dipping them into a series of solutions containing substances that would be adsorbed by the food's surface or by spraying substances onto the food surface. Adsorption happens as a result of an electrostatic attraction between substances having opposite charges. The degree of a substance's adsorption depends on the nature of the food's surface as well as the nature of the adsorbing substance. Different adsorbing substances can constitute different layers, for example polyelectrolytes (proteins and polysaccharides), charged lipids, and colloidal particles. Consequently, different types of layers could include different functional agents such as antimicrobials, antioxidants, enzymes, colors, and anti-browning agents.

NANOPACKAGING

Nanopackaging, because of its ability to improve safety and extend the shelf life of foods and beverages, is the most exciting innovation in the food industry today. Companies are already producing packaging materials based on nanotechnology, and these are already in use around the world (Brody, 2003; Baeumner, 2004; Brody, 2006a; Brody, 2006b). Numerous companies and academic institutions are developing packaging material that will be able to alert consumers if the packaged food becomes contaminated, respond to a change in environmental conditions, and self-repair holes and tears. Packaging that incorporates nanomaterials is called "smart packaging," which means that it can respond to environmental conditions, repair itself, or alert consumers to contamination or the presence of pathogens (Baeumner, 2004). Research is on-going for developing a range of smart packaging materials that will absorb oxygen, detect food pathogens such as *Salmonella* and *E.coli*, and alert consumers of spoiled food. A new nano-coated "killer paper" has been developed for food packaging to combat bacteria such as *E.coli* to extend product shelf life. Food and beverage packaging comprises 55% to 65% of the $130 billion value of packaging in the United States (Brody, 2008).

OPPORTUNITIES IN FOOD NANOTECHNOLOGIES

Understanding food materials and food processing at the nanoscale is important in order to create new and improved food products. Advancement in processes and new tools are helping researchers gain better understanding of areas such as the mechanisms of targeted delivery for both optimization of human health and novel physical, visual, and sensory effects (Paull & Lyons, 2008). Potential applications include food that can alter its color, flavor, or nutrients to suit consumers' preferences or health demands; filters that can take out toxins or modify flavors by manipulating certain molecules based on their shape and not size; and packaging that can detect when its contents are spoiling and change color to warn consumers (Mognuson, 2009).

New nanoencapsulation technologies can be used to add additional nutrients without changing a product's quality or flavor (Ross et al., 2004). This might enable producers to integrate nutrients that are not naturally occurring or could not be integrated because of their chemistry. The advantages of nanoencapsulation systems also include improved shelf life, protection of the encapsulated ingredients from the environment and undesired interaction during food processing, as well as targeting delivery of certain contents to a specific target site within the body (Augustin & Sangunsari, 2009). Nanocomposites are already available in food packaging or in plastic bottle coatings to control gas diffusion and to prolong the lifetime of various products. Nanotechnology is already being used worldwide to produce antimicrobial food contact materials commercially available as packaging or as coating on varieties of products, such as food containers, chopping boards, and refrigerators (Sozer & Kokini, 2009). Many of these materials contain nanoparticles and are regarded as safe as long as their use does not lead to the release and ingestion of these particles (Farhang, 2009). However, the long term fate and disposal of these materials, which might lead to release of nanoparticles into the environment, is a subject of great concern. These concerns have led to debate on the labeling, product approval, and regulation of these materials (Roco & Bainbridge, 2005). It is important that regulatory bodies prepare criteria for evaluating the safety of food, food packaging, and supplemental uses of nanomaterials with novel properties.

SAFETY ISSUES AND CHALLENGES

Advances in food nanotechnology offer important challenges for both academic researchers and industries. Despite rapid development in food nanotechnology, little is known about the fate and toxicity of nanoparticles. Toxicity is a major issue that is in the mind of the consumer because nanoparticles are more reactive, more mobile, and therefore likely to be more toxic. There is a possibility that nanoparticles in the body can result in increased oxidative stress, and that can generate free radicals, leading to DNA mutation, cancer, and possible death. Nanotechnology-based food ingredients, food additives, and food contact materials have been recently reported regarding their potential implications for consumer safety and regulatory controls (Kampers, 2008). International collaboration and exchange of information is being undertaken to ensure acceptance and benefits of nanotechnology products. Thus, agencies worldwide are gathering information in an effort to decide how best to proceed (Kahan et al., 2008). There is a need for specific guidelines for testing nanofoods. Regulatory authorities, such as the U.S. Food and Drug Administration (FDA), should develop guidelines for criteria to be followed in evaluating the safety of food, food packaging, and supplement uses of nanomaterial with novel properties.

NANOTECHNOLOGY IN COSMETICS APPLICATIONS

POTENTIAL COSMETICS APPLICATIONS

Applications of nanomaterials are found in cosmetics products such as moisturizers, hair care products, make-up, and sunscreen. A different class of nanoparticles is used in cosmetics and has been the subject of discussion among researchers and policy makers. However, toxicity issues are the main concerns among consumers—whether the nanomaterials are safe for dermal applications. This chapter offers a brief overview of some of the innovative nanoparticulate delivery systems for dermal applications.

LIPOSOMES. Delivery systems play an important role in the development of effective skin care products. Anti-aging treatments account for much of the growth within the skin care market. The first liposomal cosmetic product was the anti-aging cream "Capture" launched in the market in 1986. Since then varieties of products that utilize liposomal delivery capabilities have been introduced into the market. Liposomes are vesicular structures with an aqueous core surrounded by a hydrophobic lipid bilayer of phospholipids. Drug molecules in the core cannot pass through the hydrophobic bilayer, and the hydrophobic molecules can be absorbed into the bilayer, thus enabling the liposomes to carry both hydrophilic and hydrophobic molecules. The lipid bilayer of liposomes can fuse with other bilayers such as the cell membrane, which promotes release of its contents, making liposomes useful for drug delivery as well as cosmetic delivery applications. Liposomes that have vesicles in the range of nanometers are called nanoliposomes. Liposomes in general vary in size from 15 nm to several μm and can have either a single layer (unilamellar) or multilayer (multilamellar) structure. Transferosomes, a new type of liposome with improved efficiency, have been developed (Ceve, 1996) for potential applications in cosmetics and drug delivery. Transferosomes with a 200–300 nm size range can penetrate the skin with improved efficiency, better than liposomes (Thong et al., 2007).

In general, liposomes have widespread use in the cosmetic industry, and they facilitates the continuous supply of agents into the cells over a sustained period of time, making them ideal candidates for the delivery of vitamins and other active molecules (such as active ingredients and antioxidants) to regenerate the epidermis (Lautenschläger, 2006). Liposomes are also used to deliver phosphatidylcholine, one of the main components of liposomes. Phosphatidylcholines have been widely used in skin care products and shampoos for their softening and conditioning qualities of the product.

NIOSOMES. Niosomes are non-ionic surfactant-based vesicles with a similar structure to that of phospholipid vesicles' liposomes. They are formed by the self-assembly of non-ionic surfactants in aqueous media. The application of heat or physical agitation helps the process to attain a closed bilayer structure (Uchegbua & Vyas, 1998). The advantages of using niosomes in cosmetic and skin care applications include their ability to increase the stability of entrapped drugs, to improve bioavailability of poorly absorbed ingredients, and to enhance skin penetration.

Niosomes can enhance transdermal drug delivery. It is reported that noisome-encapsulated drugs have been delivered through the stratum corneum, which is considered highly impermeable (Van Hal et al., 1996). Niosomes made from a novel surfactant (Bola-surfactant) have been found very effective for percutaneous drug delivery applications (Paolino et al., 2007). Studies have shown that they improve percutaneous passage of 5-fluorouracil (5-FU) through stratum corneum and epidermis and are non-toxic (Paolino et al., 2008).

SOLID LIPID NANOPARTICLES. Nanoparticulate systems able to control release and improve targeting of skin are common both in cosmetics and pharmaceuticals. Liposomes are the best known nanoparticulate systems. As an alternative to liposomes, solid lipid nanoparticles (SLNs) were first designed for IV administration and later investigated for transdermal applications. SLNs are nanometer-sized particles with a solid lipid matrix. The mean particle size varies from 50 to 1,000 nm. They are oily droplets of lipids that are solid at body temperature and stabilized by surfactants. The process of their manufacturing is very simple, liquid lipid (oil) in nanoemulsion is exchanged with solid lipids without using any organic solvents (Müller et al., 2002). Lipids such as triglycerides, glyceryl behenate, glyceryl palmitostearate, and the wax cetyl palmitate are used for production of SLNs to be topically applied on the skin. Lipid concentration ranges from 5% to 40 %, and depending on the type and concentration of lipids used, 0.5–5% emulsifier (surfactant) has to be added for stabilization. Their appearance depends on the concentration of lipids; a lower concentration of lipids gives viscous forms to be easily applied, and a higher concentration of lipids results in semisolid appearance, also acceptable in cosmetic applications.

SLNs offer a number of potential advantages for cosmetic products. They can be used for the controlled delivery of cosmetic agents over a prolonged period of time and have been found to improve the penetration of active compounds into the stratum corneum. They can protect the encapsulated ingredients from degradation. SLN-containing formulations were found more efficient in skin hydration than placebo (Wissing & Müller, 2003). They have also been found to show UV-resistant properties, which were enhanced by incorporating molecular sunscreen (Wissing & Müller, 2001).

ETHOSOMES. Ethosomes are similar to liposomes, which are constituted of phospholipids, but are produced in the presence of ethanol. Ethosomes are produced (in the same way as liposomes but with slight modification in the process) by adding an aqueous phase to an ethanol solution (20 to 45% v/v) of soy phosphatidylcholine (5% w/w) and an active ingredient, such as azelaic acid (0.2% w/w), and mechanically stirred. Liposomes are also obtained by the same composition but in the absence of ethanol. Ethosomes are characterized by a prolonged physical stability compared to liposomes. The use of ethosomal carriers results in delivery of a high concentration of active ingredients to skin or through the skin regulated by system composition and their physical characteristics. The basic properties and the in vitro release rate kinetics of azelaic acid, alternatively vehiculated in different phospholipids based vesicles, such as ethosomes and liposomes, were reported (Esposito et al., 2004).

NANOEMULSION. Nanoemulsions are dispersions of nanoscale droplets of one liquid within another. The structure of nanoemulsions can be manipulated based on the method of preparation to give different type of products (Mason et al., 2006; Sonneville-Aubrun et al., 2004). Nanoemulsions have good applications in cosmetics because of increased shelf life of products containing those active ingredients (Tadros et al., 2004). Several cosmetic products are available on the market that use nanoemulsions. They are either transparent or translucent and have large surface area due to the small particle size. Due to the small emulsion size, stability will be higher and be more suitable to carry active ingredients.

NANOSTRUCTURE LIPID CARRIERS. Nanostructure lipid carriers (NLCs) are prepared by mixing solid lipids with liquid lipids. The high loading capacity and long term stability of the NLCs make them superior to SLNs in many cosmetic applications. This is because the NLCs have distorted structures, which makes their matrix structure imperfect and creates more spaces compared to SLNs to accommodate active compounds. NLCs are also capable of preventing chemical degradation of active compounds. NLCs show the properties of controlling occlusion without altering the properties, for example, increasing the occlusion of day cream without the glossiness of night creams. It has also been found that the release profile of active compounds can be manipulated by changing the matrix structure of nanoparticles. Lipid nanoparticles have been found to increase the penetration capabilities of active compounds compared to microparticles (Müller & Dingler, 1998). The lubricating effect and mechanical barrier of lipid nanoparticles are important in cosmetic applications for reducing irritation and allergic reactions.

NANOENCAPSULATION. Encapsulation technology has been well known for a long time in the pharmaceutical industry for drug delivery. Its cosmetic applications are also emerging. Different types of nanocapsules are required depending on the nature of the material (such as hydrophobic or hydrophilic) to be incorporated. It is also possible to functionalize these materials to target specific molecules. Polymers are generally used to create nanocapsules, which are functionalized for various applications. For cosmetic applications, polymeric nanoparticles are produced by hydrophobically modifying polyvinylalchol 10,000 (PVA) with fatty acids (FA) to obtain PVA-FA derivative (Luppi et al., 2004). PVAs are substituted for saturated fatty acids to give sufficient liophilicity to the polymer and are used for the percutaneous absorption of benzophenone-3 (BZP) or oxybenzone, widely used in many cosmetic formulations such as sunscreen lotions or emulsions, shampoos, and hair sprays. The nature of the vehicle used can enhance or block the percutaneous absorption of UV filters. Sunscreen filters are meant to work on the periphery of the skin, and percutaneous absorption is highly discouraged. Studies showed that PVC nanoparticles can limit the adsorption of BZP, the nanoparticles with a high degree of substitution, and prevent absorption more efficiently.

OPPORTUNITIES AND CHALLENGES

Delivery systems play an important role in the development of effective cosmetic products. Among delivery technologies known are lipid systems, nanoparticles, nanocapsules, polymers, and films. These nanotechnologies are loaded into such vehicles as creams, gels, patches, etc. However, only a handful of drug delivery technologies have made the leap from exclusively pharmaceuticals to cosmetics. Although there are many products available on the market using nanomaterials, opportunities still exist to exploit the benefits of nanotechnology in the cosmetic industry. Lipid based nanoparticles (SLNs, NLCs) and nanocapsules need improvements in drug loading efficiency. The production process of SLNs needs improvement to increase loading capability and stop expulsion of the contents during storage. Conditions for the formation of SLNs and NLCs and the effect of surfactants used for modifications need further studies. Better understanding of how the lipid nanoparticles affect the drug penetration into the skin and how they interact with lipids of the stratum corneum need further research (Müller et al., 2002), as well as in vivo studies on the effect of cosmetics that contain nanomaterials.

There is a need for in-depth studies into the toxicity effects of nanomaterials used in cosmetic formulations. This research, along with better regulation and reporting, will help consumers to choose products with confidence. Further, this is a challenge for cosmetic industries to understand the benefits of these nanotechnologies for the long term while retaining customer confidence.

CONCLUSION

Nanotechnology is seen by many of us as a growth area that will transform tomorrow's world. Food nanotechnology is an area of emerging interest and opens up a whole universe of new possibilities for the food industry. Nanotechnology holds great promise to provide benefits not just within food products but also around food products. It can be utilized in the food industry to detect bacteria in packaging; produce strong food flavors, textures, and color; reduce fat content; and encapsulate nutrients, such as vitamins, to ensure they do not degrade during shelf life; and improve safety by increasing barrier properties. Nanomaterials can be used to make packaging that keeps the product inside fresher for a longer time. Incorporating nanosensors in food packaging could alert consumers when a product is no longer safe to eat. In fact, nanotechnology introduces a new horizon for innovation in the food industry, but at the same time, uncertainty and health concerns are also emerging. Thus, mandatory testing of nanomodified foods is desirable before they are allowed to enter the market. Standardized test procedures to study the impact of nanoparticles on living cells are needed for the evaluation of potential hazards to humans. Improving the safety and quality of foods is a priority, and we expect that nanotechnology-derived food products will be available to consumers worldwide in the near future.

Similarly, nanotechnology in cosmetic and skin care products offers advantages and improved characteristics compared with traditional formulations. Many of the newer technologies being investigated for drug delivery may also have applications in cosmetics. However, there are concerns over their toxicity, and more research is required before the materials can be deemed safe. Further, the development of new regulations for reporting and assessing the use of nanomaterials in cosmetics is also important.

REFERENCES

Augustin, M.A.; Hemar, Y. Nano- and micro-structured assemblies for encapsulation for food ingredients. Chem. Soc. Rev. 2009, 38, 902–912.

Augustin, M.A.; Sanguansari, P. Nanostructured materials in the food industry. Adv. Food Nutr. Res. 2009, 58,183–213.

Azeredo, H.C.; Mattoso, L.H.; Wood, D.F.; Williams, T.G.; Avena-Bustillos, R.D.; Mc Hugh, T.H. Nanocomposite edible films from mango puree reinforced with cellulose nanofibers. J. Food Sci. 2009, 74, N31–N35.

Baeumner, A. Nanosensors identify pathogens in food. Food Technol. 2004, 58, 51–52.

Brody, A.L. Packaging by the numbers. Food Technol. 2008, 62, 89–91.

Brody, A.L. Food packaging climbs to summit. Food Technol. 2006a, 60, 73–75.

Brody, A.L. Nano and food packaging technologies converge. Food Technol. 2006b, 60, 92–94.

Brody, A.L. "Nano, Nano" food packaging technology. Food Technol. 2003, 57, 52–54.

Cagri, A.; Ustunol, Z.; Ryser, E.T. Antimicrobial edible films and coatings. J. Food Protect. 2004, 67, 833–848.

Cevc, G. Transfersomes, liposomes and other lipid suspensions on the skin: permeation enhancement, vesicle penetration, and transdermal drug delivery. Crit. Rev. Ther. Drug Career Syst. 1996, 13, 257–388.

Chen, H.; Weiss, J.; Shahidi, F. Nanotechnology in nutraceuticals and functional foods. Food Technol. 2006, 60, 30–36.

Esposito, E.; Menegatti, E.; Cortesi, R. Ethosomes and liposomes as topical vehicles for azelaic acid: a preformulation study. J. Cosmetic Sci. 2004, 55, 253–264.

Farhang, B. Nanotechnology and applications in food safety. Global issues in food science and technology; Barbosa-Canovas, G., Mortimer, A., Lineback, D., Spiess, W., Buckle, K., and Colonna, P., Eds.; Elsevier: U.S., 2009; pp. 401–410.

Graveland-Bikkerand, J.F.; de Kruif, C.G. Food nanotechnology. Trends Food Sci. Technol. 2006, 17, 196–203.

Gupta, A.K.; Gupta, M. Synthesis and surface engineering of iron oxide nanoparticles for biomedical applications. Biomaterials, 2005, 26, 3995–4021.

Huang, Q.; Given, P.; Qian, M., Eds.; Micro/nano encapsulation of active food ingredients. Oxford University Press: Washington D.C., 2009.

Kahan, D.M.; Braman, D.; Slovic, P.; Gastil, J.; Cohen, G. Cultural cognition of the risks and benefits of nanotechnology. Nat. Nanotechnol. 2008, 4, 87–94.

Kampers, F.W.H. Opportunities for bionanotechnology in food and the food industry. In Bionanotechnology: Global prospects; Reisner, D.E.; Bronzino, J.D., Eds.; CRC Press, Boca Raton, FL, 2008; pp. 79–90.

Kang, S.; Pinault, M.; Pfefferle, L.D.; Elimelech, M. Single-walled carbon nanotubes exhibit strong antimicrobial activity. Langmuir. 2007, 23, 8670–8673.

Lautenschläger, H. Liposomes, Handbook of cosmetic science and technology. CRS Press Taylor & Francis Group: Boca Raton, Florida, 2006; pp. 155–163.

Livney, Y.D. Milk proteins as vehicles for bioactives. Current Opin. Colloid Interface Sci. 2010, 15, 73–83.

Luppi, B.; Cerchiara, T.; Bigucci, F.; Basile, R.; Zecchi, V. Polymeric nanoparticles composed of fatty acids and polyvinyl alcohol for topical applications of sunscreens. J. Pharm. Pharmacology 2004, 56, 407–411.

Mason, T.G.; Wilking, J.N.; Meleson, K.; Chang, C.B.; Graves, S.M. Nanoemulsions: formation, structure, and physical properties. J. Phys. Condens. Matter 2006, 18, R635–R666.

McClements, D.J.; Decker, E.A. Lipid oxidation in oil-in-water emulsions: impact of molecular environment on chemical reactions in heterogeneous food systems. J. Food Sci. 2000, 65, 1270–1282.

Mognuson, B.A. Nanoscale materials in foods: existing and potential sources. ACS Symposium Series 1020, American Chemical Society- DOI: 10.1021/bk-2009-1020. 2009; chapter 4.

Morillon, V.; Debeaufort, F.; Blond, G.; Capelle, M.; Voilley, A. Factors affecting the moisture permeability of lipid-based edible films: a review. Crit. Rev. Food Sci. Nutr. 2002, 42, 67–89.

Mozafari, M.R.; Johnson, C.; Hatziantoniou, S.; Demetzos, C. Nanoliposomes and their applications in food nanotechnology. J. Liposome Res. 2008, 18, 309–327.

Müller, R.H.; Dingler, A. The next generation after the liposomes: solid lipid nanoparticles (SLN, Lipopearls) as dermal carrier in cosmetics. Eurocosmetics. 1998, 7, 8, 18–26.

Müller, R.H.; Radtke, M.; Wissing, S.A. Solid lipid nanoparticles (SLN) and nanostructured lipid carriers (NLC) in cosmetic and dermatological preparations. Adv. Drug Deliv. Rev. Suppl. 2002, 54, S15–S131.

National Science and Technology Council Committee on Technology. The National Nanotechnology Initiative (NNI): Research and Development Leading to a Revolution in Technology and Industry. Wasington(DC): Office of Science and Technology Policy. 2006.

Nickols-Richardson, S.M.; Piehowski, K.E. Nanotechnology in nutritional sciences. Minerva Biotechnol. 2008, 20, 17–126.

Paolino, D.; Cosco, D.; Muzzalupo, R.; Trapasso, E.; Picci, N.; Fresta, M. Innovative bola-surfactant niosomes as topical delivery systems of 5-fluorouracil for the treatment of skin cancer. Int J. Pharmaceutics 2008, 353, 233–242.

Paolino, D.; Muzzalupo, R.; Ricciardi, A.; Celia, C.; Picci, N.; Fresta, M. In vitro and in vivo evaluation of Bola-surfactant containing niosomes for transdermal delivery. Biomed. Microdevices 2007, 9, 421–433.

Paull, J.; Lyons, K. Nanotechnology: the next challenge for organics. J. Organic Syst. 2008, 3, 3–22.

Riley, T.; Govender, T.; Stolnik, S.; Xiong, C.D.; Garnett, M.C.; Illum, L.; Davis, S.S. Colloidal stability and drug incorporation aspects of micellar-like PLA-PEG nanoparticles. Colloids Sur B. 1999, 16, 147–159.

Ritzoulis, C.; Scoutaris, N.; Papademetriou, K.; Stavroulias, S.; Panayiotou, C. Milk protein-based emulsion gels for bone tissue engineering. Food Hydrocolloids 2005, 19, 575–581.

Roco, M.C.; Bainbridge, W.S. Societal implications of nanoscience and nanotechnology: maximizing human benefit. J. Nanopart. Res. 2005, 7, 1-13.

Ross, S.A.; Srinivas, P.R.; Clifford, A.J.; Lee, S.C.; Philbert, M.A.; Hettich, R.L. New technologies for nutrition research. J. Nutr. 2004,134, 681–685.

Shrivastava, S.; Dash, D. Agrifood Nanotechnology: a tiny revolution in food and agriculture. J. Nano Res. 2009, 58, 183–213.

Sonneville-Aubrun, O.; Simonnet, J.-T.; L'Alloret, F. Nanoemulsions: a new vehicle for skincare products. Adv. Colloid Interface Science 2004,108, 145–149.

Sozer, N.; Kokini, J.L. Nanotechnology and its application in the food sector. Trends in Biotechnology 2009, 27, 82–89.

Tadros, T.; Izquierdo, P.; Esquena, J.; Solans, C. Formation and stability of nanoemulsions. Adv. Colloid Interface Sci. 2004, 109, 303–318.

Tarver, T. Food Nanotechnology. Food Technol. 2006, 60, 22–26.

Thong, H.-Y.; Zhai, H.; Maibach, H.I. Percutaneous penetration enhancers: an overview. Skin Pharmacol. Physiol. 2007, 20, 272–282.

Tiede, K.; Boxall, A.B.A.; Tear, S.P. Detection and characterization of engineered nanoparticles in food and the environment. Food Addit. Contam. 2008, 25, 795–821.

Uchegbua, I.F.; Vyas, S.P. Non-ionic surfactant based vesicles (niosomes) in drug delivery. Int J. Pharmaceutics 1998, 172, 33–70.

Van Hal, D.; Vanrensen, A.; Devringer, T.; Junginger, H.; Bouwstra, J. Diffusion of estradiol from non-ionic surfactant vesicles through human stratum-cirneum in vitro. STP Pharm. Sci. 1996, 6, 72–78.

Vernikov, V.M.; Arianova, E.A.; Gmoshinskii, I.V.; Khotimchenko, S.A.; Tutl'ian, V.A. Nanotechnology in food production: advances and problems. Vopar Pitan. 2009, 78, 4–17.

Weiss, J.; Decker, E.A.; McClements, J.; Kristbergsson, K.; Helgason, T.; Awad, T.S. Solid lipid nanoparticles as delivery systems for bioactive food components. Food Biophys. 2008, 3, 146–154.

Weiss, J.; Takhistov, P.; McClements, D.J. Functional materials in food nanotechnology. J. Food Sci. 2006, 71, 107–116.

Wissing, S.A.; Müller, R.H. A novel sunscreen system based on tocopherol acetate incorporated into solid lipid nanoparticles (SLN). Int J. Cosm. Sci. 2001, 23, 233–243.
Wissing, S.A.; Müller, R.H. Cosmetic applications for solid lipid nanoparticles (SLN). Int J. Pharmaceutics 2003, 254, 65–68.

2

SURFACE ACTIVE LIPIDS AS ENCAPSULATION AGENTS AND DELIVERY VEHICLES

Ling-Zhi Cheong[1], Zheng Guo[1], Bena-Marie Lue[1], Rikke Miklos[2], Shuang Song[1, 3], Worawan Panpipat[1], and Xuebing Xu[1]
[1]Aarhus University, Department of Molecular Biology, Gustav Wieds Vej 10, DK-8000, Aarhus C, Denmark; [2]University of Copenhagen, Department of Food Science, Rolighedsvej 30, DK-1958, Frederiksberg C, Denmark; and [3]Beijing University of Chemical Technology, College of Life Science and Technology, No.15 of Beisanhuan East Road, 100029, Beijing, China.

INTRODUCTION

Surface active lipids such as partial acylglycerols, phospholipids, and glycolipids are amphiphilic molecules that lower either the surface tension of the medium in which they are dissolved or the interfacial tension between phases at which they are adsorbed. Due to their amphiphilicity, surface active lipids are able to self-assemble into various unique supramolecular structures ranging from micro- to nanoscales. These unique supramolecular structures are found to have enhanced physical properties ranging from increased gelation capability, guest molecules loading capacity, and in-vivo stability to targeted delivery and molecular recognition. Thus, surface active lipids have garnered huge interest as encapsulation agents and delivery vehicles in various industries including food, pharmaceutical, and cosmetics. In fact, work has also been directed towards synthesis of novel compounds and structural modification of existing surface active lipids. This chapter aims to provide a review on current state-of-the art design, synthesis, and purification of surface active lipids, namely partial acylglycerols, phospholipids, glycolipids, aminolipids and lipopeptides, phytosterol surfactant, and antioxidant esters. Physical properties of such lipids including self-assembling and formation of various supramolecular structures are elucidated. Some potential applications of the surface active lipids are also detailed.

PARTIAL ACYLGLYCEROLS

SYNTHESIS OF PARTIAL ACYLGLYCEROLS

Partial acylglycerols, namely monoacylglycerols (MAGs) and diacylglycerols (DAGs), are esters of glycerol with either one or two of their hydroxyl groups esterified with fatty acids, respectively. They exist in various isomers, namely *sn*-1 and 2 MAG and *sn*-1, 2 and 1, 3 DAG. At equilibrium, the ratio of *sn*-1 to 2 MAG and *sn*-1, 2 to 1, 3 DAG are 9 to 1 and 6 to 4, respectively (Nakajima et al., 2004). With hydrophilic hydroxyl groups in the structure, partial acylglycerols show interfacial properties and surface activity that renders them suitable for use as emulsifiers in various industries, particularly food and cosmetics (Macierzanka et al., 2009). DAGs also generated interest in the food industry as functional food with beneficial health effects including reducing postprandial lipid level, increasing β-oxidation of fat, and managing and preventing obesity (Flickinger & Matsuo, 2003).

Commercially, partial acylglycerols are produced either through chemical glycerolysis at high reaction temperatures (220–260°C) using inorganic alkaline catalysts (Jacobs et al., 2003; Yang et al., 2004) or two-steps steam hydrolysis and enzymatic esterification (Yamada et al., 2001). High temperature chemical glycerolysis is unsuitable for producing heat-sensitive polyunsaturated fatty acids enriched partial acylglycerols. It also causes development of off-flavors in the final product (Guo & Xu, 2005). In comparison, enzymatic synthesis of partial acylglycerols offers a number of advantages including reduced energy consumption, environmental-friendly operation, higher selectivity, and improved product quality. There are a number of enzymatic synthetic routes for partial acylglycerols, namely esterification, acidolysis, glycerolysis, alcoholysis, partial hydrolysis, and interesterification. Among these, extensive research has been done on glycerolysis (Damstrup et al., 2005; Kristensen et al., 2005; Yang et al., 2005), esterification (Weber & Mukherjee, 2004; Lo et al., 2007) and partial hydrolysis (Cheong et al., 2007; Babicz et al., 2010).

Enzymatic synthesis of partial acylglycerols has been extensively reviewed on several occasions (Lai et al., 2004; Lo et al., 2008). This section highlights the enzymatic glycerolysis that has the favorable product yield. Stoichiometrically, two moles of oils react with one mole of glycerol to yield three moles of partial acylglycerols. In order to produce a high yield of partial acylglycerols, phase homogeneity between the hydrophobic oil and hydrophilic glycerol is essential. Solvents such as *n*-hexane, *n*-heptane, acetone, *tert*-butanol and diethyl ether are sometimes used to improve the miscibility of reaction mixtures and, subsequently, the overall product yield. Tertiary alcohols, namely *tert*-butanol, *tert*-pentanol, or mixtures of these with hexane, were efficient reaction media for high yield of MAG (Damstrup et al., 2005). Recently, ionic liquids, which are considered eco-friendly solvents due to their negligible vapour pressure, have emerged as new reaction media. The benefits include adjustable solubility, enhanced lipases stability, positive effects on enzymes' specificity, and facilitation of recoverability and recyclability of lipases. Binary ionic liquids system of TOMA.Tf_2N/Ammoeng 102 served as a good reaction medium with at least 70% DAG formation (Kahveci et al., 2009). AOT surfactant (sodium di-2-ethylhexyl sulfosuccinate) is another example of "green" solvents which produced more than 50% yield of MAG (Fiametti et al., 2008). Despite the high efficiency, usage of solvents in the development of food-based products is generally deemed undesirable. Solvent-free enzymatic glycerolysis is found to be able to produce more than 50% DAG by using a stirred tank batch reactor, which constantly stirred the reaction substrate for phase homogeneity (Cheong et al., 2009; Kristensen et al., 2005). Another consideration for enzymatic glycerolysis is the selection of enzymes. It is best to select an enzyme that is immobilized on a hydrophobic carrier in order to prevent glycerol adsorption on the lipase, which will subsequently decrease the reaction rate. Lipases from *Rhizomucor meihei* and *Candida antarctica* immobilized on macroporous resin and acrylic resin, respectively, have been often used in enzymatic glycerolysis (Cheong et al., 2009; Kristensen et al., 2005).

The final product from enzymatic synthesis usually contains a mixture of glycerol, free fatty acids (FFA), MAG, DAG, and TAG. Thus, purification of the resultant product is necessary to obtain partial acylglycerols of high purity. Short path distillation (SPD) is a purification method where compounds with different boiling points are separated under vacuum, low evaporating temperature, and minimal residence time. This method is especially suitable for the separation of thermosensitive compounds such as polyunsaturated fatty acids enriched partial acylglycerols. In fact, SPD has been used industrially to purify partial acylglycerols up to a purity of more than 95% (Xu et al., 2002). In spite of this, other problems persist in regards to coloration and possible development of carcinogenic glycidol fatty acid esters during the refining process. The coloration of oil can be easily solved through adsorption or a bleaching process (Minoru, 2006). Thus, work is currently directed towards investigation of other purification methods or elimination of the glycidol fatty acid esters (Strijowski et al., 2010).

PHYSICAL PROPERTIES OF PARTIAL ACYLGLYCEROL AND ITS POTENTIAL APPLICATIONS

Despite the minor differences in structure, partial acylglycerols have considerably different physical properties from TAG. Physical properties of partial acylglycerols have been previously reviewed (Nakajima et al., 2004; Lo et al., 2008). Generally, MAG has the highest melting point followed by DAG and TAG. The causes of these melting point differences are the strength of hydrogen bonding of the hydroxyl groups and the fatty acid chain arrangement of the isomers (Nakajima et al., 2004). When added in low concentration to oils and fats, DAGs are found to have different effects on the melting point of the bulk oil depending on its isomers. *Sn*-1, 2 DAGs caused melting point reduction. *Sn*-1, 3 DAGs showed different effects depending on the fatty acid composition. Dipalmitoylglycerol caused melting point increment; meanwhile, dioleoylglycerol and palmitoyloleoylglycerol caused melting point reduction (Siew, 2002). This implies that partial acylglycerols have different crystal polymorphisms. TAGs exhibited three forms of crystals structure, namely the α, β' and β forms. *Sn*-1, 2 DAGs exhibited the α and β'; meanwhile, *Sn*-1, 3 DAGs exhibited the $\beta 1$ and $\beta 2$ forms (Nakajima et al., 2004). At a low degree of supercooling, DAGs exerted different effects on crystallization of bulk oil depending on their concentration. Low concentration of DAGs delayed nucleation and crystal growth; meanwhile, high concentration of DAGs promoted nucleation and crystal growth in the bulk oil (Cheong et al., 2009).

Both MAGs and DAGs have emulsifying capability due to the presence of hydroxyl groups. In fact, they have been used widely as emulsifiers in food, cosmetic, and pharmaceutical products (Kralova & Sjoblom, 2009; Macierzanka et al., 2009).

The specific emulsifying capacity depends on the fatty acid composition. Long fatty acid chains, due to stronger van der Waals forces, allow the molecules to pack closely in an ordered structure at the interface (Nakajima, 2004). DAGs having at least one long-chain unsaturated fatty acid have been reported to retain and emulsify water up to a volume fraction of 20% (Shimada & Ohashi, 2003; Nakajima, 2004).

Partial acylglycerols self-assemble into various supramolecular structures in aqueous solution. For example, MAGs first self-assemble into lamellar crystalline phase, which transforms into lamellar liquid crystalline phase, and eventually reversed bicontinuous cubic phase with increased concentration of water. All of these phases are usually on the nanoscale level, which gives high surface area for encapsulation of a significant amount of compounds (Sagalowicz et al., 2006; Leser et al., 2006). Despite the viscous gel-like property of the reversed bicontinuous cubic phase, it has been used in delivery of peptide drugs (Ericsson et al., 1991) and other drugs such as cezafolin (Sadhale & Shah, 1998), bupivacaine (Shah et al., 2001), diazepam, rifampicin, and propofol (Boyd, 2003). Such a phase is found to confer protection against degradation and controlled release of drugs (Sagalowicz et al., 2006). In addition, the lamellar liquid crystalline and reversed bicontinuous cubic phases are also used as reactors for creation of flavor compounds. The water–MAG interface served as a host for Maillard reaction precursors for formation of flavor compounds such as 2-furfurylthiol (FFT). Production of FFT showed increments of 2- to 7-fold in liquid crystalline phases as compared to the conventional aqueous phase. Similar to drug delivery, such phases are also found to contribute to controlled aroma release (Vauthey et al., 2000). It is interesting to note that the liquid crystalline phase may transform into different phases following introduction of other guest molecules. For example, when used to encapsulate sugar molecules such as glucose, the reversed bicontinuous cubic phase transformed to a reversed hexagonal phase. The sugar molecules, due to their hydrophilicity, competed with MAGs for water molecules resulting in changes in the packing parameter (Mezzenga et al., 2005). The possibility of altering both the phase behavior and number density of channels in reversed bicontinuous cubic phases by simply manipulating the position, nature, and number of double bonds in MAGs has been demonstrated (Kaasgaard & Drummond, 2006; Kulkarni et al., 2010). There seems to be a correlation between chain length and degree of unsaturation in phase behavior. Nevertheless, further studies are required before unambiguous conclusions can be drawn.

Due to their high viscosity, liquid crystalline phases, such as the reversed bicontinuous cubic phase, are difficult to handle. Attempts have been made to disperse the liquid crystalline phase into water in the form of small particles (Gustafsson et al., 1996; Gustafsson et al., 1997). Similar to the liquid crystalline in bulk phase, the dispersed liquid crystalline particle phase demonstrated phase transition with introduction of guest molecules. The dispersed liquid crystalline particle phase transformed from reversed biocontinuous cubic phase to hexagonal and micellar cubic phases and eventually lamellar liquid crystalline phase when used to encapsulate lipophilic compounds (Nakano et al., 2002; Yaghmur et al., 2005).

Unlike MAGs, DAGs have not been studied widely in terms of self-assembling properties. Patino & Dominguez (2000) studied the interfacial characteristics of DAGs' monolayers at the air and aqueous interfaces. Types of isomers are found to have effects on the formation of monolayers. *Sn*-1, 2 distearin formed a liquid-condensed monolayer that upon compression at high surface pressures transformed into a solid phase. Presence of two hydrocarbon chains in the same molecule situated at two contiguous carbon atoms contributed to hydrophobic interactions, which caused formation of stable solid film. *Sn*-1, 3 distearin, on the other hand, formed a liquid-expanded monolayer. *Sn*-1, 3 distearins also have two hydrocarbon chains, but they are separated by a hydroxyl group that favors a more expanded structure. Mixture of both isomers at a ratio of 1 to 1 resulted in a liquid expanded structure, which then transformed into a liquid condensed structure. This transition is mainly due to rearrangement of the isomers at the interface (Patino & Dominguez, 2000).

PHOSPHOLIPIDS

STRUCTURAL MODIFICATION OF PHOSPHOLIPIDS

Phospholipids (PLs) are amphiphilic molecules containing both hydrophilic head groups (negatively charged phosphate bonded to polar groups such as choline and inositol) and hydrophobic tails (long chain fatty acids). PLs can be classified into three main classes, namely glycerophospholipids, sphingolipids, and ether phospholipids (Hawthorne & Ansell, 1982). As a major component of cell membranes, PLs provide structural assurance for membrane integrity, assist in membrane functions (Lemmon, 2008; Schmidt et al., 2006), and regulate basic biological processes (Kölsch et al., 2008; Aoyama, 2009). PLs possess surface activity and thus are able to self-assemble into various supramolecular structures such as bilayer, capsule, and liposome. These structures, specifically liposome, can be used to encapsulate active ingredients for enhanced delivery and bioavailability (Tran et al., 2009; Hillaireau et al., 2009). To date, structural modification of PLs has been done intensively.

PLs can be structurally modified through chemical or enzymatic means. Chemically, PLs can be modified through *N*-acetylation, hydrolysis, hydroxylation, and hydrogenation. In *N*-acetylation, PLs containing free amino groups such as phosphatidyl ethanolamine (PE) are acetylated using acetic anhydride (Doig & Diks, 2003; Gunther, 1984). The *N*-acetylated PE is insoluble in ethanol; thus, it can be easily separated from a mixture containing ethanol-soluble phosphatidyl choline. This method, known as batch ethanol fractionation, is used commercially to produce PC of varying purity (Doig & Diks, 2003).

Chemical hydrolysis of PLs can be performed using acid or basic catalysts to produce lysophospholipids (LPLs) and eventually glycerophospho compounds (Zuidam & Crommelin, 1995). Recently, chemical hydrolysis of PLs in a liposomal environment has been studied for drug delivery. Several factors, namely temperature, pH, charge of the liposomes, and types of buffer used, were found to affect chemical hydrolysis of PLs in a liposomal environment (Zuidam & Crommelin, 1995; Vernooij et al., 2002). Despite the lower cost, chemical catalysts possess shortcomings such as formation of complex reaction products due to non-specificity of the catalysts and undesirable coloration of the reaction product.

Both hydroxylation and hydrogenation of PLs occur at the carbon-carbon double bonds of the unsaturated fatty acids in PLs. In hydroxylation, hydroxyl groups are attached to the double bonds under acidic environments to yield hydroxylated PLs (Doig & Diks, 2003; Vandana et al., 2003), which have lighter color, lower iodine value, higher hydrophilic capacity, and improved dispersion in cold water (Vandana et al., 2003). Hydrogenation of PLs increases the degree of saturation, which subsequently improves the stability of the resultant saturated PLs. As natural PLs contain impurities, it is important to select chemical catalysts with high resistance towards poisoning such as palladium on carbon and platinum (Brois et al., 1992; Schneider, 1997). Hydrogenation of PLs using palladium on carbon is only limited to applications in pharmaceutical and non-food industrial products due to the high cost of such catalysts and restriction in food applications (Guo et al., 2005).

Enzymatic modification of PLs involves either modification of the acyl groups or the head groups. Phospholipase A_1 (PLA_1) (EC 3.1.1.32) and phospholipase A_2 (PLA_2) (EC 3.1.1.4) hydrolyze or synthesize the *sn*-1 and *sn*-2 acyl ester bonds of PLs, respectively. It has been found that 1, 3-specific lipase works in a similar manner as PLA_1. Meanwhile, phospholipase B (PLB) (EC 3.1.1.5) hydrolyzes both positional ester bonds. Despite the similar catalytic function, lipases surpass phospholipases in terms of practical industrial application due to the cheaper price, easier availability, and higher stability. Phospholipase C (PLC) (EC 3.1.4.3) and phospholipase D (PLD) (EC 3.1.4.4) are used to modify the head groups of PLs. The former modifies the phosphorus-oxygen bond between phosphate and glycerol, and the latter modifies the phosphorus-oxygen bond between phosphate and head group (Hirche & Ulbrich-Hofmann, 2000; De Maria et al., 2007).

Generally, modifications of the acyl groups occur either through a two-step hydrolysis and esterification (Svensson et al, 1992; Adlercreutz et al., 2002) or a single step interesterification (acidolysis/transesterification) (Peng et al., 2002; Vikbjerg et al., 2005). Interesterification is simpler and easier to perform but the two-step hydrolysis and esterification yields products of higher purity (Adlercreutz et al., 2001). Some of the important parameters affecting the aforementioned reactions include types of enzyme, water activity, acyl donor, reaction media (solvent or solvent-free), time, and temperature (Guo et al., 2005). Solvents are usually used due to the viscous reaction system. They are selected by taking into consideration the log *P* value (polarity), solubility of substrates and products, and volatility and toxicity of the solvents (Park et al., 2000; Senanayak & Shahidi, 2002). Water activity is important to maintain enzyme activity and prevent hydrolysis of the reaction products. It is controlled either by equilibrating the reaction system with saturated salt solution or adding salt hydrates with a definite water activity value (Hosokawa et al., 1995).

Among the reactions to modify the head groups, PLD-catalyzed transphosphatidylation is considerably the most interesting. It produces novel structured PLs such as PL-terpenes (Yamamoto et al., 2008), PL-sterol (Hossen & Hernandez, 2004), PL-serinol, PL-Tris, PL-diethanolamine/triethanolamine, and PL-BisTris (Dippe et al., 2008). Nevertheless, these processes are unlikely to be commercialized except for production of high-added value product in the pharmaceutical industry due to the high cost of PLD.

Parameters affecting PLD-catalyzed transphosphatidylation include the source of PLD, nature of the alcohol acceptor, and reaction media (Hirche & Ulbrich-Hofmann, 2000). PLD from *Streptomyces* sp., which catalyzes transphosphatidylation more than the competitive hydrolysis, is most frequently used (Uesugi & Hatanaka, 2009). In terms of alcohol acceptors, hydrophilic alcohols with small molecular weight and primary hydroxyl group are more easily incorporated. Biphasic system comprised of buffer with solvent such as ethyl acetate, diethyl ether, or dichloromethane are frequently used as reaction media to solubilize substrates differing in polarity (PLs are hydrophobic, whereas alcohol acceptors are hydrophilic). In some cases, benzene or a mixture of hexane and 1-octanol has also been used (Ulbrich-Hofmann, 2003; Hirche & Ulbrich-Hofmann, 2000). To minimize solvent usage, enzymatic transphophatidylation in a pure aqueous suspension system with high reaction efficiency has been developed (Iwasaki et al., 2003).

Hydrolysis of the reaction product is a major concern in transphosphatidylation. Polarity of the alcohol acceptors and solvent systems is found to have significant effect. Transphosphatidylation is faster than hydrolysis in a polar solvent when amphiphilic alcohol acceptors are used. Meanwhile, transphosphatidylation is faster than hydrolysis in polar solvents such as diethyl ether when hydrophilic alcohol acceptors are used. In addition, hydrolysis of secondary alcohols is more apparent when using PLD from *Streptomyces* sp. as compared to PLD from cabbage (Hirche & Ulbrich-Hofmann, 2000).

Structured PLs can be purified using various techniques such as liquid-liquid extraction, ultrafiltration using polysulfone membrane on polyester support, and column chromatography (Vikbjerg et al., 2006a; Vikbjerb et al., 2006b). Although these processes purified the structured PLs to a purity of more than 90%, they are time consuming and labor intensive. Further investigation and optimization are required in order for the purification process to be economically feasible.

PHYSICAL PROPERTIES OF PHOSPHOLIPIDS, STRUCTURED PHOSPHOLIPIDS, AND THEIR POTENTIAL APPLICATIONS

Structured PLs possess different physicochemical properties. The hydrophile-lipophile balance (HLB), for example, increased from 4-5 in deoiled lecithin to 8-9 in hydrolyzed lechitin and 9-10 and 10-12 in acetylated and hydroxylated lecithin, respectively (Schneider, 1997; Sinram, 1991). In addition, fatty acids with different acyl chain length (C2-C18) and degree of unsaturation are found to affect emulsification properties and oxidation stability (Pedersen, 2001). For example, emulsions prepared using structured PLs containing caproic acid are more stable and less prone to phase separation as compared to emulsions prepared using pure soybean PC (Trotta et al., 2002). Similar findings are reported by Vikjberg and colleagues (2006c), in which stable emulsions are formed using structured PLs containing caprylic acids as compared to pure soybean PC.

Due to their amphiphilic property, PLs have the ability to absorb at the surfaces or interface between two immiscible boundaries, for example between solid-liquid and two immiscible liquid-liquids. Through adsorption at the interfaces, PLs lower the free energy between the phase boundaries (surface tension) and thus increase their miscibility (Holmberg et al., 2002). PLs self-assemble into a range of supramolecular structures at the interfaces. Some of the known factors affecting the self-assembled structures include the molecular structures and concentration of PLs, temperature, and pH of the phase boundaries. Self-assembling properties of PLs have been carefully reviewed and elucidated on several occasions (Holmberg et al., 2002; Sagar et al., 2007). Some of the basic self-assembling properties of PLs, and also nanosized structures formed by structured PLs, are elucidated in the following section.

At a low PL concentration (above the critical micelle concentration), PLs first self-assemble into micelles in oil-in-water. In a micelle, the hydrophobic tails of PLs are directed towards the oily interior of the cluster and the hydrophilic head groups are directed towards the surrounding aqueous solution. Micellar structures grow into different systems depending on the type of PLs, degree of concentration, temperature, presence of counter ion, and cosolute (Holmberg et al., 2002; Lee et al., 2010). For PLs with long chain fatty acids, the spherical micelle may grow into polydisperse cylindrical linear rod-like micelle from ten to hundreds of nanometers in length. The linear growth may lead to branched structures which, at high enough concentrations, may form bicontinuous structures such as hexagonal, lamellar, and cubic phases. The hexagonal micellar phase consists of long cylindrical micelles arranged in hexagonal patterns. Meanwhile, the lamellar phase is built up of bilayers of PL molecules alternating with water layers. The PL bilayers can range from stiff and planar to flexible and undulating. Except for the lamellar phase, all the aforementioned structures have a reversed counterpart in water-in-oil. For example, a reversed hexagonal phase is built up of hexagonally packed water cylinders with the polar heads directed towards the interior cluster while the non-polar tails are directed towards the continuum oily phase (Holmberg et al., 2002).

In addition to the aforementioned conventional structures in both water-in-oil and oil-in-water, PLs' natural occurring lecithin can grow from spherical reverse micelles in the presence of nonpolar organic solvent to an organogel with the addition of polar additive such as water. In the nonpolar organic media alone, lecithin self-assembled into spherical reverse micelles. Addition of the polar additive provided a polar bridge (hydrogen bonding) to bind two adjacent lecithin molecules. This grew further to form flexible, long tubular micelles, which upon reaching a critical length began to overlap and entangle forming a viscous jelly-like phase. Being thermodynamically stable, it possessed prolonged shelf life. As a result, lecithin organogels are found to be suitable for topical drug delivery (Kumar & Katare, 2005).

Structured PLs such as those conjugated with polymer have been synthesized and used to form polymeric micelles as nanoscale carriers for poorly soluble drugs. For example, poly(ethylene glycol)-phosphatidylethanolamine (PEG-PL) formed micelles of approximately 15 nm, which efficiently encapsulated similar size water-soluble anticancer drugs such as doxorubicin hydrochloride and vinorelbine tartrate (Wang et al., 2010). Aside from encapsulation, PLs conjugate with polymer, namely phospholipid-polyethyleneimine, and are found to form micelle-like nanoparticles suitable for systemic gene and siRNA delivery. The micelle-like nanoparticles have high in vivo stability, which is often an issue in conventional gene delivery (Ko et al., 2009; Navarro et al., 2011).

Liposomes or lipid vesicles represent another important supramolecular structure formed from self-assembling of PLs. In water phase, surface active lipids such as PLs arrange themselves by exposing their polar head groups toward the water phase and hydrophobic hydrocarbon moieties adhere together in the bilayer. They form closed vesicles with an aqueous core and an internal hydrophobic domain that is trapped between the lipid bilayers (Barani & Montazer, 2008). Liposomes can be divided into three main classes, namely multilamellar vesicles (MLV), which contain more than a single bilayer membrane with a size range of 0.1–6.0μm; small unilamellar vesicles (SUV); and large unilamellar vesicles (LUV), which both contain only a single bilayer membrane with size ranges from 0.02 to 0.05μm and 0.06μm, respectively (Henna Lu et al., 2011). The "rigidity" or "fluidity" of a liposome is determined by degree of chain length and unsaturation. Liposomes consisting of saturated PLs with long acyl chains, such as dipalmitoylphosphatidylcholine, have more rigid and impermeable structures than unsaturated monomers with short acyl chains (Crommelin et al., 2003). In the last decades, nanosized liposomes (nanosomes) have been synthesized either using natural PL or a combination of different PLs for encapsulation of antimicrobial peptide. The liposomes encapsulated in antimicrobial peptide are found to lower the viable counts of lactococci in cheese without affecting the rheological and sensory characteristics. Thus, it is suitable for controlling spoilage and pathogenic bacteria in food (Malheiros et al., 2010). Phospholipids nanosomes can also be used to encapsulate and deliver protein macromolecule anticancer and antiHIV drugs (Castor, 2005).

The surface of liposomes can be modified for additional functionalities. "Homing" devices which have targetability such as antibodies, saccharides, and peptide can be attached covalently to the surface of the liposomes forming "targeted" liposomes (Crommelin et al., 2003). This approach has been used to form immunoliposomes for delivery of antibody specifically to tumor cells (Mizoue et al., 2002). Another approach is to coat the surface of liposomes with hydrophilic chains such as poly (ethylene glycol) (PEG) and poly (hydroxyethyl L-asparagine) (PHEA) to form "stealth" liposomes, which have increased circulation time. Increase in circulation time of "stealth" liposomes prevented phagocytosis in vivo and increased liposome concentration accumulated on target cells (Metselaar, 2003). "Stealth" liposomes loaded with doxorubicin are currently marketed for treatment of solid tumor (Crommelin et al., 2003).

In addition to modifying the surface of liposomes, novel methods can be developed to direct self-assembly of liposomes. Hong and colleagues (2010) adapted a technique known as controlled microfluidic mixing and nanoparticle determination to manufacture liposome-hydrogel hybrid nanoparticles. Unlike liposomes and hydrogel, which have their own advantages and disadvantages, liposome-hydrogel hybrid nanoparticles utilize the strengths of each material. The liposome-hydrogel hybrid nanoparticle is envisioned to revolutionize drug delivery, in that it can target specific cells such as tumors, pass through the target cell, and slowly release the drug content (Hong et al., 2010).

GLYCOLIPIDS

Glycolipids refer to compounds containing one or more monosaccharide residues bound by a glycosidic linkage to a hydrophobic moiety such as acylglycerol, sphingoid, ceramide, or prenyl phosphate. Depending on their lipid moieties, glycolipids can be generally classified as glycoglycerolipids, glycophosphatidylinositols, or glycosphingolipids (Chester, 1997). Having both hydrophobic and hydrophilic moieties, the amphiphilic glycolipids possess surface activity, which renders them possible "green" alternatives to the commonly used non-degradable and highly toxic ethylene-oxide-containing surfactants (Faivre & Rosilio, 2010). Glycolipids also play a vital role in molecular recognition mechanisms in which they are the receptors for various pathogenic agents such as bacteria and viruses (Garelli-Calvet et al., 1992; Blanzat et al., 1999). In addition, glycolipids have recently been demonstrated to self-assemble into meso- and nanostructures (John et al., 2001; Jung et al., 2002; Roussel et al., 2006). Due to these unique properties, glycolipids have found applications as soft nanomaterials, which can be used as encapsulation agents and delivery vehicles in numerous industries ranging from nutrition to pharmaceuticals and cosmetics (Vemula et al., 2006).

SYNTHESIS OF GLYCOLIPIDS

Chemical synthesis of glycolipids, particularly alkyl glucosides or polyglycosides, has a long history tracing back to more than 100 years ago. It began with Hermann Emil Fisher's work on reacting a monosaccharide with an alcohol in the presence of an acid catalyst. The basis of Fisher glycosylation has been successfully applied in industrial production of alkyl polyglycoside surfactants reaching an estimated annual capacity of between 30,000 and 50,000 tons (Behler et al., 2001). To date, chemical glycosylation has been used to synthesize various glycolipids such as cardanol-based glycolipids (John et al., 2001), long-chain phenyl glucoside (Jung et al., 2002), monosaccharide esters with diamino-aromatic linkers (John et al., 2004), and alginate derived alkyl oligosaccharides (Roussel et al., 2005; Roussel et al., 2006).

The chemical glycosylation process can be either a direct synthesis (Hill et al., 1988; John et al., 2001), a two-steps transacetalization (Bergfeld & Seifert, 1994; Roussel et al., 2005), or a stereoselective substitution reaction on suitably activated carbohydrate substrates (Jung et al., 2002). Direct synthesis is employed when a monosaccharide such as glucose is used as the acyl acceptor. The monosaccharide is suspended in an excess acyl donor such as fatty alcohol in the presence of an acidic catalyst, for example, sulfonic acid, Et_3N or $BF_3.OEt_2$, at 100–120°C. Solvents such as anhydrous dichloromethane or tetrahydrofuran are sometimes used to facilitate the solvation of both acyl donor and acceptor (Hill et al., 1988; Rybinski & Hill, 1998; John et al., 2001). Two-step transacetalization, on the other hand, is applied when oligo- or polysaccharide is used as the acyl acceptor. In the first step, the carbohydrate reacts with a short-chain alcohol such as butanol in the presence of an acidic catalyst at 100–140°C to yield a short-chain alkyl glycoside. Following that, the short-chain alkyl glycoside is transacetalized with a long-chain alcohol to form the final alkyl glycoside (Bergfeld & Seifert, 1994; Rybinski & Hill, 1998; Roussel et al., 2005). In stereoselective substitution reaction, the anomeric carbon of the carbohydrates is first activated by leaving groups such as halides, sulfonium, trichloroacetimidate, or nitrate groups. Alcohol of various chain lengths can then be stereoselectively attached to the activated carbon (Rybinski & Hill, 1998; Jung et al., 2002).

Chemical glycosylation is terminated by alkaline neutralization using, for example, sodium hydroxide or bicarbonate. Both direct synthesis and two-step transacetalization yield a complex mixture of residual glucose (<1%), excess fatty alcohol, alkyl mono-, alkyl oligo-, and alkyl polyglycosides. The composition of the alkyl glycosides depends largely on the ratio of the carbohydrates to alcohol which can be easily adjust to yield, for example, a predominant alkyl monoglycoside (>60%) (Rybinski & Hill, 1998). A stereoselective substitution reaction may result in the formation of exclusive products rather than of complex reaction mixtures. In all cases, excess fatty alcohol can be effectively removed through short path distillation resulting in high purity alkyl glycosides (98–99%). For purified products containing alkyl mono-, alkyl oligo-, and alkyl polyglycosides, a further purification can be achieved through column chromatography to obtain the different components (John et al., 2001). Nevertheless, this is not usually done industrially as it is not cost-efficient. In fact, the mixture of different alkyl glycosides is usually applied directly, for example, as detergents used in the bottle cleaning industry (Rybinski & Hill, 1998).

One of the major concerns in chemical glycosylation is the formation of secondary products such as glucose polymer, ethers, and colored impurities due to the usage of acid catalysts and distillation at prolonged high temperature. This could be controlled and reduced by optimizing the reaction duration, using hydrophobic acids such as alkyl benzene sulfonic acids, and minimizing the residence time of thermal stressing through short path distillation (Rybinski & Hill, 1998). In addition, the non-specificity of chemical catalysts and presence of multiple hydroxyl groups in carbohydrates led to low yield of specific isomers. As a result, there is a growing interest in the usage of biocatalysts including lipases and peptidases in synthesis of glycolipids (Chang & Shaw, 2009).

The early work in enzymatic synthesis of glycolipids involved the use of different biocatalysts including lipases and proteases and various solvent systems to form regioselective glycosidic (glycosylation), ester (esterification), or amide bonds between the carbohydrate and acyl donor (Chang & Shaw, 2009). Table 2-A shows a compilation of enzymatic synthesis of various glycolipids in different solvent systems during the last decade. Several enzymes, particularly lipase from *Candida antarctica*, has shown high catalytic efficiency (John et al., 2006; Chaiyaso et al., 2006; Vemula et al., 2006).

The major challenge in enzymatic synthesis of glycolipids is selection of a suitable solvent system to dissolve both the hydrophilic sugar and hydrophobic lipid substrates without causing denaturation or inactivation of the enzyme. The solvent system must be cheap and possess low toxic potential in order for application of resultant glycolipids in food, pharmaceutical, and cosmetics industries. Acetone, ethyl methylketone ,and *n*-hexane have been shown in some instances to provide reasonable solubility of the substrates and also retain the enzymatic activity (Yan et al., 2001; John et al., 2006; Vemula et al., 2006). The aforementioned solvents are regarded as either class 2 or class 3 solvents with low toxic potential by the United States Food and Drug Administration (U.S. FDA, 2003).

As ionic liquids are gaining recognition as "green" solvents in the field of biocatalysis, attempts have been made to use ionic liquids as reaction media for enzymatic synthesis of glycolipid. Liu and colleagues (2005) found ionic liquids based on the dicyanimide anion to dissolve high amounts of carbohydrates from glucose to starch and cellulose. Despite the excellent sugar dissolution property, this group of ionic liquids caused dissolution of lipase *Candida antarctica* B, which is notable for its excellent stability in various solvent and reaction systems. It is postulated that the dicyanamide anion interacts strongly with the enzyme's protein, resulting in breakage of the intermolecular protein bonds and subsequently causing dissolution and irreversible enzyme inactivation (van Rantwijk et al., 2006). Ganske and Bornscheuer (2005) were among the pioneers to synthesize glucose fatty acid esters in high yield in a reaction media containing a mixture of ionic liquid ([BMIM] [BF_4] and [BMIM] [PF_6]) with *tert*-butanol. Following that, other techniques have been developed to synthesize glycolipids in high yield in pure ionic liquid media such as employing a supersaturated carbohydrate solution in ionic liquid (Lee et al., 2008a; Lee et al., 2008b) and ultrasonic irradiation (Ha et al., 2010).

The cost of ionic liquids and stringent rules regarding their usage in food processing has garnered research towards solvent-free enzymatic synthesis of glycolipids. Saccharide-fatty acid esters, specifically fructose-oleic acid esters (FOE), have been shown to dissolve fructose and oleic acid in a solvent-free environment (Dang et al., 2005). First, a mixture of FOE and oleic acid can be pumped into a fructose desorption column to facilitate the solvation of fructose then a reaction mixture (fructose, FOE and oleic acid) is pumped into reactors (stirred tank batch reactor and packed

Table 2-A. Enzymatic Synthesis of Glycolipids.

Substrate					
Acyl acceptor	**Acyl donor**	**Enzymes**	**Solvents**	**Products**	**Refs**
Glucose Fructose α-D-methylglucopyranoside Maltose	Myristic acid	Novozyme 435 (*Candida antarctica*)	*tert*-butanol:pyridine (55:45 v/v)	Myristyl glucose (23.61ug/ml/min) Myristyl fructose (22.3umol/g/min) Myristyl α-D-methylglucopyranoside (26.9umol/g/min) Myristyl maltose (1.9 umol/g/min)	Degn & Zimmermann (2001)
Glucose	Caprylic acid Palmitic acid Stearic acid	Chirazyme L-2 (*Candida antarctica*)	Ethyl methylketone	Caprylyl glucose (31.8 wt %) Palmitoyl glucose (62.5 wt %) Stearoyl glucose (79.2 wt %)	Yan et al. (2001)
Maltose Glucose Sucrose	Vinyl laurate Vinyl palmitate	*Thermomyces lanuginosus* Novozyme 435 (*Candida antarctica*)	2-methyl-2-butanol:dimethylsulfoxide	6-*O*-acylsucrose 6,6′-di-*O*-acylsucrose 6,1′-di-*O*-acylsucrose 6′-*O*-acylmaltose 6′-*O*-acylglucose	Ferrer et al. (2005)
Trehalose Sucrose Maltose Lactose	Vinyl stearate Vinyl butyrate	Lipase B from *Candida antarctica*	Acetone	Trehalose-6, 6′-distearate (>50 wt %) Trehalose-6, 6′-dibutyrate (>50 wt %)	John et al. (2006b)
D-Glucose L-ascorbic acid D-Fructose D-Galactose D-Galactose α-methyl-D-glucose	Palm fatty acid distillates	(Novozyme 435) *Candida antarctica*	Acetone 2-methyl-2-butanol *tert*-butanol MTBE *n*-hexane	Glucose fatty acid esters (76 wt %) L-ascorbic acid fatty acid esters (62 wt %) D-Fructose fatty acid esters (39 wt %) D-Galactose fatty acid esters (7 wt %) D-Galactose fatty acid esters (18 wt %) α-methyl-D-glucose fatty acid esters (76 wt %)	Chaiyaso et al. (2006)
D-Amygdalin	Fatty acid vinyl ester (C4, C14, C18)	(Novozyme 435) *Candida antarctica*	Acetone	D-Amygdalin fatty acid esters (>90%) (C4, C14, C18)	Vemula et al. (2006)
D-(+)-Mannitol D-(+)-Sorbitol	Fatty acid vinyl ester (C4, C8)	(Novozyme 435) *Candida antarctica*	Acetone	D-(+)-Mannitol fatty acid esters D-(+)-Sorbitol fatty acid esters	Jadhav et al. (2010)

bed bioreactor) containing enzyme for reaction. This solvent-free reaction produced high yield of saccharide fatty acid esters (FOE), although the reaction kinetics is 2- to 3-fold slower than a solvent reaction (Pyo & Hayes, 2008; 2009). Recently, Ye and colleagues (2010) found enzymatic synthesis of saccharide-fatty acid esters using suspensions of saccharide crystals to be comparable in terms of both product yield and reaction kinetics.

PHYSICOCHEMICAL PROPERTIES AND POTENTIAL APPLICATIONS

Glycolipids are widely used industrially as surfactants in personal care and household products such as shampoo, shower gel, toothpaste, and detergents due to their interfacial properties. Nickel and colleagues (1992) found the critical micelle concentration (CMC) of alkyl polyglycosides to decrease with alkyl chain length. Similar findings are found in the newly designed alginate-derived alkyl oligosaccharides. Alginate-derived alkyl oligosaccharides with medium chain length (C12 and C14) had CMC values that are comparable with commercial neutral surfactants (polyethylene glycol—PEG). They have the capability of reducing surface tension comparable to that of PEG. In addition, alginate-derived alkyl oligosaccharides can be designed with different alkyl chain lengths to give rise to a series of surfactants with a broad range of hydrophile-lipophile balance (HLB). Short and medium alkyl chain lengths produced surfactants with high HLB, whereas, long alkyl chain length produced surfactants with low HLB. The series of surfactants with broad range of HLB enabled alginate-derived alkyl oligosaccharides to be applied for formation of different emulsions, be it water-in-oil or oil-in-water emulsion (Benvegnu & Sassi, 2010). Sucrose esters with a broad range of HLB values have been studied and evaluated in microencapsulation of protein for parenteral delivery. Despite the lower encapsulation capacity, a lesser amount of sucrose esters is required to encapsulate protein compared to the control polyvinyl alcohol (Youan et al., 2003). This implies glycolipids can be designed with slight modifications to the hydrophilic head group and hydrophobic alkyl chain length to produce a series of surfactants as encapsulation agents for different compounds.

In addition, glycolipids also possess good foaming behavior, which is essential in cosmetics and detergent industries. Alkyl polyglycosides, particularly short and medium chain alkyl polyglycosides, foamed considerably better with higher foam volume than the conventional surfactant fatty alcohol ethoxylates. Alkyl polyglyocoside foam is also finer, more creamy, and more stable than the fatty alcohol ethoxylates foam (Busch et al., 1993). Alginate-derived alkyl oligosaccharide is also found to have superior foaming properties than the commercially available sodium dodecyl sulfate (SDS). Higher foam volume per minute can be achieved using 10 times less alginate-derived alkyl oligosaccharides compared to SDS (Benvegnu & Sassi, 2010).

A number of synthesized glycolipids have recently been shown to self-assemble into various meso- and nanostructures, which have potential applications as encapsulation agents and delivery vehicles. Cardanol-based glycolipids, both with or without double bonds in the alkyl chain, formed a white fluffy cotton-like material when dispersed in aqueous solution at room temperature. This material was observed as having coiled or twisted nanosized tubular structures under a transmission electron microscope. Despite the similar structures, only saturated cardanol-based glycolipids possessed thermoreversible gelation capability. This implies the degree of saturation in cardanol-based glycolipids affects the molecular orientation and packing. Despite the lack of gelation property, unsaturated cardanol-based glycolipids formed vesicles at a temperature above 46°C (John et al., 2001). Although related literature on further practical application of cardanol-based glycolipids is absent, its self-assembling properties propose possible application as encapsulation agents. In fact, it has prompted synthesis of several glycolipids such as monosaccharide esters with diamino-aromatic linkers (John et al., 2004), trehalose-based diesters (John et al., 2006), amygdalin-based esters (Vemula et al., 2006), alginate derived alkyl oligosaccharides (Roussel et al., 2006), and dialkanoate derivatives of sugar alcohols (Jadhav et al., 2010). Among the aforementioned glycolipids, amygdalin-based esters and dialkanoate derivatives of sugar alcohols were studied comprehensively from synthesis to self-assembling properties and practical applications.

Long chain amygdalin-based ester was found to form thermoreversible nanosized hydrogel in a broad range of solvents. The self-assembling property of the hydrogel can be briefly described as a highly interdigitated bilayer structure with the alkyl chain tilted with respect to the normal layer plane in which the hydrophilic parts face inside the assembly and the hydrophobic chains are exposed to the outer side of the assembly (Vemula et al., 2006). The hydrogel structure is stabilized by the presence of several interactions, namely hydrogen bonding within the sugar, π-π stacking of the phenyl ring, and van der Waals interactions within the alkyl chain. This robust hydrogel is proven suitable for drug delivery through encapsulation of the hydrophobic curcumin in its hydrophobic pockets within the gel. The curcumin-encapsulated gel was stable and remained intact over a broad range of temperatures. Release can be triggered by incubating the curcumin-encapsulated gel at 37°C in the presence of lipase. Lipase hydrolyzed the long chain esters attached to the amygdalin, thus disrupting the gel and releasing the curcumin (Vemula et al., 2006; John & Vemula, 2006).

Dialkanoate derivatives of sugar alcohols are shown to selectively gel or solidify the oil phase from an oil-water mixture at room temperature. The oil-contained organogels formed were thermoreversible and stable for several months. Both alkyl chain length and the structure of the sugar alcohol affect the gelling ability in which longer alkyl chain length demonstrated higher gelling ability. In terms of rheological property, the oil-contained organogels were viscous and stiff, and could be easily removed from the water. Similar to long chain amygdalin-based ester, the alkyl chain of the organogel tilted with the hydrophilic parts faced inside the assembly and the hydrophobic chains exposed to the outer side of the assembly. The hydrophobic tails on either side of the sugar head ensure the molecules remain in the non-polar oil layer (Jadhav et al., 2010). Due to the aforementioned property, dialkanoate derivatives of sugar alcohols appear to be promising model candidates for oil-spill remediation.

Despite the lack of practical application study, monosaccharide esters with di-amino-aromatic linkers possessed unique properties worth noting. They possessed similar chemical composition to cardanol-based glycolipids except for the presence of two secondary amine groups in the former. They also self-assembled into nano-sized tubular structures in water. The unique property of this compound is the di-aminopyridine head group, which formed 3-point hydrogen bonding with artificial receptors including thymidine analogues uracil and the anticancer compound 5-fluorouracil. This property is useful in molecular recognition and gene delivery (John et al., 2004; John & Vemula, 2006).

AMINOLIPID AND LIPOPEPTIDE

Aminolipid and lipopeptide are molecular conjugates of lipid- and protein-based moieties linked together either through amide or ester bonds. They are found to occur naturally in the inner and outer membranes of some bacteria species. For example, N-[3-(hexadecanoyloxy) hexadecanoyl]-ornithine (Fig. 2.1), which has structural similarity to ceramides, is a type of aminolipid present in various bacteria's membranes (Kawai et al., 1999). Some of these conjugates, such as N-arachidonoyl serine (Fig. 2.1), which are isolated from bovine brain are also found in plants and mammalian tissues (Milman et al., 2006). Marine organisms have recently been shown to be rich in lipopepides with unusual fatty acids and amino acid derivatives. Mololipids (Fig. 2.1) isolated from sponges consist of bis-amides between a bromo-tyrosine derivative (formerly known as moloka'iamine) and long chain fatty acids ranging from C14 to C20 (Ross et al., 2000). Naturally occurring aminolipid and lipopeptide usually exist at low concentrations, but they have important biological activities such as possessing immunological effect, involvement in molecular signaling, and displaying anti-HIV activity. Besides that, aminolipid and lipopeptide also possess surface activity due to their amphiphatic properties. In fact, Daptomycin cyclic lipopeptide (a seven-amino-acid ring structure coupled to one molecule of 3-hydroxy-13-methyl-tetradecanoic acid) (Fig. 2.1) produced by *B. subtilis* is able to lower the surface tension from 72 to 27.9 mN/m at concentrations as low as 50 ppm, making it one of the most powerful biosurfactant known today (Marahiel et al., 1977). Therefore, there is a growing interest in designing and developing amino-lipids and lipopepides for industrial application.

SYNTHESIS OF AMINOPEPTIDE AND LIPOPEPTIDE

Molecular engineering is one of the approaches used to produce new or modified microbial-based lipopeptides. The first step in molecular engineering is to determine goals such as modulation of the bioactivity of lipoprotein according to specific requirements, elimination of residues causing undesired side effects, particularly in medical applications, and general optimization of the emulsifying, solubilization, and foaming properties. Based on the goals, a proper strategy can be determined

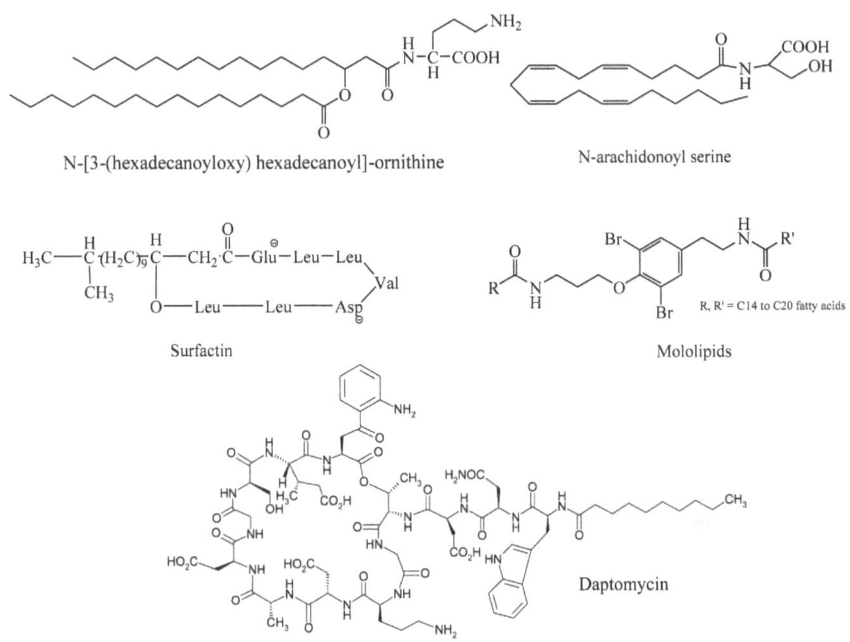

Fig. 2.1. Molecular structures of some natural existing aminolipids and lipopeptides.

to design the lipopeptide either by changing the type and sequence of amino acid residues in the peptide backbone or the nature, length, and branching of the fatty acid chain moiety. Specific directed mutagenesis can then be used to modify the enzymatic activity of distinct domains such as the substrate specificity of A-domains or of the terminal thioesterase I (TEI)-domain responsible for cyclization and release of the lipopeptide (Koglin et al., 2010).

Synthesis of lipidated peptides can also be obtained either by modification of natural or existing peptides or by de novo synthesis. The three most common types of lipidation of peptides or proteins are myristoylation of the N-terminal amino group, S-prenylation (farnesyl- and geranylgeranyl groups) of cysteine residues at or close to the C-terminus, and S-palmitoylation of cysteines throughout the chain (Marilyn, 1999) (Fig. 2.2A). Lability of the thioester under basic conditions and towards nucleophiles poses considerable problems in protein fatty acylation. In addition, under acidic conditions, addition of the acid to the prenyl group double bonds occurs readily. Thus, protecting groups that can be removed under extremely mild and preferably neutral conditions is required for the synthesis of lipopeptides embodying an acid-labile farnesyl thioether and a base-labile thioester. Enzyme labile-protecting groups such as acetyl choline esterase (AChE) and butyryl choline esterase (BChE) are employed for the C-terminus of the peptide; meanwhile, lipase and

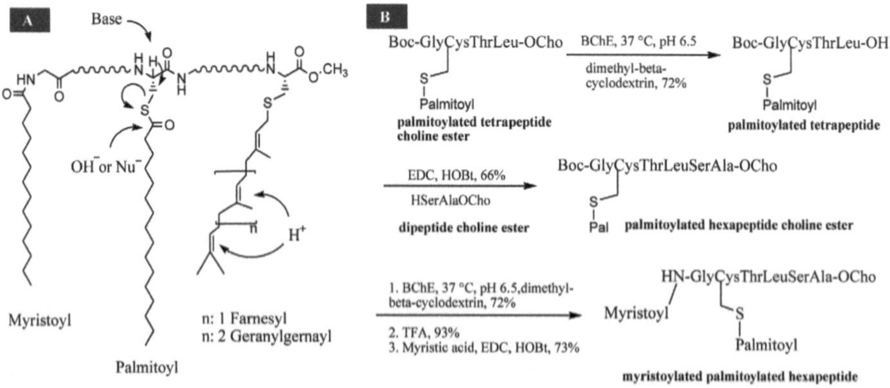

Fig. 2.2.A. Lipid modification of peptides; **B.** Choline ester strategy for the synthesis of a myristoylated and palmitoylated hexapeptide. TFA: trifluoroacetic acid, EDC: 1-ethyl-3-(dimethylamino)-propylcarbodiimide hydrochloride, HOBt: 1-hydroxybenzotriazole, Cho: choline, BChE: butyryl choline esterase, Boc: tert-butoxycarbonyl.

esterase are used for the N-terminus of the peptide (Kadereit & Waldman, 2000). Fig. 2.2B presents an example of using choline esterase to synthesize a myristoylated and palmitoylated hexapeptide.

De novo synthesis of lipopeptides includes synthesis of peptides with amino acids as building blocks and coupling fatty acid with peptide by using or not using linkers. Solid-phase peptide synthesis (SPPS) has been well developed and documented (Albericio, 2000). Generally, SPPS consists of two stages, the first stage consisting of peptide chain assembly with protected amino acid derivatives on a polymeric support, and the second stage consisting of concurrent cleavage of the peptide from the resin support and all side chain protecting groups to yield the new crude peptide. In the first stage, the C-terminal amino acid residue of the target peptide is attached to an insoluble support via its carboxyl group. Any functional groups in the amino acid side chains must be masked with "permanent" protecting groups that are not affected by the reaction conditions during peptide chain assembly. The "temporary" protecting group masking the α-amino group during the initial resin loading is removed. An excess of the second amino acid is introduced, with the carboxyl group of this amino acid being activated for amide bond formation through generation of an activated ester or by reaction with a coupling agent. After coupling, excess reagents are removed by washing and the protecting group is removed from the N-terminal of the dipeptide prior to the addition of the third amino acid. This process is repeated until the desired peptide sequence is assembled. Finally, in the second stage, the peptide formed is released from the solid support and all the "permanent" protecting groups on the side chain are removed. The superiority of this technique lies in the ability to perform wash cycles after each reaction, removing excess reagent with all of the growing peptide of interest remaining covalently attached to the insoluble resin (Chan & White, 2004). There are two major forms of SPPS, namely the Fmoc-based SPSS and t-Boc-based SPSS. In Fmoc-based SPPS, 9-fluorenylmethyloxycarbonyl (Fmoc) is used as the α-amino protecting group and pyridine is used as the deprotection reagent, which removes the Fmoc group for reaction of the α-amino group with an incoming activated amino acid. Meanwhile, t-Boc-based SPPS uses *tert*-butoxycarbonyl (t-Boc) to temporarily protect the α-amino group, which is removed with acids such as trifluoroacetic acid (TFA). This forms a positively-charged amino group, which is simultaneously neutralized and coupled to the incoming activated amino acid. Both types of reactions are driven to completion by using an excess (2- to 4-fold) of activated amino acid. The advantages and disadvantages of both methods have been documented elsewhere (Fields & Noble, 1990). Fig. 2.3 describes the Fmoc-based SPPS synthetic procedure of an amphiphilic lipopeptide (ALP), composed of amphiphilic oligopeptide moieties (6a-c) with alternating hydrophobic and hydrophilic amino acid residues, (Leu-Glu)n interlinked by a succinyl moiety to the phospholipid tails, 1,2-dioleoyl-sn-glycero-3-phosphoethanolamine (DOPE). The peptidic part with variable length was synthesized using a standard Fmoc-based SPPS protocol. After removal of the Fmoc group, the 1, 2-dioleoyl-sn-glycero-3-phosphoethanolamine-N-succinic acid was coupled to the anchored peptides. The acid-labile Sieber amide resin permitted selective cleavage of the fully protected lipopeptides from the solid support under mild conditions. Then t-Butyl protecting groups were removed under acidic conditions yielding the target ALPs that could be purified by RP-HPLC (Cavalli et al., 2006).

SELF-ASSEMBLING PROPERTIES OF AMINOPEPTIDE AND LIPOPEPTIDE AND THEIR POTENTIAL INDUSTRIAL APPLICATIONS

Amino acids are natural building blocks of protein and vary in their structures. Proteinogenic or standard amino acids refer to the 22 amino acids that are naturally incorporated into polypeptides. Amino acids can be divided into those with electrically charged side chains such as arginine and glutamic, those with polar uncharged side chains such as serine and cysteine, and those with hydrophobic side chains such as phenylalanine and leucine. Therefore, amino acids are excellent building blocks for lipid-like and surfactant-like peptides, whereby the hydrophobic amino acids act as the tail and one or two hydrophilic amino acids act as the head. These designer peptides usually have the ability to self-assemble into various supramolecular structures such as peptide β-sheet, peptide nanofibers, and peptide tube (Zhang, 2003).

Despite having different compositions and sequences, surfactant-like peptides share a common feature with the hydrophilic heads consisting of one to two charged amino acids and the hydrophobic tails consisting of four or more consecutive hydrophobic amino acids. Based on this simple rule, different types of surfactant-like peptides can be designed. They can be either negatively charged, such as acetyl-AAAAAAD (A6D), acetyl-VVVVVVD (V6D), acetyl-VVVVVVDD (V6D2), and acetyl-LLLLLLDD (L6D2), or positively charged, such as acetyl-AAAAAAK-NH2

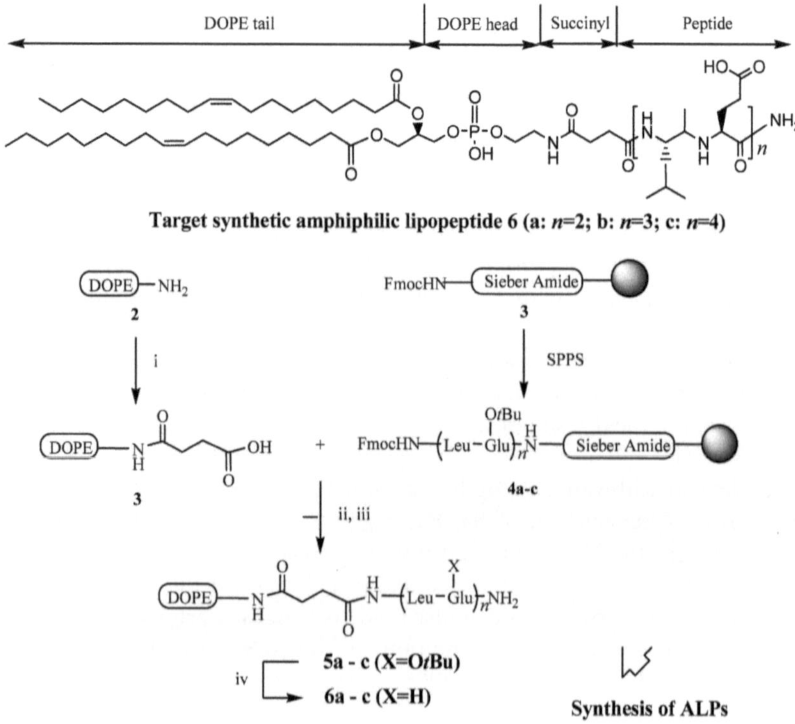

(i) Succinic anhydride, TEA in DCM. Solid phase peptide synthesis (SPPS): (a) (Fmoc removal) 20% piperidine in NMP; (b) (coupling) PyBOP, Fmoc-Glu (O*t*Bu)-OH or Fmoc-Leu-OH, DiPEA in NMP; (c) (capping) 5% Ac_2O, 5% 2, 6-lutidine in NMP; iterated *n* times (*n* = 2, **4a**; *n* = 3, **4b**; *n* = 4, **4c**). (ii) (a) (Fmoc removal) 20% piperidine in NMP; (b) (coupling) PyBOP, DiPEA in DCM. (iii) Cleavage from the resin: 1% TFA in DCM. (iv) *t*-Bu removal: 90% TFA in H2O. Compounds: **1.** DOPE, 1,2-dioleoyl-*sn*-glycero-3-phosphoethanolamine; **2.** 1,2-dioleoyl-*sn*-glycero-3-phosphoethanolamine-*N*-succinic acid; 3. acid-labile Sieber amide resin; **4.** Fmoc peptides on solid support; **5.** t-butyl protecting ALPs; **6.** target ALPs. Abbrs.: ALP, Amphiphilic lipopeptide; TEA, triethylamine; DCM, dichloromethane; Fmoc, 9-fluorenylmethyloxycarbonyl; NMP, N-Methylpyrrolidone; PyBOP, benzotriazol-1-yl-oxytripyrrolidinophosphonium hexafluorophosphate; Glu, Glutamic acid; Leu, leucine; DiPEA, diisopropylethylamine;

Fig. 2.3. Synthesis of amphiphilic lipopeptides.

(A6K), acetyl-VVVVVVK-NH2 (V6K), acetyl-VVVVVVH-NH2 (V6H), acetyl-VVVVVVKK-NH2 (V6K2), acetyl-LLLLLLKK-NH2 (L6K2), and so on (Table 2-B). These surfactant-like peptides self-assemble in water to form various structures including nanotubes and nanovesicles with an average diameter of 30–50 nm (Zhao, 2009).

As lipidated-peptides fuse easily into the cell membrane and interact with proteins and enzymes, lipopeptides often show different bioactivities. To date, there

Table 2-B. Natural Amino Acids Mostly Used for Synthesis of Lipid-like Peptides.

Property	Full name (abbrev.)	One letter code
Hydrophobic	Glycine (Gly)	G
	Alanine (Ala)	A
	Valine (Val)	V
	Leucine (Leu)	L
	Isoleucine (Ile)	I
Hydrophilic (negatively charged)	Aspartic acid (Asp)	D
	Glutamic acid (Glu)	E
Hydrophilic (positively charged)	Lysine (Lys)	K
	Histidine (His)	H
	Arginine (Arg)	R

is limited knowledge about the biological activities of lipopeptides and potential risks for cells and tissues when exposed to large dosages of lipopeptides. Therefore, industrial application of aminopeptides and lipopeptides, such as fermented biosurfactants like cyclic lipopeptide surfactin, is limited to antiadhesives, biomedical, and therapeutic applications. Applications in detergent, cosmetic, and food industries are few and remain unexplored (Nitschke & Costa, 2007).

In addition to self-assembling properties, lipopeptides possess broad antibiotic properties. To date, lipopeptides have been applied in the pharmaceutical industry as potent antifungal, antimicrobial, and antiviral agents. An example is Daptomycin (Cubicin) (Fig. 2.1), which is a lipopeptide antibiotic approved in the United States in 2003 for the treatment of skin and skin structure infections caused by Gram-positive pathogens (Baltz et al., 2005). Lipopeptide analogues are also applied as vaccines and adjuvants. Highly purified antigenic moieties offer potential advantages over traditional vaccines, because they have a high degree of safety and the capacity of eliciting highly specific immune responses. In spite of this, they tend to be poorly immunogenic in vivo and require coadministration of adjuvant to enhance their cellular immunity. Synthetic analogs of bacterial lipopeptides, which are safe and easy to handle and administer, can be used as adjuvants.

Designer lipopeptides are structurally simple and affordable to produce on a large scale. The self-assembling properties of lipopeptides into various nanosized structures such as nanofibers and nanotubes provide potential application as growth templates for a range of inorganic materials such as silver, gold, and semiconducting materials. They can also be used for the development of electrochemical biosensing platforms (Baltz et al., 2005). Lipopeptide has also been found to enhance the catalytic abilities of lipase by encapsulating the enzyme in peptide nanotubes. The enhanced catalytic ability is mainly due to the conformational change into an "active" structure following encapsulation (Scanlon & Aggeli, 2008). In addition, surfactant or lipid-like peptides can also be used as targeting delivery systems for genes, drugs, and SiRNA.

PHYTOSTEROL SURFACTANTS

Sterols are found in plants and animals. Sterols from animal origins such as cholesterol are collectively known as zoosterol. Meanwhile, sterols originating from plants are called phytosterols. Some of the common phytosterols are β-sitosterol, campesterol, stigmasterol, and ergosterol. Generally, the chemical structure of cholesterol consists of a C27-backbone of dimethylsterane and a branched aliphatic side-chain. Phytosterol differs in terms of chemical structure from cholesterol by the presence of an extra methyl or ethyl group on the cholesterol side-chain at the C-24 position (Marangoni & Poli, 2010).

Phytosterols can be found in vegetable oils, nuts, grains, and grain-derived products (Marangoni & Poli, 2010). However, only vegetable oils (including byproducts of seed oil production) and tall oil that contains up to 80% β-sitosterol are used as raw materials for large scale isolation of phytosterols (Fernandes & Cabral, 2007; Marangoni & Poli, 2010). Vegetable oil deodorizer distillates are also found to contain free sterols and phytosterol esters (Torres et al., 2009). Phytosterols have beneficial health effects including anti-cholesterol, anti-athrogenicity, anti-cancer and anti-oxidative properties (Nashed et al., 2005). Nevertheless, phytosterols are almost insoluble in water and slightly soluble in edible oils. In order to increase their solubility and to facilitate incorporation into lipid-based foods, phytosterols are often formulated in esterified forms (Fernandez & Vega-López, 2005). Recently, there is a growing interest in developing phytosterol-based surfactants that have low toxicity such as ethoxylated phytosterols, hydrophilic and lipophilic phytosterol, and phospholipids-phytosterol conjugates for pharmaceutical and cosmetics applications.

SYNTHESIS, PHYSICAL PROPERTIES AND APPLICATIONS OF PHYTOSTEROLS-BASED SURFACTANTS

The synthesis and application of ethoxylated phytosterols have been reviewed by Folmer (2003). Commercial ethoxylation of phytosterols is usually performed using base catalysts such as potassium hydroxide. Upon completing ethoxylation, the product is neutralized with acid. This reaction yields ethoxylated phytosterol with a polydisperse oxyethylene head group (Fig. 2.4). For example, commercial ethoxylated phytosterol produced by Nikko Chemicals, Japan, consists of a hydrophobic phytosterol with a hydrophilic polyoxyethylene chain of a range of 5 to 30 units. The main applications of this product are as emulsifier, emollient, dispersant, and solubilizer (Folmer & Nyden, 2008). Formulas containing sterol ethoxylates for moisturizing lotions, lipsticks, shampoos and hair rinses have been reported (Cognis, 2011). In fact, Cognis GmbH has commercialized ethoxylated phytosterol derived from rapeseed sterols and polyethylene glycol (GENEROL® R E5 and GENEROL® R

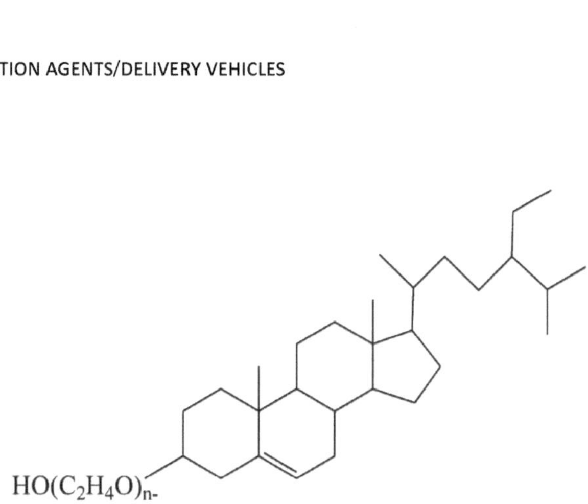

Fig. 2.4. Chemical structure of β-sitosterol ethoxylate (n =30).

E10) as coemulsifiers for the preparation of cosmetic emulsions (Product data sheet of Cognis Care Chemical, 2011).

Physical properties, particularly self-assembling properties, of ethoxylated phytosterol depend largely on the size of the oxyethylene units. At low concentration, ethoxylated phytosterol with small hydrophilic head group of 5 oxyethylene units are insoluble in water. With increasing concentration, they are able to form a lamellar liquid crystalline region. Nevertheless, as the hydrophilicity is still low, the lamellar liquid crystalline region exists as a separate phase from the bulk water phase. For ethoxylated phytosterol containing 10 to 20 oxyethylene units, a micellar solution is formed at low temperature, which eventually transforms into hexagonal liquid crystal. Hydrophilic ethoxylated phytosterol with 30 oxyethylene units also forms a micellar solution, which transforms into a micellar cubic phase before a final hexagonal phase at a higher concentration (Folmer et al., 1999). Similar findings are observed by Naito and colleagues (2005). In addition, Naito and colleagues (2005) observed that the addition of another lipophilic nonionic co-surfactant, such as polyoxyehytlene dodecylether, to the micellar solution of water and ethoxylated phytosterol (10–20 oxyethylene units) caused dilute micellar solutions to become highly viscoelastic solutions of worm-like micelle. The higher viscoelastic solutions may have useful application as surfactants and thickeners in personal care products.

Ethoxylated phytosterol (30 oxythylene units) is also found to form a lamellar structure similar to those of human intercellular stratum corneum lipids when used as co-surfactants together with polyoxypropylene dimethyl ether in an ultra fine emulsion composed of polar oil and water. Small angle scattering measurements indicated that the lamellar layers composed of the surfactants and oily ingredients stacked on the surface of the emulsified particle (Oka et al., 2008). Although the mechanisms on the effect of such structure on skin has not been clearly elucidated, there are reports on formulation containing similar structure, giving excellent improvement on dry and scaly skin (Sone et al., 1994; Kim, 2007).

Nanosized dispersion of hexosome mesophase composed of ternary mixtures of glycerol monooleate, tricaprylin, and water can be used as delivery vehicles for peptide drugs (Amar-Yuli & Garti, 2005). Libster and colleagues (2009) found that a low concentration of ethoxylated phytosterol with 30 oxyethelene units was able to stabilize the hexagonal mesophases, which co-existed with small vesicles and disordered particles. Increasing concentration of ethoxylated phytosterol led to phase transition from hexagonal mesophases to unilamellar and multilamellar phases. When evaluated as delivery vehicles for peptide drugs, these ethoxylated phytosterol-stabilized hexosome mesophases provided sustained release during the first 10 hours with drastically increased release in the 10–24 hour range. The bulky and rigid properties of ethoxylated phytosterol affected the critical packing parameters, which at low concentrations stabilize the nanosized hexagonal liquid crystals for applications as drug delivery vehicles, especially when sustained and controlled release are required. In fact, phytosterol ethoxylates (15–50 ethylene glycol units) can be used as a surface active agent for delivery of drugs with poor water solubility such as felodipine, probucol, and oxazepam (Soderlind, 2003). Despite having high capability to solubilize drugs, ethoxylated phytosterols possess very low cell damaging activity as hemolysis or cell lysis comparable to those of typical non-ionic surfactant. In addition, ethoxylated phytosterols can be used to produce an aqueous solution of the sparingly soluble substances to prevent cell damage (Soderlind, 2004).

In addition to ethoxylated phytosterols, other phytosterol-derived surfactants such as hydrophilic and lipophilic phytosterols and phospholipid-phytosterol conjugates have also been synthesized. Chung and Choi (2007) synthesized a hydrophilic derivative of β-sitosterol by esterification with poly (ethylene glycol) in a two-step reaction. The first step involved synthesis of an intermediate (carboxyethyl-β-sitosterol, CES), which gave carboxylic functionality to β-sitosterol by using succinic anhydride. The second step was the coupling of the hydrophilic PEG to CES in the presence of a basic catalyst (4-dimethylaminopyridine) and dehydrating agents (1, 3 dicyclohexylcarbodiimide). The water solubility of synthesized hydrophilic β-sitosterol significantly depends on the degree of substitution (DS, average molar number of β-sitosterol moieties per molecule of hydrophilic β-sitosterol). Hydrophilic β-sitosterol with a DS of 1.08 is soluble in water at concentrations of up to 0.15 g/ml at 25°C.

Chung and colleagues (2008) also developed a chemical synthetic route for lipophilic β-sitosterol. Newly synthesized lipophilic β-sitosterol is an oil-soluble compound that is esterified with oleic acid in the presence of a basic catalyst (4-dimethylaminopyridine) and dehydrating agents (1,3 dicyclohexylcarbodiimide) at 25°C. The product is purified by recrystallization in ethyl alcohol. Synthesized lipophilic β-sitosterol has solubility of up to 0.05 g/ml in edible oil at 25°C. The solubility of lipophilic β-sitosterol derived from oleic acid is found to have 3-fold higher solubility in edible oil than that of their stearic acid counterpart (Chung & Cho, 2006). Both hydrophilic and lipophilic β-sitosterol derivatives also demonstrated comparable effects to original β-sitosterol in lowering blood cholesterol levels in rats. Nevertheless, they possess superiority over β-sitosterol in terms of solubility (Chung et al., 2008). To date, few if any studies have been found on self-assembling properties for both hydrophilic and lipophilic derivatives of β-sitosterol.

Phospholipid-phytosterol conjugates can be synthesized through the transphosphatidylation reaction between phosphatidylcholine and β-sitosterol using PLD from *Streptomyces* sp. in a biphasic system of an organic solvent with an acetate buffer (Hossen & Hernandez, 2004). The phosphatidyl choline-β-sitosterol conjugate has higher polarity than β-sitosterol alone and can be easily separated from the bulk reaction mixture using a thin layer chromatography. This modified phytosterol is characterized as an amphiphile molecule and can be used for beverage applications.

ANTIOXIDANT ESTERS

DESIGN, SYNTHESIS, AND PURIFICATION OF ANTIOXIDANT ESTERS

Conjugation of antioxidants with lipidic moieties results in surface active lipids with a wide range of uses. To date, naturally occurring antioxidants such as polyphenols, ascorbic acid, and tocopherols have been successfully conjugated to lipidic moieties, including fatty acids, fatty alcohols, and triacylglycerides (Villeneuve, 2007; Karmee, 2009; Choo & Birch, 2009). Structural features of the substrates, namely size, presence of primary or secondary alcohol groups, and additional substituents on the antioxidants, as well as the length and degree of unsaturation of the lipidic moieties, determine the yield of the antioxidant esters.

Solubility differences between antioxidant and lipidic substrates, general low solubility, and reactivity of some substrates of interest, such as polyphenols, represent the main hurdles in synthesis of antioxidant esters. Thus, solvent selection is of the utmost importance, especially in enzymatic synthesis. While non-polar organic solvents are generally compatible with enzymes, it is well established that polar organic solvents tend to deactivate enzymes such as lipase by stripping essential water from its surface (Klibanov, 1997). Intermediate polarity solvents such as tertiary alcohols are thus often employed in order to maintain lipase activity while allowing for solubilization of both hydrophilic and hydrophobic substrates. For example, acylation of rutin with varying chain length fatty acids (C4:0–C18:3) in 2-methylbutan-2-ol resulted in 27–62% conversion after 168 hours (Viskupicova et al., 2010). Higher rutin ester yields of70–77% after 96 hours were also reported in acetone (Lue et al., 2010a). Ionic liquids have also been employed to solve the substrate solubility issue. For instance, esters of phenolic acids, flavonoids, and ascorbic acid have all been successfully produced via enzymatic reactions in ionic liquid systems (Katsoura et al., 2009; Hu et al., 2009; Chen et al., 2008; van Rantwijk & Sheldon, 2007). Nevertheless, more research is needed to address issues such as enzyme compatibility, ionic liquid toxicity, product separation, and overall economic feasibility.

Reaction systems involving chemo-enzymatic set-ups have also been used successfully for synthesis of antioxidant esters. Chemical synthesis of a chlorogenic acid methyl ester followed by lipase-catalyzed transesterification with selected fatty alcohols yielded 61–93% conversion (López-Giraldo et al., 2007). Feruloylated monoacyl- and diacylglycerols were also produced following a two-step process whereby ferulic acid was chemically esterified with glycerol and then enzymatically esterified with oleic acid (Sun et al., 2009).

Regarding purification of these compounds, currently used methods include liquid-liquid extraction techniques or chromatographic separation such as TLC and HPLC (López-Giraldo et al., 2007; Burham et al., 2009). For example, one efficient system for isolation of gram quantities of rutin ester products (>97% pure) followed a two-step solvent extraction procedure employing heptane, water, and ethyl acetate (Lue et al., 2010a). Yet in reality, most purification methods reported in the literature are often time consuming and labor intensive, therefore not suitable or realistic for large quantity production. Thus, existing methodologies need to be streamlined to make them more efficient in terms of time, cost, and waste generation. In addition, care must be taken not to subject the products to excessive heat or other conditions whereby premature oxidation or degradation of the products can occur.

PHYSICAL PROPERTIES OF THE ANTIOXIDANT ESTERS

Physiochemical properties of antioxidant-containing surface active lipids vary based on their structural characteristics. For instance, chlorogenic acid derivatives possessing longer hydrocarbon chains are more soluble in lipidic systems than derivatives with shorter chains (Sasaki et al., 2010). In addition to increased lipid solubility, chlorogenic acid derivatives with longer hydrocarbon chains (C8, C12) also exhibit increased surface activity. As such, these compounds are more likely to concentrate at the oil-water interface of emulsion droplets. Conjugation with longer chain length acyl groups also results in increasingly lipophilic rutin esters, which undoubtedly partition differently into more hydrophilic and hydrophobic media (Lue et al., 2010b; Viskupicova et al., 2010). Stearyl ferulate was previously employed in the preparation of solid lipid nanoparticles (SLNs) measuring approximately 150–200 nm in size (Trombino et al., 2009). In addition to being smaller than nanoparticles prepared from stearic acid, these stable spherical particles were capable of entrapping higher levels of drug compounds, making them very interesting candidates for topical formulations. Based on cytotoxicity studies using RAT-1 immortalized fibroplast cells, stearyl-ferulate-based SLNs may be tolerated at levels up to 5 μM.

Similarly, the addition of saturated acyl L-ascorbates such as caproyl ascorbate to linoleic acid produces emulsions with decreased droplet sizes and improved oxidative stability following encapsulation (Watanabe et al., 2002). Thus, these antioxidant-containing surface active lipids worked both as emulsifiers and as antioxidants towards the encapsulated linoleic acid. All in all, as amphiphilic compounds, acyl L-ascorbates can form supramolecular self-assembled aggregates in aqueous media whereby an outer hydrophilic surface can protect an inner lipophilic core (Karmee,

2009). Solubilization of lipophilic drugs within the core of the assembly may confer a significant level of protection. Still, more in-depth studies are needed to adequately assess and characterize the physiochemical properties of these diverse antioxidant-containing lipidic compounds.

POTENTIAL APPLICATIONS AS ENCAPSULATING AGENTS OR DELIVERY VEHICLES

To date, the application of diverse antioxidant esters has been directed mainly towards the food, pharmaceutical, and cosmetics industries. For example, flavonoid esters of n-substituted fatty acids (C6-C22) were patented for use in cosmetics, pharmaceutical formulations, and nutritional products based on their antioxidant and skin protecting properties against UV-radiation (Moussou et al., 2005). Accordingly, flavonoid esters could be directly dissolved in the oil phase of the formulation and totally or partially incorporated in liposomes or microcapsules to control the release of the active compound. Conjugation of flavonoids with lipidic compounds could also make them more suitable as encapsulating materials or delivery vehicles, as they exhibit good chemical stability.

This good chemical stability was also noted by Trombino and colleagues (2009) following the preparation of stearyl ferulate-based SLNs for encapsulation and stabilization of oxygen, heat and light sensitive β-carotene and α-tocopherol. Prepared via a microemulsion technique at around 70–75°C, stearyl ferulate-based SLNs were found to have a much better encapsulation efficiency than stearic acid-based SLNs. As delivery vehicles, stearyl-based SLNs were able to stabilize β-carotene and α-tocopherol and prevent their oxidation and degradation to a higher extent than stearic acid-based SNLs.

Finally, even more recent was the patent by Chung and colleagues (2010) detailing the use of aldehyde-conjugated flavonoid preparations to form delivery vehicles for increased dosing of flavonoids or additional bioactive agents. It is clear that the use of these types of compounds is expanding, largely based on their enhanced physicochemical properties.

CONCLUSION

Molecular self-assembly is a key concept in supramolecular chemistry since assembly of the molecules is directed through noncovalent interactions (e.g., hydrogen bonding, metal coordination, hydrophobic forces, van der Waals forces, π-π interactions, and/or electrostatic), as well as electromagnetic interactions. Common examples include the formation of micelles, vesicles, liquid crystal phases, and Langmuir monolayers by surfactant molecules. Further examples of supramolecular assemblies demonstrate that a variety of different shapes and sizes can be obtained using molecular self-assembly. There has been huge interest recently concerning the use of molecular self-assembly to form the required physical structures in food, pharmaceutical, and cosmetic applications. This includes the encapsulation technology, delivery vehicle, and nanoparticle formation.

The use of surface active lipids as encapsulating agents and delivery vehicles of various compounds has a number of advantages. For example, antioxidant esters can help protect encapsulated material through a variety of mechanisms, including acting as a physical barrier, exhibiting radical-scavenging activity, and chelating pro-oxidative metals in the system just to name a few. In contrast, problems are also encountered when these materials are used, including instability due to pro-oxidant activity, degradation by harsh conditions such as high temperatures, and the possibility of leakage of vehicle contents under certain conditions. Thus, work is still ongoing to develop novel compounds and modifying structure of existing surface active lipids for enhanced properties. However, it is clear that surface active lipids have the potential to enhance a wide range of formulations and products within the food, pharmaceutical, and cosmetic industries, and progress continues to be made.

REFERENCES

Adlercreutz, D.; Budde, H.; Wehtje, E. Synthesis of phosphatidylcholine with defined fatty acid in the sn-1 position by lipase-catalyzed esterification and transesterification reaction. Biotechnol. Bioeng. 2002, 78, 403–411.

Adlercreutz, P.; Virto, C.; Persson, M.; Vaz, S.; Adlercreutz, D.; Svensson, I.; Wehtje, E. Enzymatic conversion of polar lipids, principles, problems and solutions. J. Mol. Catal. B-Enzym. 2001, 11, 173–178.

Albericio, F. Solid-phase synthesis: a practical guide; CRC Press: London, 2000; pp. 848.

Amar-Yuli, I.; Garti, N. Transitions induced by solubilized fat into reverse hexagonal mesophases. Colloids Surf., B 2005, 43, 72–82.

Aoyama, T. Phospholipid signaling in root hair development. Plant. Cell monographs 2009, 12, 171–189.

Babicz, I.; Leite, S.G.F.; de Souza, R.O.M.A.; Antunes, O.A.C. Lipase-catalyzed diacylglycerol production under sonochemical irradiation. Ultrason. Sonochem. 2010, 17, 4–6.

Baltz, R.H.; Miao, V.; Wrigley, S.K. Natural products to drugs: daptomycin and related lipopeptide antibiotics. Nat. Prod. Rep. 2005, 22, 717–741.

Barani, H.; Montazer, M. A review on applications of liposomes in textile processing. J. Lipos. Res. 2008, 18, 249–262.

Behler, A.; Biermann, M.; Hill, K.; Raths, H-C.; Victor, M-E.S.; Uphues, G. Industrial surfactant syntheses. Reactions and syntheses in surfactant systems; CRC Press: London, 2001; pp. 1–45.

Benvegnu, T.; Sassi, J-F. Oligomannuronates from seaweeds as renewable sources for the development of green surfactants. Top. Curr. Chem. 2010, 294, 143–164.

Bergfeld, M.J.; Seifert, J. Method for the Continuous Production of Alkylglycosides. EP 0617045 (1994).

Blanzat, M.; Perez, E.; Rico-Lattes, I.; Lattes. A. Synthesis and anti-HIV activity of catanionic analogs of galactosylceramide. New J. Chem. 1999, 23, 1063–1065.

Boyd, B.J. Characterisation of drug release from cubosomes using the pressure ultra filtration method. Int. J. Pharm. 2003, 260, 239–247.

Brois, S.J.; Ryer, J.; Femino, F.; Deen, H.E.; Gutierrez A. Hydrogenated Lecithin for Friction and Flow Properites. US 5135669 (1992).
Burham, H.; Rasheed, R.A.G.A.; Noor, N.M.; Badruddin, S.; Sidek, H. Enzymatic synthesis of palm-based ascorbyl esters. J. Mol. Catal. B: Enzym. 2009, 58, 153–157.
Busch, P.; Hensen, H.; Tesmann, H. Alkyl polyglycoside–a new generation of surfactants for cosmetics. Tenside Surfactants Deterg. 1993, 30,116–121.
Castor, T.P. Phospholipid nanosomes. Curr. Drug Deliv. 2005, 2, 4, 329–340.
Cavalli, S.; Handgraaf, J.W.; Tellers, E.E. Two-dimensional ordered β-sheet lipopeptide monolayers. J. Am. Chem. Soc. 2006, 128, 13959–13966.
Chaiyaso, T.; H-kittikun, A.; Zimmermann, W. Biocatalytic acylation of carbohydrates with fatty acids from palm fatty acid distillates. J. Ind. Microbiol. Biotechnol. 2006, 33, 338–342.
Chan, W.C.; White, P.D. Fmoc solid phase peptide synthesis: a practical approach. Oxford: United Kingdom, 2004, pp. 216.
Chang, S.W.; Shaw, J.F. Biocatalysis for the production of carbohydrate esters. New Biotechnol. 2009, 26, 3, 109–116.
Chen, B.; Guo, Z.; Let, M.B.; Lue, B.-M.; Xu, X. Preparation of CLA ascorbyl ester with improved volumetric productivity by an ionic liquid-based reaction system. Org. Biomol. Chem. 2008, 6, 3196–3201.
Cheong, L.Z.; Tan, C.P.; Long, K.; Yusoff, M.S.A.; Arifin, N.; Lo, S.K.; et al. Production of a diacylglycerol-enriched palm olein using lipase-catalyzed partial hydrolysis, optimization using response surface methodology. Food Chem. 2007, 105, 1614–1622.
Cheong, L.Z.; Zhang, H.; Xu, Y.; Xu, X. Physical characterization of lard partial acylglycerols and their effects on melting and crystallization properties of blends with rapeseed oil. J. Agric. Food Chem. 2009, 57, 11, 5020–5027.
Chester, M.A. Nomenclature of glycolipids. Pure Appl. Chem. 1997, 69, 2475–2487.
Choo, W.-S.; Birch, E.J. Radical scavenging activity of lyophilized products from lipase-catalyzed transesterification of triolein with cinnamic acid and ferulic acids. Lipids 2009, 44, 145–152.
Chung, D-W.; Cho, Y.T. Study on the synthesis and characteristics of lipophilic derivatives of b-sitosterol. J. Korean Ind. Eng. Chem. 2006, 17, 375–380.
Chung, D-W.; Choi, Y.T. Synthesis and solubility of hydrophilic derivatives of β-sitosterol. J. Ind. Eng. Chem. 2007, 13, 367–372.
Chung, D-W.; Kim, W.D.; Noh, S.K.; Dong, M.S. Effects of hydrophilic and lipophilic-sitosterol derivatives on cholesterol absorption and plasma cholesterol levels in rats. J. Agric. Food Chem. 2008, 56, 6665–6670.
Chung, J.E.; Kurisawa, M.; Yang, Y.Y.; Zhuo, L. Aldehyde Conjugated Flavonoid Preparations. US 7858080 (2010).
Cognis. Product data sheet of Cognis Care Chemical. http://www.cognis.com/company/Businesses/Care+Chemicals/Product+Catalog/ProduktePDS2.htm, Jan. 2011.
Crommelin, D.J.A.; Bosand, G.W.; Storm, G. Liposomes—successful carrier systems for targeted delivery of drugs. Drug Delivery Peptides, Proteins & Liposomes, Business Briefing, Pharmatech 2003, 209–213.
Damstrup, M. L.; Jensen, T.; Sparsø, F. V.; Kiil, S.; Jensen, A. D.; Xu, X. Solvent optimization for efficient enzymatic monoacylglycerol production based on a glycerolysis reaction. J. Am. Oil Chem. Soc. 2005, 82, 559–564.
Dang, H.T.; Obiri, O.; Hayes, D.G. Feed batch addition of saccharide during saccharide-fatty acid esterification catalyzed by immobilized lipase, time course, water activity, and kinetic model. J. Am. Oil Chem. Soc. 2005, 82(7), 487–493.

Degn, P.; Zimmermann, W. Optimization of carbohydrate fatty acid ester synthesis in organic media by a lipase from Candida Antarctica. Biotechnol. Bioeng. 2001, 74(6), 483–491.

De Maria, L.; Vind, J.; Oxenbøll, K. M.; Svendsen, A.; Patkar, S. Phospholipases and their industrial applications. Appl. Microbiol. Biot. 2007, 74(2), 290–300.

Dippe, M.; Mrestani-Klaus, C.; Schierhorn A.; Ulbrich-Hofmann R. Phospholipase D-catalyzed synthesis of new phospholipids with polar head groups. Chem. Phys. Lipids 2008, 152(2), 71–77.

Doig, S.D.; Diks, R.M.M. Toolbox for modification of the lecithin head group. Eur. J. Lipid Sci. Technol. 2003, 105, 368–376.

Ericsson, B.; Eriksson, P. O.; Löfroth, J. E.; Engström, S. Cubic phases as delivery systems for peptide drugs. Polymeric drugs and drug delivery system, Dunn R.L.; Ottenbritte, R.M. American Chemical Society, Washington, DC, 1991. 251–265.

Faivre, V.; Rosilio, V. Interest of glycolipids in drug delivery, from physicochemical properties to drug targeting. Expert Opin. Drug Deliv. 2010, 7(9), 1031–1048.

Fernandes, P.; Cabral, J.M.S. Phytosterols, applications and recovery methods. Bioresour. Technol. 2007; 98, 2335–2350.

Fernandez, M.L.; Vega-López, S. Efficacy and safety of sitosterol in the management of blood cholesterol levels. Cardiovasc. Drug Rev. 2005; 23, 57–70.

Ferrer, M.; Soliveri, J.; Plou, F.J.; L'opez-Cort'es, N.; Reyes-Duarte,D.; Christensen, M.; Copa-Patino, J.L.; Ballesteros, A. Synthesis of sugar esters in solvent mixtures by lipases from *Thermomyces lanuginosus* and *Candida Antarctica* B and their antimicrobial properties. Enz. Microb. Technol. 2005, 391–398.

Fiametti, K.G.; Rovani S.; de Oliveira D.; Corazzo M.L.; Treichel H.; Oliveira J.V. Assessment of variable effects on solvent-free monoacylglycerol enzymatic production in AOT surfactant. Eur. J. Lipid Sci. Technol. 2008, 110, 510–515.

Fields, G. B.; Noble, R. L. Solid phase peptide synthesis utilizing 9-fluorenylmethoxycarbonyl amino acids. Int J Pept Protein Res. 1990, 35(3), 161–214.

Flickinger, B. D.; Matsuo, N. Nutritional Characteristics of DAG Oil. Lipids. 2003; 38, 129–132.

Folmer, B.M. Sterol surfactants, from synthesis to applications. Adv. Colloid Interface Sci. 2003; 103, 99–119.

Folmer, B.M.; Nyden, M. Structure and dynamics of micelles and cubic phase structures with ethoxylated phytosterol surfactant in water. Langmuir. 2008, 24, 6441–6446.

Folmer, B.M.; Svensson, M.; Holmberg, K.; Brown, W. The physicochemical behavior of phytosterol ethoxylates. J. Colloid Interface Sci. 1999, 213, 112–120.

Ganske, F.; Bornscheuer, U.T. Optimization of lipase-catalyzed glucose fatty acid ester synthesis in a two-phase system containing ionic liquid and t-BuOH. J. Mol. Catal. B Enzym. 2005, 36, 40–42.

Garelli-Calvet, R.; Latge, P.; Rico, I.; Lattes, A.; Puget, A. A new surfactant series, the N-alkyl-amino-1-deoxylactitols: application for extraction of 'op' opiate receptors from frog brain. Biochim. Biophys. Acta. 1992, 1109, 55–58.

Gunther, B-R. Process for the Separation of Acylated Phospholipids from Phosphatidylcholine Products Containing the Same. US 4443378 (1984).

Guo, Z.; Xu, X. New opportunity for enzymatic modification of fats and oils with industrial potentials. Org. Biomol. Chem. 2005, 3, 2615–2619.

Guo, Z.; Vikbjerg, A. F.; Xu, X. Enzymatic modification of phospholipids for functional applications and human nutrition. Biotechnol. Adv. 2005, 23, 203–259.

Gustafsson, J.; Ljusberg-Wahren, H.; Almgren, M.; Larsson, K. Cubic lipid–water phase dispersed into submicron particles. Langmuir. 1996, 12, 20, 4611–4613.

Gustafsson, J.; Ljusberg-Wahren, H.; Almgren, M.; Larsson, K. Submicron particles of reversed lipid phases in water stabilized by a nonionic amphiphilic polymer. Langmuir, 1997, 13, 6964–6971.

Ha, S.H.; Hiep, N.M.; Koo Y.M. Enhanced production of fructose palmitate by lipase-catalyzed esterification in ionic liquids. Biotechnol. Bioprocess Eng. 2010, 15, 126–130.

Hawthorne, J.N.; Ansell, G.B. Phospholipids, New comprehensive biochemistry. Hawthorne, J.N., Ed.; Elsevier Biomedical Press: Amsterdam, 1982.

Henna Lu, F.S.; Nielsen, N.S.; Timm-Heinrich, M.; Jacobsen, C. Oxidative stability of marine phospholipids in the liposomal form and their applications. Lipids. 2011, 46, 3–23.

Hill, K.; Wuest, W.; Wollmann, J.; Biermann, M.; Rossmaier, H.; Eskuchen, R.; Bruuns, A.; Hellmann, G.; Ott, K.H.; Winkle, W.; Wollman, K. Differences Between Alkyl Polyglucosides of Natural Alcohol and Oxo-alcohol. EP 0437460 (1988).

Hillaireau, H.; Couvreur, P. Nanocarriers' entry into the cell: relevance to drug delivery. Cell Mol. Life Sci. 2009, 66, 2873–2896.

Hirche, F.; Ulbrich-Hofmann, R. The interdependence of solvent, acceptor alcohol, and enzyme source in transphosphatidylation by phospholipase D. Biocatal. Biotransfor. 2000, 18, 343–353.

Holmberg, K.; Janssen, B.; Kronberg, B.; Lindman, B. Surfactants and polymers in aqueous solution. John Wiley & Sons, Ltd.: West Sussex, England, 2002.

Hong, J.S.; Stavis, S.M.; Lacerda, S.H.D.; Locascio, L.E.; Raghavan, S.R.; Gaitan, M. Microfluidic directed self-assembly of liposome-hydrogel hybrid nanoparticles. Langmuir. 2010, 26, 13, 11581–11588.

Hosokawa, M.; Takahashi, K.; Miyazaki, N.; Okamura, K.; Hatano, M. Application of water mimics on preparation of eicosapentaenoic and docosahexaenoic acid containing glycerolipids. J. Am. Oil Chem. Soc. 1995, 72, 421–425.

Hossen, M.; Hernandez, E. Phospholipase D-catalyzed synthesis of novel phospholipid-phytosterol conjugates. Lipids 2004, 39, 777–782.

Hu, Y.; Guo, Z.; Lue, B.-M.; Xu, X. Enzymatic synthesis of esculin ester in ionic liquids buffered with organic solvents. J. Agric. Food Chem. 2009. 57, 3845–3852.

Iwasaki, Y.; Mizumoto, Y.; Okada, T.; Yamamoto, T.; Tsutsumi, K.; Yamane, T. An aqueous suspension system for phospholipase D - mediated synthesis of PS without toxic organic solvent. J. Am. Oil Chem. Soc. 2003, 80, 653–657.

Jacobs, L.; Lee, I.; Poppe, G. Chemical Process for the Production of 1,3-diglyceride Oils. US20030104109 (2003).

Jadhav, S.R.; Vemula, P.K.; Kumar, R.; Raghavan S.R.; John, G. Sugar-derived phase-selective molecular gelators as model solidifiers for oil spills. Angew. Chem. Int. Ed. 2010, 49, 7695–7698.

John, G.; Masuda, M.; Okada, Y.; Yase, K.; Shimizu, T. Nanotube formation from renewable resources via coiled nanofibers. Adv. Mater. 2001, 13, 10, 715–718.

John, G.; Mason, M.; Ajayan, P.M.; Dordick, J.S. Lipid-based nanotubes as functional architectures with embedded fluorescence and recognition capabilitites. J. Am. Chem. Soc. 2004, 126, 15012–15013.

John, G.; Vemula, P.K. Design and development of soft nanomaterials from biobased amphiphiles. Soft Matter. 2006a, 2, 909–914.

John, G.; Zhu, G.; Li, J.; Dordick, J.S. Enzymatically derived sugar-containing self-assembled organogels with nanostructured morphologies. Angew. Chem. Int. Ed. 2006b, 45, 4772–4775.

Jung, J.H.; John, G.; Yoshida, K.; Shimizu, T. Self-assembling structures of long-chain phenyl glucoside influenced by the introduction of double bonds. J. Am. Chem. Soc. 2002, 124, 10674–10675.

Kaasgaard, T.; Drummond, C.J. Ordered 2-D and 3-D nanostructured amphiphilic self-assembly materials stable in excess solvent. Phys. Chem. Chem. Phys. 2006, 8, 4957–4975.

Kadereit, D.; Waldmann, H. Chemoenzymatic Synthesis of Lipidated Peptides. Monatshefte für Chemie. 2000, 131, 571–584.

Kahveci, D.; Guo, Z.; Ozcelik, B; Xu, X. Optimization of enzymatic synthesis of diacylglycerols in binary medium systems containing ionic liquids. Food Chem. 2009, 119, 880–885.

Karmee, S.K. Biocatalytic synthesis of ascorbyl esters and their biotechnological applications. Appl. Microbiol. Biotechnol. 2009. 81, 1013–1022.

Katsoura, M.H.; Polydera, A.C.; Tsironis, L.D.; Petraki, M.P.; Rajačić, S.K.; Tselepis, A.D.; Stamatis, H. Efficient enzymatic preparation of hydroxycinnamates in ionic liquids enhances their antioxidant effect on lipoproteins oxidative modification. New Biotechnol. 2009, 26, 83–91.

Kawai, Y.; Nakagawa, Y.; Matuyama, T.; Akagawa, K.; Itagawa, K.; Fukase, K.; Kusumoto, S.; Nishijima, M.; Yano, I. A typical bacterial ornithine-containing lipid Nalpha-(D)-[3-(hexadecanoyloxy) hexadecanoyl]-ornithine is a strong stimulant for macrophages and a useful adjuvant. FEMS Immunol Med Microbiol. 1999, 23, 1, 67–73.

Klibanov, A.M. Why are enzymes less active in organic solvents than water? Trends Biotechnol. 1997, 15, 97–101.

Kim, I. A new mimic structure of intercellular lipids based on O/LC/W emulsions for highly strengthening skin barrier function. Fragrance J. 2007, 35, 84–86.

Ko, Y.T.; Kale, A.; Hartner,W.C. Self-assembling micelle-like nanoparticles based on phospholipid-polyethyleneimine conjugates for systemic gene delivery. J. Control Release. 2009, 133, 132–138.

Kölsch, V.; Charest, P.G.; Firtel, R.A. The regulation of cell motility and chemotaxis by phospholipid signaling. J. Cell Sci. 2008, 121, pt. 5, 551–559.

Kono, K.; Takagishi, T. Temperature-sensitive liposomes. Methods Enzymol. 2004, 387, 73–82.

Koglin, A.; Doetsch, V; Bernhard, F. Molecular engineering aspects for the production of new and modified biosurfactants. Adv. Exp. Med. Biol. 2010, 672, 158–169.

Kralova, I.; Sjoblom, J. Surfactants used in food industry: a review. J. Disper. Sci. Technol. 2009, 30, 1363–1383.

Kristensen, J.B.; Xu, X.; Mu, H. Diacylglycerol synthesis by enzymatic glycerolysis, screening of commercially available lipases. J. Am. Oil Chem. Soc. 2005, 82, 329–334.

Kulkarni, C.V.; Tang, T.Y.; Seddon, A.M.; Seddon, J.M.; Ces, O.; Templer, R.H. Engineering bicontinuous cubic structures at the nanoscale—the role of chain splay. Soft Matter. 2010, 6, 3191–3194.

Kumar, R.; Katare, O. Lecithin organogels as a potential phospholipid-structured system for topical drug delivery: a review. AAPS Pharm. Sci. Tech. 2005, 6, 298–310.

Lai, O.M; Lo, S.K.; Akoh, C.C. Patent review on lipid technology. Healthful lipid, Lai, O.M., and Akoh, C.C., Eds.; AOCS Press: Champaign, Illinois, 2004; pp. 433–507.

Lasic, D.D.; Vallner, J.J.; Working, P.K. Sterically stabilized liposomes in cancer therapy and gene delivery. Curr. Opin. Mol. Ther. 1999, 1, 2, 177–185.

Lee, S.H.; Dang, D.T.; Ha, S.H.; Chang, W.J.; Koo, Y.M. Lipase-catalyzed synthesis of fatty acid sugar ester using extremely supersaturated sugar solution in ionic liquid. Biotechnol. Bioeng. 2008a, 99, 1, 1–8.

Lee, S.H.; Ha, S.H.; Hiep, N.M.; Chang, W.C.; Koo, Y.M. Lipase-catalyzed synthesis of glucose fatty acid ester using ionic liquids mixtures. J. Biotechnol. 2008b, 133, 486–489.

Lee, H-Y; Diehn, K.K.; Ko, S.W.; Tung, S-H; Raghavan S.R. Can simple salts influence self-assembly in oil? Multivalent cations as efficient gelators of lecithin organosols. Langmuir. 2010, 26, 17, 13831–13838.

Lemmon, M.A. Membrane recognition by phospholipid-binding domains. Nat. Rev. Mol. Cell Biol. 2008, 9, 99–111.

Leser, M.E.; Sagalowicz, L.; Michel, M.; Watzke, H.J. Self-assembly of polar food lipids. Adv. Colloid Interface Sci. 2006, 125–136.

Libster, D.; Aserin, A.; Yariv, D.; Shonham, G.; Garti, N. Soft matter dispersion with ordered inner structure, stabilized by ethoxylated phytosterols. Colloids Surf. B: Biointerfaces. 2009, 74, 202–215.

Liu, Q.; Janssen, M.H.A.; van Rantwijk, F.; Sheldon, R.A. Room-temperature ionic liquids that dissolve carbohydrates in high concentration. Green. Chem. 2005, 7, 39–42.

Lo, S.K.; Cheong, L.Z.; Arifin, N.; Tan, C.P.; Long, K.; Yusoff, M.S.A.; et al. Diacylglycerol and triacylglycerol as responses in a dual response surface-optimized process for diacylglycerol production by lipase-catalyzed esterification in a pilot packed-bed enzyme reactor. J. Agr. Food Chem. 2007, 55, 5595–5603.

Lo, S.K.; Tan, C.P.; Long, K.; Yusoff, M.S.A.; Lai, O.M. Diacylglycerol oil—properties, processes and products: a review. Food Bioprocess Technol. 2008, 1, 223–233.

López-Giraldo, L.J.; Laguerre, M.; Lecomte, J.; Figueroa-Espinoza, M.C.; Barouh, N.; Baréa, B.; Villeneuve, P. Lipase-catalyzed synthesis of chlorogenate fatty esters in solvent-free medium. Enzyme Microb. Technol. 2007. 41, 721–726.

Lue, B.-M.; Guo, Z.; Glasius, M.; Xu, X. Scalable preparation of high purity rutin fatty acid esters following enzymatic synthesis. J. Am. Oil Chem. Soc. 2010a, 87, 1, 55–61.

Lue,.B.-M.; Nielsen, N.S.; Jacobsen, C.; Hellgren, L.; Guo, Z.; Xu, X. Antioxidant properties of modified rutin esters by DPPH, reducing power, iron chelation and human low density lipoprotein assays. Food Chem. 2010b, 123, 221–230.

Macierzanka, A.; Szelag, H.; Szumala, P.; Pawlowicz, R.; Mackie, A.R.; Ridout, M.J. Effect of crystalline emulsifier composition on structural transformations of water-in-oil emulsions, emulsification, and quiescent conditions. Colloid Surface A. 2009, 334, 40–52.

Malheiros, P.S.; Daroit, D.J.; Brandelli, A. Food applications of liposome-encapsulated antimicrobial peptides. Trends Food Sci. Technol. 2010, 21, 284–292.

Marangoni, F.; Poli, A. Phytosterols and cardiovascular health. Pharmacol. Res. 2010, 61, 193–199.

Marahiel, M.; Denders, W.; Krause, M.; Kleinkauf, H. Biological role of gramicidin-S in spore functions. Studies on gramicidin-S-negative mutants of Bacillus brevis ATCC 9999. Eur. J. Biochem. 1977, 99, 49–52.

Metselaar, J.M.; Bruin, P.; de Boer, L.W.; de Vringer, T.; Snel, C.; Oussoren, C.; Wauben, M.H.M.; Crommelin, D.J.A.; Storm, G.; Hennink, W.R. A novel family of L-amino acid-based biodegradable polymer-lipid conjugates for the development of long-circulating liposomes with effective drug-targeting capacity, Bioconjug Chem. 2003, 14, 1156–1164.

Mezzenga, R.; Grigorov, M.; Zhang, Z.D.; Servais, C.; Sagalowicz, L.; Romoscanu, A.; et al. Polysaccharide-induced order-to-order transitions in lyotropic liquid crystals. Langmuir. 2005, 21,14, 6165–6169.

Milman, G.; Maor, Y.; Abu-Lafi, S.; Harowitz, M.; Gallily, S.; Mo, F.M.; Offertaler, L.; Pacher, P.; Kunos, G.; Mechoulam, R. N-arachidonoyl L-serine, an endocannabinoid-like brain constituent with vasodilatory properties. Proc. Natl. Acad. Sci. USA 2006, 103, 7, 2428–2433.

Minoru, K.; Keiji, S.; Toshiteru, K. Process for Producing Fat or Oil. EP1717318 (2006).

Mizoue, T.; Horibe, T.; Maruyama, K.; Takizawa, T.; Iwatsuru, M.; Kono, K.; Yanagie, H.; Moriyasu, F. Targetability and intracellular delivery of anti-BCG antibody-modified, pH-sensitive fusogenic immunoliposomes to tumor cells. Int. J. Pharm. 2002, 237, 129–137.

Moussou, P.; Falcimaigne, A.; Ghoul, M., Danous, L.; Pauly, G. Esters of Flavonoids with Substituted C6-C22 Fatty Acids. WO 2005/000831 A1 (2005).

Naito, N.; Acharya, D.P.; Tanimura J.; Kunieda, H. Phase behavior of polyoxyethylene phytosteryl ether/polyoxyethylene dodecyl ether/water systems, J. Oleo Sci. 2005, 54, 7–13.

Nakajima, Y. Water-retaining ability of diacylglycerol. J. Am. Oil Chem. Soc. 2004, 81, 907–912.

Nakajima, Y.; Fukasawa, J.; Shimada, A. Physicochemical properties of diacylglycerols. Diacylglycerol Oil; Katsuragi, Y., Yasukawa, T., Matsuo, N., Flickinger, B.D., Tokimitsu, I., and Matlock, M.G. AOCS Press: Champaign, Illinois, 2004; pp. 182–196.

Nakano, M.; Teshigawara, T.; Sugita, A.; Leesajakul, W.; Taniguchi, A.; Kamo, T. et al. Dispersions of liquid crystalline phases of the monoolein/oleic acid/pluronic F127 system. Langmuir. 2002, 18, 9283–9288.

Nashed, B.; Yeganeh, B.; Hayglass, K.T.; Moghadasian, M.H. Antiatherogenic effects of dietary plant sterols are associated with inhibition of proinflammatory cytokine production in Apo E-KO mice. J. Nutr. 2005, 135, 2438–2444.

Navarro, G.; Sawant, R.R.; Essex, S.; ILarduya, C.T.; Torchilin, V.P. Phospholipid-polyethylenimine conjugate-based micelle-like nanoparticles for siRNA delivery. Drug Deliv. Transl. Res. 2011, 1, 25–33.

Nickel, D.; Nitsch, C.; Kurzendofer, P.; von Rybinski, W. Interfacial properties of surfactant mixtures with alkyl polyglycosides. Progr. Colloid Polym. Sci. 1992, 89, 249–252.

Nitschke, M.; Costa S.G.V.A.O. Biosurfactants in food industry. Trends Food Sci. Technol. 2007, 18, 252–259.

Oka, T.; Miyahara, R.; Teshigawara, T.; Watanabe, K. Development of novel cosmetic base using sterol surfactant. I. preparation of novel emulsified particles with sterol surfactant. J. Oleo Sci. 2008, 57, 567–575.

Park, C.W.; Kwon, S.J.; Han, J.J.; Rhee, J.S. Transesterification of phosphatidylcholine with eicosapentaenoic acid ether ester using phospholipase A2 in organic solvent. Biotechnol. Lett. 2000, 22, 147–150.

Patino, J.M.R.; Dominguez, M.R. Interfacial characteristics of diglyceride monolayers at the air/aqueous phase interface. Colloid Surface A 2000, 168, 35–43.

Pedersen, K.B. Interesterification of Phospholipids. US 6284501 (2001).

Peng, L.; Xu, X.; Mu, H.; Høy, C-E; Adler-Nissen J. Production of structured phospholipids by lipase-catalyzed acidolysis, optimization using response surface methodology. Enzyme Microb. Tech. 2002, 31, 4, 523–532.

Pyo, S.H.; Hayes, D.G. Desorption of fructose from a packed column to an oleic acid/fructose oleate mixture for employment in a bioreactor system. J. Am. Oil Chem. Soc. 2008, 85, 1033–1040.

Pyo, S.H.; Hayes, D.G. Designs of bioreactor systems for solvent-free lipase catalyzed synthesis of fructose-oleic acid esters. J. Am. Oil Chem. Soc. 2009, 86, 521–529.

Resh, M.D. Fatty acylation of proteins: new insights into membrane targeting of myristoylated and palmitoylated proteins, Biochim Biophysic Acta Mol. Cell Res.1999, 1451, 1, 1–16.

Ross, S.A.; Weete J.D.; Schinazi, R.F.; Wirtz, S.S.; Tharnish, P.; Scheuer, P.J.; Hamann, M.T. Mololipids, a new series of anti-HIV bromotyramine-derived compounds from a sponge of the order verongida. J. Nat. Prod. 2000, 63, 4, 501–503.

Roussel, M.; Benvegnu, T.; Lognone, V.; Deit, H.L.; Soutrel, I.; Laurent, I.; Plusquellec, D. Synthesis and physic-chemical properties of novel biocompatible alkyl D-mannopyranosiduronate surfactants derived from alginate. Eur. J. Org. Chem. 2005, 3085–3094.

Roussel, M.; Lognone, V.; Plusquellec, D.; Benvegnu, T. Monolayer lipid membrane-forming dissymmetrical bolaamphiphiles derived from alginate oligosaccharides. Chem. Commun. 2006, 3622–3624.

Rybinski, W.V.; Hill, K. Alkyl polyglycosides—properties and applications of a new class of surfactants. Angew. Chem. Int. Ed. 1998, 37, 1328–1345.

Sadhale, Y.; Shah, J.C. Glyceryl monooleate cubic phasegel as chemical stability enhancer of cefazolin and cefuroxime. Pharm. Dev. Technol. 1998, 3, 4, 549–556.

Sagalowicz, L.; Leser, M.E.; Watzke, H.J.; Michel, M. Monoglyceride self-assembly structures as delivery vehicles. Trends Food Sci. Tech. 2006, 17, 204–214.

Sagar, G.H.; Arunagirinathan M.A.; Bellare, J.R. Self-assembled surfactant nano-structures important in dmg delivery: a review. Indian J. Exp. Biol. 2007, 45, 2, 133–159.

Sasaki, K.; Alamed, J.; Weiss, J.; Villeneuve, P.; López Giraldo, L.J.; Lecomte, J.; Figueroa-Espinoza, M.-C.; Decker, E.A. Relationship between the physical properties of chlorogenic acid esters and their ability to inhibit lipid oxidation in oil-in-water emulsions. Food Chem. 2010, 118, 830–835.

Scanlon, S.; Aggeli, A. Self-assembling peptide nanotubes. Nano Today. 2008, 3, 3–4, 22–30.

Schmidt, D.; Jiang, Q-X; MacKinnon, R. Phospholipids and the origin of cationic gating charges in voltage sensors. Nature, 2006, 444, 775–779.

Schneider, M. Phospholipids. Lipid technologies and applications. Gunstone, F.D., and Padley, F.B. CRC Press: London, 1997; p. 62.

Schneider, M. Phospholipids for functional food. Eur. J. Lipid Sci. Technol. 2001, 103, 98–101.

Senanayak, S.P.N.J.; Shahidi, F. Enzyme-catalyzed synthesis of structured lipids via acidolysis of seal (Phoca groenlandica) blubber oil with capric acid. Food Res. Int. 2002, 35, 745–752.

Shah, J.C.; Sadhale, Y.; Dakshina, M.C. Cubic phase gels as drug delivery systems. Adv. Drug. Deliver. Rev. 2001, 47, 229–250.

Shimada, A.; Ohashi, K. Interfacial and emulsifying properties of diacylglycerol. Food Sci. Technol. Res. 2003, 9, 142–147.

Siew, W.L. Understanding the interactions of diacylglycerols with oils for better product performance. Palm Oil Dev. 2002, 6–12.

Sinram, R.D. The added value of specialty lecithins. Oil Mill Gazetteer, 1991, 22–26.

Soderlind, E. New Use of Ethoxylated Phytosterols and Phytostanols. WO 03/039511 A1 (2003).

Soderlind, E. Use of Ethoxylated Phytosterols and Phytostanols. US 20040254129A1 (2004).

Sone, T.; Hanamizu, T.; Yoshikawa, S.; Ichikawa, M.; Yokokura, T. Moisturizing effect of lipid forming lamellar structures. J. Jpn. Cosmet. Sci. Soc. 1994, 31, 34–39.

Strijowski, U.; Heinz, V.; Franke, K. Removal of 3-MCPD esters and related substances after refining by absorbent material. Eur. J. Lipid Sci. Technol. 2011, 113, 387–392.

Sun, S.; Yang, G.; Bi, Y.; Xiao, F. Chemoenzymatic synthesis of feruloylated monoacyl- and diacylglycerols in ionic liquids. Biotechnol. Let. 2009. 31, 1885-1889.

Svensson, I.; Adlercreutz, P.; Mattiasson, B. Lipase-catalyzed acidolysis of phosphatidylcholine at controlled water activity. J. Am. Oil Chem. Soc. 1992, 69, 986–991.

Torres, C.F.; Fornari, T.; Torrelo, G.; Señoráns, F.J.; Reglero, G. Production of phytosterol esters from soybean oil deodorizer distillates. Eur. J. Lipid Sci. Technol. 2009, 111, 459–463.

Tran, M.A.; Watts, R.J.; Robertson G.P. Use of liposomes as drug delivery vehicles for treatment of melanoma. Pigment Cell Melanoma Res. 2009, 22, 388–399.

Trombino, S.; Cassano, R.; Muzzalupo, R.; Pingitore, A.; Cione, E.; Picci, N. Stearyl ferulate-based solid lipid nanoparticles for the encapsulation and stabilization of beta-carotene and alpha-tocopherol. Colloids Surf. B 2009. 72, 181–187.

Trotta, M.; Pattarino, F.; Ignonia, T. Stability of drug-carrier emulsions containing phosphatidylcholine mixtures. Eur. J. Pharm. Biopharm. 2002, 53, 2, 203–208.

Uesugi, Y.; Hatanaka, T. Phospholipase D mechanism using Streptomyces PLD. Biochim. Biophysic. Acta Mol. Cell Biol. Lipids 2009, 1791, 9, 962–969.

Ulbrich-Hofmann, R. Enzyme-catalyzed transphosphatidylation. Eur. J. Lipid Sci. Technol. 2003, 105, 305–308.

Ulbrich-Hofmann, R.; Lerchner, A.; Oblozinsky, M.; Bezakova, L. Phospholipase D and its application in biocatalysis. Biotechnol. Let. 2005, 27, 8, 535–544.

U.S. Department of Health and Human Services. Guidance for Industry Q3C Tables and Lists. U.S. Food and Drugs Administration, 2003.

van Rantvijk, F.; Secundon, F.; Sheldon, R.A. Structure and activity of Candida Antarctica lipase B in ionic liquids. Green. Chem. 2006, 8, 282–286.

van Rantwijk, F.; Sheldon, R.A. Biocatalysis in ionic liquids. Chem. Rev. 2007. 107, 2757–2785.

Vandana, V.; Karuna, M.S.L.; Prasad P.S.S.; Prasad, R.B.N. Process for the Preparation of Hydroxylated Lecithin from Crude Soybean Lecithin. US 6638544 (2003).

Vauthey, S.; Milo, C.; Frossard, P.;Garti, N.; Leser, M.E.; Watzke, H.J. Structured fluids as microreactors for flavor formation by the Maillard reaction. J. Agr. Food Chem. 2000, 48, 10, 4808–4816.

Vemula, P.K.; Li, J.; John, G. Enzyme catalysis: tool to make and break amygdalin hydrogelators from renewable resources, a delivery model for hydrophobic drugs. J. Am. Chem. Soc. 2006, 128, 8932–8938.

Vernooij, E.A.A.M.; Kettenes-van den Bosch, J.J.; Underberg, W.J.M.; Crommelin D.J.A. Chemical hydrolysis of DOTAP and DOPE in a liposomal environment. J. Control. Release 2002, 79, 299–303.

Vikbjerg, A.F.; Mu, H.; Xu X. Lipase-catalyzed acyl exchange of soybean phosphatidylcholine in n-hexane: a critical evaluation of both acyl incorporation and product recovery. Biotechnol. Progr. 2005, 21, 2, 397–404.

Vikbjerg, A.F.; Jonsson, G.; Mu, H.; Xu, X. Application of ultra filtration membranes for purification of structured phospholipids produced by lipase-catalyzed acidolysis. Sep. Purif. Technol. 2006a, 50, 2, 184–191.

Vikbjerg, A.F.; Rusig, J-Y; Jonsson, G.; Mu, H.; Xu. X. Strategies for lipase-catalyzed production and the purification of structured phospholipids. Eur. J. Lipid Sci. Technol. 2006b, 108, 802–811.

Vikbjerg, A.F.; Rusig, J.Y.; Jonsson, G.; Mu, H.; Xu, X. Comparative evaluation of the emulsifying properties of phosphatidylcholine after enzymatic acyl modification. J. Agric. Food Chem. 2006c, 54, 9, 3310–3316.

Villeneuve, P. Lipases in lipophilization reactions. Biotech. Adv. 2007, 25, 515–536.

Viskupicova, J.; Danihelova, M.; Ondrejovic, M.; Liptaj, T.; Sturdik, E. Lipophilic rutin derivatives for antioxidant protection of oil-based foods. Food Chem. 2010, 123, 45–50.

Wang, Y.; Wang, R.; Lu, X.; Lu, W.; Zhang, C.; Liang, W. Pegylated phospholipids-based self-assembly with water-soluble drugs. Pharmaceut. Res. 2010, 27, 361–370.

Watanabe, Y., Fang, X., Minemoto, Y., Adachi, S., Matsuno, R. Suppressive effect of saturated acyl L-ascorbate on the oxidation of linoleic acid encapsulated with maltodextrin or gum arabic by spray-drying. J. Agric. Food Chem. 2002, 50, 3984–3987.

Weber, N.; Mukherjee, K.D. Solvent-free lipase-catalyzed preparation of diacylglycerols. J. Agr. Food Chem. 2004, 52, 5347–5353.

Xu, X.; Jacobsen, C.; Nielsen, N.S.; Heinrich, M.T.; Zhou, D.Q. Purification and deodorization of structured lipids by short path distillation. Eur. J. Lipid Sci. Technol. 2002, 104, 745–755.

Yaghmur, A.; de Campo, L.; Sagalowicz, L.; Leser, M.E.; Glatter, O. Emulsified microemulsions and oil-containing liquid crystalline phases. Langmuir. 2005, 21, 569–577.

Yamamoto, Y.; Hosokawa, M.; Kurihara, H.; Miyashita, K. Preparation of phosphatidylated terpenes via phospholipase D-mediated transphosphatidylation. J. Am. Oil Chem. Soc. 2008, 85, 313–320.

Yamada, Y.; Shimizu, M.; Suguira, M.; Yamada, N. Process for Producing Diglycerides. US6261812 (2001).

Yan, Y.; Bornscheuer, U.T.; Stadler, G.; Lutz-Wahl, S.; Otto, R.T.; Reuss, M.; Schmid, R.D. Regioselective lipase-catalyzed synthesis of glucose ester on a preparative scale. Eur. J. Lipid Sci. Technol. 2001, 103, 583–587.

Yang, T.; Zhang, H.; Mu, H.; Sinclair, A.J.; Xu, X. Diacylglycerols from butterfat: production by glycerolysis and short-path destillation and analysis of physical properties. J. Am. Oil Chem. Soc. 2004, 81, 979–987.

Yang, T.; Rebsdorf, M.; Engelrud, U.; Xu, X. Enzymatic production of monoacylglycerols containing polyunsaturated fatty acids through an efficient glycerolysis system. J. Agr. Food Chem. 2005, 53, 1475–1481.

Ye, R.; Pyo, S.H.; Hayes, D.G. Lipase-catalyzed synthesis of saccharide-fatty acid esters using suspensions of saccharide crystals in solvent-free media. J. Am. Oil Chem. Soc. 2010, 87, 281–293.

Youan, B.C.; Hussain, A.; Nguyen N.T. Evaluation of sucrose esters as alternative surfactants in micro encapsulation of proteins by the solvent evaporation method. Pharm. Sci. 2003, 5, 2, 1–9.

Zhao, X. Design of self-assembling surfactant-like peptides and their applications. Curr. Op. Colloid Interface Sci. 2009, 14, 340–348.

Zhang, S. Building from the bottom up. Materials Today, 2003, 6, 5, 20–27.

Zuidam, N.J.; Crommelin, J.A. Chemical hydrolysis of phospholipids. J. Pharm. Sci. 1995, 84, 1113–1119.

3

MILK PHOSPHOLIPIDS: A NANOCARRIER SYSTEM FOR DELIVERY OF BIOACTIVE COMPOUNDS

Bita Farhang and Milena Corredig
Department of Food Science, University of Guelph, Guelph, Ontario, N1G 2W1, Canada

INTRODUCTION

Development of new functional foods requires technologies for incorporating health-promoting compounds into matrices without reducing their bio-functionality, protecting them from degradation, and maintaining their bioavailability. The production of nanoliposomes is considered to be an effective technology for the encapsulation and controlled release of nutraceuticals and bioactive compounds, as well as for enhancing their stability and bioavailability. However, application of nanoliposome technology in foods has yet to be fully exploited. Liposomes are often prepared with methodologies requiring the use of non-food-grade materials, such as the drying and reconstituting method. On the other hand, microfluidics and extrusion can be employed to prepare food grade unilamellar phospholipid vesicles of about 100 nm in diameter, also called nanoliposomes.

Although liposomes are generally prepared with phospholipids from soy or egg, in recent years there has been a growing interest in the health benefits and functional properties of milk fat globule membrane (MFGM)-derived phospholipid extracts as these ingredients have become commercially available. The unique composition of the MFGM-derived phospholipid fractions and the low cost of some of the sources, such as buttermilk, may offer some advantages in the manufacture of nanoliposomes for the entrapment of bioactive compounds in food systems.

In this chapter, the preparation, structure, and physiochemical properties of nanoliposomes prepared with phospholipids from milk are described, as are their potential function as delivery systems for bioactive compounds.

OVERVIEW OF ENCAPSULATION OF BIOACTIVE COMPOUNDS

The obesity epidemic, the increased incidence of cardiovascular disease, diabetes, and other chronic conditions are increasing consumer awareness of how dietary and lifestyle intervention can affect quality of life. For this reason, food products are no longer just a source of nutrition, but they are becoming a platform for delivery of functions related to health and well-being. Food scientists are then challenged to develop the next generation of foods, foods containing bioactive compounds tailored to particular biological functions in our bodies. Providing new platforms for functional foods holds many challenges. For example, high levels of plant polyphenols would affect the taste as well as the structure and texture of a milk-based drink; vitamins and polyunsaturated fatty acids would oxidize during storage; and nutritionally significant levels of certain minerals, fiber, or astringent molecules in a drink would cause major problems in stability, texture, taste, and mouthfeel. These examples illustrate why an increased level of sophistication in product design is needed—to make sure that the next generation of processed foods can really deliver the promised functionality.

Optimal delivery systems in foods need not only to be able to protect the active compounds against degradation and prevent interactions with the food components during processing and storage but also need to control the release rate of the incorporated compounds (Kirby et al., 1991; Liu & Park, 2009; Mozafari et al., 2008). Moreover, the delivery matrix should be compatible with the environment in the gut and should not only protect but optimize the bioavailability of the encapsulated compound.

Liposomes are lipid vesicles containing an aqueous core and a bilayer membrane mostly composed of amphiphilic molecules, phospholipids. These liquid crystal vesicles are formed spontaneously in the presence of water, and they can entrap solutes. Because of the lipid bilayer, liposomes have the unique property, compared to other delivery systems such as emulsion droplets for example, to carry and compartmentalize both water and lipid soluble bioactives within a single vesicle (Brgles et al., 2008; Lasic, 1993). Phospholipids that can form liposome structures include, but are not limited to, phosphatidylcholine, sphingomyelin, phosphatidylserine, and phophatidylinositol. Depending on the amount of charged lipid present in the bilayer, the space between lamellae will change, affecting the physical properties of the vesicles (Tyrrell et al., 1976). The degree of saturation and the length of the fatty acid chains present on the lipid molecules also change the physical properties of the bilayer membrane, namely the curvature, stability, and permeability (Tyrell et al., 1976). Changes in permeability of the membrane can also be obtained by the addition of cholesterol molecules, as the presence of this molecule affects the fluidity of the bilayer (Demel & Kruyff, 1976). As liposomes have a similar composition to that of cell membranes, they are well compatible with tissues, and they have been widely studied for intracellular drug uptake (Frkanec et al., 2003; Gregoriadis et al., 1971; Rahman et al., 1974; Takeuchi et al., 2005; Zhang et al., 2005).

It has been shown that liposomes are rapidly removed from circulation in a rat model, and they are trapped mainly in the liver (Gregoriadis & Ryman, 1972). The physical properties, such as particle size and charge, affect the association with the cells and the rate of clearance from the blood stream (Juliano & Stamp, 1975).

Because of the encouraging results from pharmaceutical and medical studies (Mozafari et al., 2008; Taylor et al., 2005) and because phospholipids have been associated with several health benefits (Benton et al., 2001; Hellhammer et al., 2010; Kivinen et al., 1992; Saito & Ishihara, 1977; Van Hooijdonk et al., 2000), food scientists have begun to suggest the application of liposome technology for controlled delivery of functional components in food products (Lariviere et al., 1991; Thompson et al., 2009). Liposomes are well suited as they can easily be made with food-grade ingredients. There are many potential applications for liposomes in the food industry, ranging from protecting sensitive ingredients to increasing the efficacy of food additives to confining undesirable flavors (New, 1990; Taylor et al., 2005).

During preparation or storage, liposomes may undergo physical changes or chemical degradation. Oxidation of fatty acid chains of the phospholipid molecules can cause an increase in the permeability of the bilayers. The oxidation can be minimized by applying different strategies, such as using high quality lipids, avoiding high temperatures for the preparation and storage, using various antioxidants, or employing phospholipids with more saturated fatty acid chains (Jan Zuidam et al., 2003).

The properties of liposomes to protect molecules by encapsulation have been widely demonstrated. For example, it has been shown that liposomes containing ascorbic acid protect it from what would be otherwise rapid oxidation: encapsulated ascorbic acid retained more than 50% activity after 50 days storage at 4°C while free ascorbic acid lost all activity after 18 days of storage under the same conditions (Kirby et al., 1991). Egg phospholipids have been employed to entrap β-galactosidase in liposomes, to prevent the immediate hydrolysis of lactose in milk (Rodrıguez-Nogales & Delgadillo Lopez, 2006), and to encapsulate bioactive molecules and proteins into nanoliposomes (Brgles et al., 2008; Taylor et al., 2007; Xia et al., 2006). The entrapment of lipid soluble molecules such as carotenoids, retinol, or phytosterols in liposome bilayer membranes has also been reported (Chan et al. 2004; Junghans et al., 2000; Lee et al., 2005). However, in most of these studies the common technique of solvent dehydration and reconstitution in water was used for preparation of the liposomes and, in a few cases, physical and chemical stability of the liposomes was tested. The preparation method, the permeability and fluidity of the lipid bilayer (or bilayers), and the size of the liposomes have a tremendous effect on the stability and the encapsulation properties of the liposomes. Encapsulation is usually quantified as encapsulation efficiency, that is, the ratio between the amount of material present in the core of the phospholipid vesicles and the total amount of added material. It is important to note that the entrapment is a function of the internal aqueous phase of the liposomes (Tyrrell et al., 1976).

LIPOSOMES AND NANOLIPOSOMES: SIZE, STRUCTURE, AND STABILITY

Phospholipids have defined polar head groups and non-polar hydrocarbon tails, and the molecules organize themselves into ordered aggregates where the non-polar regions are oriented towards the interior, away from the water phase. This is why both hydrophobic and hydrophilic molecules can be entrapped in these vesicles. The aqueous core can be surrounded by one or more bilayers, (uni- or multi-lamellar, respectively), depending on the preparation method. Their size can vary between 30 and 40 nm and a few μm. Liposomes are usually divided into three classes: large multilamellar vesicles (MLV), large unilamellar vesicles (LUL), and small unilamellar vesicles (SUL). There are many different strategies for the preparation of liposomes (Lasic, 1993; New, 1990; New, 1994; Mozafari et al., 2008).

The size of the liposome and the lamellarity depend on the level of mechanical stress during the dispersion process. One of the most widespread methods to prepare liposomes involves the evaporation of the lipid in a thin film, followed by addition of a buffer and then sonication or other mechanical treatments. Simple agitation causes the formation of MLVs. MLVs have limited industrial applications because of their large diameters, heterogeneity of size, multiple internal compartments, and inconsistencies from preparation to preparation. High levels of mechanical stress, such as high-pressure homogenization, sonication or extrusion, lead to small and unilamellar liposomes (Barnadas-Rodrıguez & Sabes, 2001; Lasch et al., 2003; Lasic, 1993; Mui & Hope, 2007; Thompson & Singh, 2006). By reducing the size, the macroscopic appearance of the liposome dispersion changes and the stability to sedimentation or creaming is significantly reduced. It has been shown that nanoliposomes with a size between 70 and 200 nm and with good size homogeneity are of advantage for prolonged circulation in the bloodstream and for higher surface area availability for interactions (Lasch et al., 2003). It is claimed that multilamellar liposomes will release material more gradually than unilamellar vesicles, as their concentric membranes degrade more slowly (New, 1990).

Encapsulation efficiencies as high as 50% (Chan et al., 2004; Kirby et al., 1991) have been reported for hydrophilic materials prepared with a standard thin-film liposome preparation method, with repeated freezing and thawing of the dispersion. This practice increases the encapsulation efficiency. In general, lower encapsulation efficiencies are reported for hydrophilic molecules incorporated in unilamellar nanoliposomes, high-pressure homogenization where liposomes have a monomodal distribution of diameters of about 100 nm (Alexander et al., 2011; Thompson et al., 2006). Because of the regulatory requirements, to prepare liposomes to be incorporated in foods, microfluidization or other high energy input techniques are more feasible than the dehydration and hydration technique.

Nanoliposomes such as those prepared by microfluidization are of particular interest for the food industry, as their small size and substantial surface area provide unique functionalities. Indeed, structures at the nano scale have potential because of the greatly increased surface area available for interactions with target molecules and tissues. In addition, the small size of these liposomes makes them less susceptible to physical destabilization. However, a better understanding of the physicochemical characteristics of these nanometer-size particles and how they can change the structure, texture, and quality of the food matrix is still needed. In particular, it is not well known how the encapsulation efficiency, the stability of the membranes, and the absorption and bioavailability of the components are affected in a small unilamellar vesicle prepared with microfluidization or high-pressure homogenization, compared to what is known in the literature from the numerous studies using liposomes prepared with solvents such as chloroform.

It is claimed that large, unilamellar or multilamellar vesicles show higher encapsulation efficiency. Hydrophobic compounds, in particular, are better incorporated in MLVs' bilayers. As the MLVs have large size, they can also incorporate more hydrophilic compounds; however, their stability is significantly lower than that of SUVs. LUVs also show high encapsulation efficiency, as the core volume of the liposome is large. These types of liposomes can be prepared by sequential extrusion of MLVs through a series of filters of reducing pore size under low pressure (Lasch et al., 2003; Mui & Hope, 2007).

In extrusion or membrane homogenization, the dispersed phase containing large liposomes is forced to pass through small pores, and this causes the formation of a homogeneous population of small vesicles (Taylor et al., 2005). Although these methods also can be carried out using food-grade solvents, the most up-scalable method of preparation of liposomes is microfluidization or high-pressure homogenization. In a microfluidizer, the size reduction is caused by the collision of two high-velocity fluid streams of the lipid dispersion in an interaction chamber (Barnadas-Rodrıguez et al., 2003; Taylor et al., 2005). Within the interaction chamber, cavitations, along with shear and impact, reduce the size of the particles.

Fig. 3.1 shows the size distribution of different soy phospholipid dispersions (10, 15, and 25% w/v), as measured using dynamic light scattering. The average apparent diameter of these nanoliposomes was between 100 and 130 nm, made by high-pressure homogenizer (Alexander et al., 2011). This has been previously shown by others. For example, liposomes with an average diameter of 95 ± 5 nm can be obtained with soy or milk phospholipids by microfluidization (Thompson et al., 2006). The encapsulation efficiency of the three liposome dispersions shown in Fig. 3.1 was measured by determining the amount of ascorbic acid entrapped in the liposomes (Kirby et al., 1991). The unentrapped ascorbic acid was separated using column chromatography, as described by others (Thompson et al., 2009). As shown in Fig. 3.2, the encapsulation efficiency increases with the concentration of phospholipid, doubling from 15% to 30% of ascorbic acid entrapped, increasing the concentration of lipids from 10% to 25%. However, it is important to point out that the encapsulation efficiency is a function of the volume of water entrapped in the liposome's core, and this parameter remains constant to about 26%/nm^{-3} (10^{-6}) (Fig. 3.2). These results highlight the importance of taking into account the aqueous volume entrapped when looking at differences in permeability and encapsulation efficiencies of the various membranes.

Using microfluidization it is possible to obtain liposome dispersions of MFGM phospholipid with an average hydrodynamic diameter of 90 ± 5 nm. The micrographs depicted in Figs. 3.3 and 3.4 clearly illustrate the differences between vesicles prepared by microfluidization and extrusion. Cryogenic transmission electron microscopy (cryo-TEM) is usually employed to provide information on the microstructure of

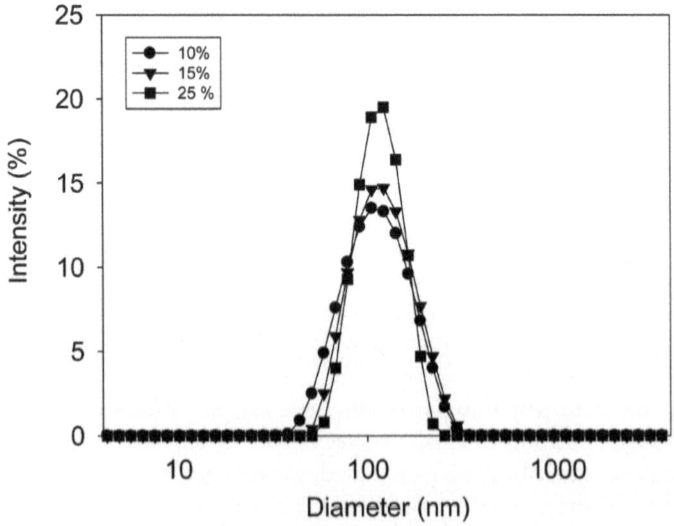

Fig. 3.1. Apparent diameter measured by dynamic light scattering of liposomes, prepared using high-pressure homogenization, with varying concentrations of soybean phospholipids (10, 15, and 25% w/v).

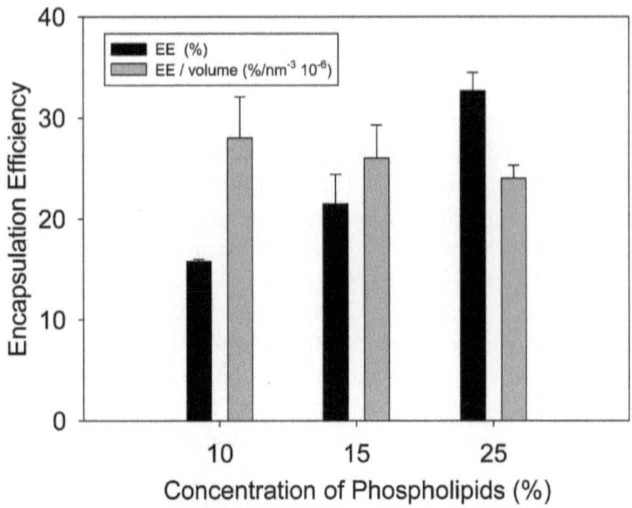

Fig. 3.2. Encapsulation efficiency (EE) of soy phospholipid nanoliposomes (average diameter 100 nm) prepared with 10, 15, and 25% (w/v). Encapsulation efficiency was calculated using ascorbic acid as hydrophilic molecule marker, and values are also adjusted by the total core volume of the liposomes.

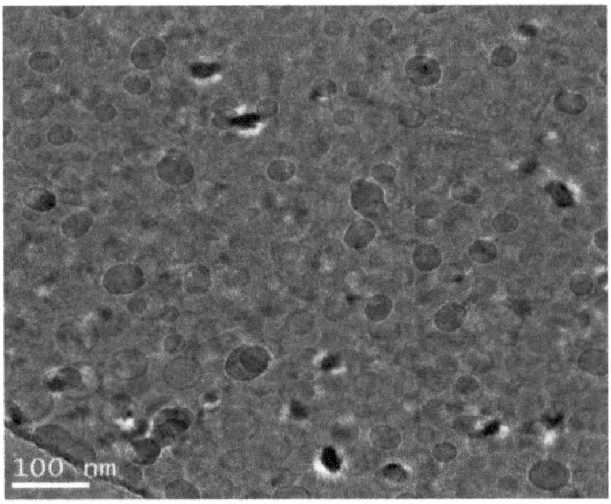

Fig. 3.3. cryo-TEM micrograph of nanoliposomes prepared with MFGM phospholipids (10% w/v) using 5 passes through the microfluidizer. Bar = 100 nm.

liposomes and other amphiphilic self-assemblies (Won, 2004). With cryo-TEM it is possible to visualize the outer membrane that surrounds the internal aqueous space, and the liposomes can be clearly identified as predominantly spherical in shape. While the microfluidized dispersions (Fig. 3.3) showed a uniform distribution of unilamellar vesicles after five passes, the liposomes prepared with five passes through the extruder, although still small in size, showed multilamellar membranes. The vesicles (10% phospholipid) had a hydrodynamic diameter of 120 ± 10 nm.

In extrusion or membrane homogenization the dispersed phase containing large liposomes is forced to pass through a membrane or a filter with a uniform pore size distribution, generating a homogeneous population of small vesicles. In the case of extrusion, usually the dispersions need a higher number of passes compared to microfluidization (or high-pressure homogenization) to obtain unilamellar nanoliposomes of similar size (data not shown).

MILK PHOSPHOLIPIDS AND THEIR POTENTIAL APPLICATIONS AS DELIVERY SYSTEMS

Most research has been reported on liposomes prepared from soy- or egg-derived phospholipids, (Lasic, 1993; New, 1990). Only recently a new extract from the milk fat globule membrane (MFGM) has become commercially available, and it has been suggested as a viable alternative to the more traditional soy or egg sources (Thompson et al., 2006; Thompson & Singh, 2006). The MFGM in native milk surrounds the fat globules and protects them from oxidation and coalescence. As MFGM derives from the mammary cell its composition is similar to that of any

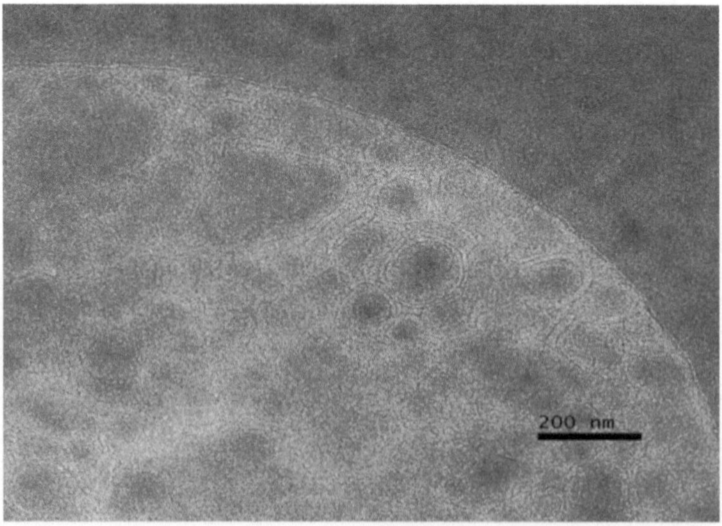

Fig. 3.4. cryo-TEM micrograph of nanoliposomes prepared with MFGM phospholipids (10% w/v) using 5 passes through the extruder. Bar = 200 nm.

biological membrane, consisting of phospholipids, glycoproteins, neutral lipids, enzymes, and other minor components (McPherson & Kitchen, 1983; Ward et al., 2006). The phospholipids from milk can be extracted from a number of by-products such as buttermilk or butter serum (Corredig & Dalgleish, 1997; Corredig et al., 2003; Gallier et al., 2010; Rombaut et al., 2006). Table 3-A illustrates the difference in composition between the three different phospholipid sources: soy, egg, or milk. The composition of polar lipids in the MFGM greatly differs from that of soy and egg phospholipids (Burling & Graverholt, 2008). While egg phospholipids are predominantly phosphatidylcholine, soy phospholipids are a mixture of phosphatidylcholine, phosphatidylethanolamine, and phosphatidylinositol (Table 3-A). The polar lipids present in the MFGM belong to two main classes: the glycerophospholipids, such as phosphatidylcholine, phosphatidylethanolamine, phosphatidylinositol, and so on, and the sphingolipids, based on sphingosine, and amino alcohol with a long unsaturated hydrocarbon chain. Among phospholipids, phosphatidylcholine and phosphatidylethanolamine are present at a high ratio in the MFGM isolates, and anionic phospholipids (phosphatidylinositol and phosphatidylserine) are present in lower amount (Rombaut & Dewettinck, 2006). The main sphingolipids found in MFGM are sphingomyelin, cerebrosides, and ganglyosides. The milk phospholipids are rich in unsaturated fatty acids, with about 25% of C18:1 and 6% of C18:2 (Fauquant et al., 2007), and they play an essential role in imparting fluidity to the membrane. Sphingomyelin has high saturated fatty acid content (McPherson & Kitchen, 1983). The MFGM phospholipids have a high amount of phosphatidylcholine and sphingomyelin, and this ratio between the two polar lipids imparts unique properties of the bilayer membrane (Barenholz & Thompson, 1980).

While the hydrophobic regions of phosphatidylcholine are composed of two fatty acids esterified to the backbone of glycerol and the two acyl chains in position 1 and 2 are almost equal in length, the hydrophobic regions of sphingomyelins are very different. The acyl chain (on average, about 20 carbons in length) is linked to the primary amino group of the sphingosine with an amide bond, and the second portion of the hydrophobic region is the paraffinic chain on the sphingosine, and it is shorter in length, leading to an asymmetric structure (Barenholz & Thompson, 1980).

The co-presence of phosphatydilcholine and sphingomyelin in the lipid bilayer will impart unique properties to the bilayer and, as a consequence, to the liposomes prepared with these fractions. Most sphingomyelins have their transition temperature in the physiological temperature range, while phosphatidylcholine have transition temperatures below 37°C. It has been shown that sphingomyelins cause a more structured gel phase in a bilayer membrane (Barenholtz & Thomson, 1980). Recently, microstructural evidence using confocal microscopy showed the presence of phase separated areas in the native MFGM, with segregated domains of sphingomyelin and cholesterol (Lopez et al., 2010). The profound difference in the composition of milk phospholipids from the other plant sources creates new opportunities to obtain vesicles with different physico-chemical characteristics. For example, the lower level of unsaturated lipids in milk phospholipids may make them less susceptible to oxidation, and the differences in transition temperatures may cause different permeability and storage stability when used as delivery systems.

Another very attractive aspect of the use of MFGM phospholipids is their recognized biological activity (Spitsberg, 2005). Phospholipids in general have been the subject of a number of nutritional studies, which have suggested that their consumption may result in health related benefits (Hellhammer et al., 2010; Imaizumi et al., 1992; Van Hooijdonk et al., 2000; Vesper et al., 1999;, Ward et al., 2006). In particular, in milk, sphingolipids and glycosphingolipids outstand. While sphingo-

Table 3-A. Relative Phospholipid Composition (%) in Different Commercial Sources (Burling & Graverholt, 2008).

	Soy	Egg	MFGM
Phosphatidylcholine	34	75	27
Phosphatidylethanolamine	21	15	25
Phosphatidylinositol	18	0.4	8
Phosphatidylserine	0.5	0	12
Sphingomyelin	0	1.5	24
Phosphatidyl acid	9	0	0
Others	17.5	8.1	4

lipids contain polar groups, such as choline, gangliosides are further derivatized on their polar heads with sugar molecules. These polar compounds are increasingly being recognized as molecules with important bioactivities (Vesper et al., 1999). Although there is no required dietary amount for the consumption of sphingolipids, it has been reported that they bring significant health benefits, for example to the development of the gut maturation (Oshida et al., 2003). MFGM phospholipids have been shown to have anticarcinogenic properties, as there is experimental evidence brought forth to their inhibition of aberrant crypt foci in mice (Dillehay et al., 1994; Schmelz et al., 1996). Sphingomyelin metabolites, ceramide, sphingosine, and others have also shown biological functions such as cell differentiation, apoptosis, and immune recognition (Gallier et al., 2010; Sanchez Juanes et al., 2009; Ward et al., 2006). In addition, it has been shown that consumption of dietary sphingolipids can reduce serum LDL-cholesterol in rats (Imaizumi et al., 1992) and that the sphingolipids from milk sources are more effective than those from egg (Noh & Koo, 2004). The molecules seem to compete with cholesterol for the loading of intestinal mixed micelles, lowering the cholesterol uptake from food, and ultimately causing a significant reduction of LDL cholesterol. Furthermore, it has been reported that sphingomyelins can increase the level of serum HDL (Schmitt, 2009). The higher inhibitory effect of sphingomyelins from milk has been attributed to the higher degree of saturation and the longer chain length of its fatty acyl group (Noh & Koo, 2004).

Phosphatidylserine has also raised significant interest as an active component of nutraceuticals. A substantial portion (15%) of the human brain phospholipids pool consists of phosphatidylserine (Cenacchi et al., 1993). As the brain ages, material is lost and replaced by inactive substances like cholesterol (Burling & Graverholt, 2008). The beneficial effects of phsophatidylserine for improved cognitive performance, decreased of aging-induced memory loss, and as a stress-release agent have been discussed (Burling & Graverholt, 2008; Helhammer et al., 2010).

The unique composition of the MFGM-derived phospholipid fractions and the low cost of some of the sources (for example, buttermilk, a waste product of butter making) may offer some advantages in the manufacture of liposomes for the entrapment of bioactive compounds in food systems. As previously mentioned, the increase in the availability of commercial fractions of MFGM lipids makes the development of functional ingredients for foods with added value much more possible. Moreover, the unique composition of the MFGM phospholipids fraction (50% saturated, 35% monounsaturated, and 15% polyunsaturated fatty acids) results in products more resistant to oxidation. In the case of food applications, therefore, liposomes prepared with MFGM may provide a number of advantages over the more traditional soy and egg sources. Compared to the characterization of soy- or egg-derived phospholipid liposomes and their encapsulation behavior for drug delivery, the information on the encapsulation properties of MFGM liposomes and their potential for application in food systems are scarce.

Thompson et al. (2006) showed that the liposomes prepared from MFGM fractions were significantly different to those produced from soy phospholipids. The

MFGM liposomes had a higher phase transition temperature, a thicker membrane (about 60 Å) and lower membrane permeability than soy liposomes (Thomson et al., 2006). As for the other phospholipid sources, changes in the method of preparation would affect the size and structure of the liposomes. Heating or thin film hydration causes the formation of larger liposomes than those prepared using a microfluidizer (Thompson et al., 2006).

Liposomes prepared with MFGM phospholipid dispersions have been characterized under a variety of conditions, including a range of pH, temperature, and time combinations that may be used for storage and processing conditions commonly encountered in the food industry (Thompson et al., 2006). In addition, the properties and stability of the liposome dispersions produced from MFGM phospholipids have been compared to those produced from nonhydrogenated soy fractions by measuring changes in average hydrodynamic diameter, peroxide value, and conjugated diene levels of samples (Thompson & Singh, 2006). The MFGM liposomes showed little change in the apparent size with pH, and the dispersions were consistently more stable at all pH values during heating compared to those prepared with soy phospholipids (Thompson et al., 2006). As mentioned above, the most significant difference between the phospholipid compositions of these two fractions is the high level of sphingomyelin present in the milk-derived fractions. Sphingomyelin has a more structured gel phase than phosphatidylcholine, and sphingomyelin membranes seem to be more stable than phosphatidylcholine bilayers (Thompson et al., 2006).

In particular, MFGM liposomes show better stability during storage at temperatures in the range of 4–35°C compared to soy phospholipid liposomes. In addition, the MFGM liposomes are less prone to physical aggregation during heating regimes typical of food processing, from temperatures of 55°C to 141°C. This is of particular importance when considering the use of MFGM liposomes as delivery systems in food matrices. It has been shown that, if the heating treatments are carried out at pH near neutral, aggregation is limited and destabilization of MFGM liposomes occurs at a lower pH compared to that of soy liposome dispersions (Thompson et al., 2006). The MFGM liposomes have also been shown to be less prone to aggregation and/or fusion than soy phospholipid liposomes in buffers with added salt (Thompson et al., 2006); however, more work needs to be carried out in complex matrices, as the ionic strength and the presence of other components may affect their stability and the permeability of the vesicles.

The stability of the MFGM phospholipid liposomes has been attributed to the presence of a thicker membrane and the different organization of the phospholipids in their bilayer compared to their soy counterpart. It has also been hypothesized that the differences in the bilayer organization would improve their encapsulation behavior, by reducing permeability (Thompson et al., 2009); however, very few studies have been conducted. Liposomes produced from the MFGM-derived phospholipids showed significantly higher entrapment efficiencies for β-carotene and potassium chromate (hydrophobic and hydrophilic tracer molecules, respectively) than those measured for soy liposomes (Thompson et al., 2009). The stability and encapsulation behavior of nanoliposomes prepared from milk phospholipids using microfluidization have also been observed using ascorbic acid as a model hydrophilic molecule. Ascorbic acid is added extensively to food products for two different purposes: as a vitamin supplement to reinforce dietary intake of vitamin C and as an antioxidant to protect the sensory and nutritive quality of the food itself. However, vitamin C is very labile and can degrade rapidly in food systems by a variety of different mechanisms. Fig. 3.5 illustrates the stability of ascorbic acid encapsulated in nanoliposomes during storage at 4°C. Encapsulation was measured as reported in the literature (Kirby et al., 1991). About 70% of encapsulated vitamin C remained after storage at 4°C and pH 7 for 7 weeks without significant increases in peroxide value and liposome size (Farhang et al., unpublished).

We have also studied the simultaneous encapsulation of ascorbic acid with β-carotene. Liposomes ideally can incorporate both hydrophobic and hydrophilic bioactive compounds. Carotenoids are unstable, insoluble in water, and their bioavailability may be improved by using liposomal formulations as the delivery systems (Shojia & Nakashima, 2004). However, it is not clear if the incorporation of the β-carotene in the bilayer would affect the permeability of the vesicles. Nanoliposomes of approximately 100 nm in diameter were prepared with 10% MFGM phospholipid dispersions containing ascorbic acid. Incorporation of β-carotene yielded larger nanoliposomes, with an apparent diameter of 200 nm, and by simultaneous incorporation of β-carotene and vitamin C there was an incorporation efficiency of

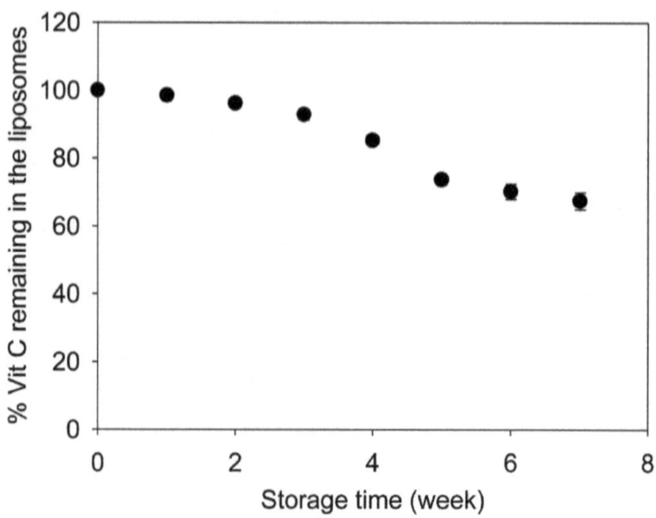

Fig. 3.5. Amount of ascorbic acid remaining encapsulated in nanoliposomes prepared with MFGM phospholipids (10% w/v), using 5 passes through the microfluidizer, as a function of storage time at 4°C.

92% for β-carotene without any significant change in vitamin C encapsulation (about 26%) (Farhang et al., unpublished data). The results would suggest that β-carotene enhanced the lamellar size of the liposomes without increasing the lumen volume. As fractions of milk phospholipids derived from the milk fat globule membrane become commercially available, there will be a demand for these ingredients for the preparation of liposome material as a delivery system, and studies on stability of these nanoparticles in relevant food matrices will become increasingly critical.

CONCLUSIONS

As more information becomes available on the health benefits associated with the consumption of MFGM components, there are increasing advantages of using MFGM-derived phospholipids in formulation of delivery systems for foods. These advantages are in addition to the differences in technological performance of these nanoliposomes compared to those prepared with soy or egg phospholidids. Therefore, as the number of opportunities to use liposomes in the food industry increases, ranging from delivery systems to incorporating bioactive molecules, antimicrobials, or active ingredients, more research is needed to better understand the differences in physicochemical properties and encapsulation behavior between MFGM and other phospholipid sources. More importantly, it will be necessary to understand liposomes' behavior in relevant food matrices and during processing and storage. The rise of customer awareness about the increasing preference for natural and healthy components should give these milk bioactive compounds an ever-increasing role in the field of nutraceuticals delivery systems. In addition, the preparation of liposomes based on MFGM material would create an all-dairy ingredient for milk based products.

REFERENCES

Alexander, M.; Acero Lopez, A.; Fang, Y.; Corredig, M. Incorporation of phytosterols in soy phospholipid nanoliposomes: encapsulation efficiency and stability. LWT-Int. J. Food Sci. Technol. Submitted.

Barnadas-Rodrıguez, R.; Sabes, M. Factors involved in the production of liposomes with a high-pressure homogenizer. Int. J. Pharm. 2001, 213, 175–186.

Barenholz, Y.; Thompson, T.E. Sphingomyelins in bilayers and biological membranes. Biochim. Biophys. Acta 1980, 604, 129–158.

Benton, D.; Donohoe, R.T.; Sillance, B.; Nabb, S. The influence of phosphatidylserine supplementation on mood and heart rate when faced with an acute stressor. Nutr. Neurosci. 2001, 4, 3, 169–178.

Brgles, M.; Jurašin, D.; Dutour Sikiri, M.; Frkanec, R.; Tomaši, J. Entrapment of ovalbumin into liposomes—Factors affecting entrapment efficiency, liposome size, and zeta potential. J. Liposome Res. 2008, 18, 3, 235–248.

Burling, H.; Graverholt, G. Milk–a new source for bioactive phospholipids for use in food formulations. Lipid Technol. 2008, 20, 10, 229–231.

Cenacchi, T.; Bertoldin, T.; Farina, C.; Fiori, M.G.; Crepaldi, G. Cognitive decline in the elderly: a double blind, placebo-controlled multicenter study on efficacy of phosphatidylserine administration. Aging (Milano) 1993, 5, 2, 123–133.

Chan, Y.H.; Chen, B.H.; Chiu, C.P.; Lu, Y.F. The influence of phytosterols on the encapsulation efficiency of cholesterol liposomes. Int. J. Food Sci. Technol. 2004, 39, 985–995.

Corredig, M.; Dalgleish, D.G. Isolates from industrial buttermilk: Emulsifying properties of material derived from the milk fat globule membrane. J. Agric. Food Chem. 1997, 45, 4595–4600.

Corredig, M.; Roesch, R.R.; Dalgleish, D.G. Production of a Novel Ingredient from Buttermilk, J. Dairy Sci. 2003, 86, 2744–2750.

Demel, R.A.; de Kruyff, B. The function of sterols in membranes. Biochim. Biophys. Acta 1976, 457,109–132.

Dillehay, D.L.; Webb, S.K.; Schmelz, E.M.; Merrill, A.H. Jr. Dietary sphingomyelin inhibits 1, 2 dimethylhydrazine-induced colon cancer in CF1 mice. J. Nutr. 1994, 124, 615–620.

Fauquant, C.; Briard-Brion, V.; Leconte, N.; Guichardant, M.; Michalski, M.C. Membrane phospholipids and sterols in microfiltered milk fat globules. Eur. J. Lipid Sci. Technol. 2007, 109, 1167–1173.

Frkanec, R.; Travaš, D.; Krstanovi, M.; a Špoljar, B.H.; Ljevakovi, U.; Vraneši, B.; Frkanec, L.; Tomaši, J. Entrapment of peptidoglycans and adamantyltripeptides into liposomes: an HPLC assay for determination of encapsulation efficiency. J. Liposome Res. 2003, 13, 3, 279–294.

Gallier, S.; Gragson, D.; Cabral, C.; Jime'Nez-Flores R.; Everett, D.W. Composition and fatty acid distribution of bovine milk phospholipids from processed milk products. J. Agric. Food Chem. 2010, 58, 10503–10511.

Gregoriadis, G.; Leathwood, P.D.; Ryman, B.E. Enzyme entrapment in liposomes. FEBS Letters 1971, 14, 2, 95–99.

Gregoriadis, G. Ryman, B.E. Fate of protein-containing liposomes injected into rats. Eur. J. Biochem. 1972, 24, 485–449.

Hellhammer, J.; Waladkhani, A.R.; Hero, T.; Buss, C. Effects of milk phospholipid on memory and phychological stress response. Br. Food J. 2010, 112, 1124–1137.

Imaizumi, K.; Tominaga, A.; Sato, M.; Sugano, M. Effects of dietary sphingolipids on levels of serum and liver lipids in rats. Nutr. Res. 1992, 12, 543–548.

Jan Zuidam, N.; van Winden, E.; de Vrueh, R.; Crommelin, D.J.A. Stability, storage, and sterilization of liposomes. In Liposomes practical approach, second edition; V.P. Torchilin, V. Weissig; Oxford University Press: New York 2003; pp. 149–155.

Juliano, R.L.; Stamp, D. The effect of particle size and charge on the clearance rates of liposomes and liposomes encapsulated drugs. Bioch. Biophys. Res. Comm. 1975, 63, 651–658.

Junghans, A.; Sies, H.; Stahl, W. Carotenoid-containing unilamellar liposomes loaded with glutathione: a model to study hydrophobic-hydrophilic antioxidant interaction. Free Radical Res. 2000, 33, 6, 801–808.

Kirby, C.J.; Whittle, C.J.; Rigby, N.; Coxon, D.T.; Law, B.A. Stabilization of ascorbic acid by microencapsulation in liposomes. Int. J. Food Sci. Technol. 1991, 26, 437–449.

Kivinen, A.; Tarpila, S.; Salminen, S.; Vapaatalo, H. Gastroprotection with milk phospholipids: a first human study. Milchwissenschaft. 1992, 47, 694–696.

Larivière, B.; El Soda, M.; Soucy, Y.; Trépanier, G.; Paquin, P.; Vuillemard, J.C. Microfluidized liposomes for the acceleration of cheese ripening. Int. Dairy J. 1991, 1, 111–124.

Lasch, J.; Weissig, V.; Brandl, M.; Preparation of liposomes. In Liposomes practical approach, second edition; V.P. Torchilin, V Weissig; Oxford University Press: New York, 2003; pp. 3–22.

Lasic, D.D. Liposomes from physics to applications; Elsevier science B.V.: Amesterdam, the Netherlands, 1993; pp. 3–7, 62–79.

Lee, S.; Lee, K.; Kim, J.J.; Lim, S.H. The effect of cholesterol in the liposome bilayer on the stabilization of incorporated retinol. J. Liposome Res. 2005, 15, 3, 157–166.

Liu N.; Park H.J. Chitosan-coated nanoliposome as vitamin E carrier. J. Microencapsulation 2009, 26, 3, 235–242.

Lopez, C.; Madec, M.-N.; Jinenez-Flores, R. Lipid rafts in the bovine milk fat globule membrane revealed by the lateral segregation of phospholipids and heterogeneous distribution of glycoproteins. Food Chem. 2010, 120, 22–33.

McPherson, A.V.; Kitchen, B.J.; Reviews of the progress of dairy science: the bovine milk fat globule membrane – its formation, composition, structure and behaviour in milk and dairy products. J. Dairy Res. 1983, 50, 107–133.

Mozafari, M.R.; Johnson, C.; Hatziantoniou S.; Demetzos C. Nanoliposomes and their applications in food nanotechnology. J. Liposome Res. 2008, 18, 309–327.

Mui, B.; Hope, M.J. Formation of large unilamellar vesicles by extrusion, in liposome preparation and related technology, Third edition; G. Gregoriadis, Ed.; Informa healthcare USA Inc: New York, 2007; pp. 55–65.

New, R.R.C. Introduction. In liposomes: a practical approach. New R.R.C., IRL Press: New York, 1990; pp. 1–32.

New, R.R.C. Influence of Liposome characteristics on their properties and fate. In liposomes as tools in basic research and industry, Philippot, J.R.; Schuber, F.; CRC Press: Florida, 1994; pp. 4–20.

Noh, S.K.; Koo, S.I. Milk sphingomyelin is more effective than egg sphingomyelin in inhibiting intestinal absorption of cholesterol and fat in rats. Am. Soc. Nutr. Sci. 2004, 2611–2616.

Oshida, K.; Shimizu, T.; Takase, M.; Tamura, Y.; Shimizu, T.; Yamashiro, Y. Effects of dietary sphingomylelin on cerntral nervous system myelination in developing rats. Pediatr. Res. 2003, 53, 589–593.

Rahman, Y; Rosenthal, M.W.; Cerny, E.A.; Moretti, E.S. Preparation and prolonged tissue retention of liposome-encapsulated chelating agents. J. Laboratory Clin. Med. 1974, 83, 4, 640–647.

Rodrıguez-Nogales, J.; Delgadillo Lopez, A. A novel approach to develop β-galactosidase entrapped in liposomes in order to prevent an immediate hydrolysis of lactose in milk. Int. Dairy J. 2006 , 16, 4, 354–360.

Rombaut, R.; Van Camp, J.; Dewettinck, K. Phospho- and sphingolipid distribution during processing of milk ,butter and whey. Int. J. Food Sci. Technol. 2006, 41, 435–443.

Rombaut, R.; Dewettinck, K. Properties, analysis and purification of milk polar lipids, Int. Dairy J. 2006, 16, 1362–1373.

Sanchez-Juanes, F.; Alonso, J.M.; Zancada, L.; Hueso, P. Glycosphingolipids from bovine milk and milk fat globule membranes: a comparative study. Adhesion to enterotoxigenic Escherichia coli strains. Bio. Chem. 2009, 390, 31–40.

Saito, H.; Ishihara, K.J. Antioxidant activity and active sites of phospholipids as antioxidants. J. Am. Oil Chem. Soc. 1997, 74, 1531–1536.

Schmitt, H. Phospholipids–Making lives healthier. Wellness Foods Europe May/June 2009, 36–41.

Schmelz, E.M.; Dillehay, D.L.; Webb, S.K.; Reiter, A.; Adams, J.; Merrill, A.H. Jr. Sphingomyelin consumption suppresses aberrant colonic crypt foci and increases the proportion of adenomas

versus adenocarcinomas in CF1 mice treated with 1,2-dimethylhydrazine: implications for dietary sphingolipids and colon carcinogenesis. Cancer Res. 1996, 56, 4936–4941.
Shojia, Y.; Nakashimaa, H. Nutraceutics and delivery systems. J. Drug Targeting 2004, 12, 6, 385–391.
Spitsberg, V.L. Invited review: bovine milk fat globule membrane as a potential nutraceutical. J. Dairy Sci. 2005, 88, 2289–2294.
Takeuchi, H.; Matsui, Y.; Sugihara, H.; Yamamoto, H.; Kawashima, Y. Effectiveness of submicron-sized, chitosan-coated liposomes in oral administration of peptide drugs. Int. J. Pharm. 2005, 303, 160– 170.
Taylor, T.; Davidson, P.; Bruce ,B.; Weiss, J. Liposomal nanocapsules in food science and agriculture. Crit. Rev. Food Sci. Nutr. 2005, 45, 587–605.
Taylor, T.M.; Gaysinsky, S.; Davidson, P.M.; Bruce, B.D.; Weiss, J. Characterization of antimicrobial-bearing liposomes by z-potential, vesicle size and encapsulation efficiency. Food Biophysics. 2007, 2, 1–9.
Thompson, A.; Singh, H. Preparation of liposomes from milk fat globule membrane phospholipids using a microfluidiser. J. Dairy Sci. 2006, 89, 410–419.
Thompson, A.; Haisman, D.; Singh, H. Physical Stability of Liposomes Prepared from Milk Fat Globule Membrane and Soya Phospholipids. J. Agric. Food Chem. 2006, 54, 6390–6397.
Thompson, A.; Couchoud, A.; Singh H.; Comparison of hydrophobic and hydrophilic encapsulation using liposomes prepared from milk fat globule-derived phospholipids and soya phospholipids. Dairy Sci. Technol. 2009, 89, 99–113.
Tyrrell, D.A.; Heath, T.D.; Colley, C.M.; Ryman, B.E. New aspects of liposomes. Biochim. Biophys. Acta 1976, 457, 259–302.
Van Hooijdonk, A.C.M.; Kussendrager, K.D.; Steijns, J.M. In vivo antimicrobial and antiviral activity of components in bovine milk and colostrums involved in non-specic defence. Br. J. Nutr. 2000, 84, S127–S134.
Vesper, H.; Schmelz, E.M.; Nikolova-Karakashian, M.N.; Dillehay, D.L.; Lynch, D.V.; Merrill, A.H. Jr. Sphingolipids in food and the emerging importance of sphingolipids in nutrition. J. Nutr. 1999, 129, 1239–1250.
Xia, S.; Xu, S.; Zhang, X. Optimization in the preparation of coenzyme Q10 nanoliposomes. J. Agric. Food Chem. 2006, 54, 6358–6366.
Ward, R.E.; German, J.B.; Corredig, M. Composition, applications, fractionational significance of milk fat globule membrane material. In Advanced Dairy Chemistry, P.F. Fox and P.L.H. McSweeney, Eds.; Springer: New York, 2006; pp. 213–244.
Won, Y. Imaging nanostructured fluids using cryo-TEM. Korean J. Chem. Eng. 2004, 21, 1, 296–302.
Zhang, N.; Ping, Q.N.; Huang, G.H.; Xu, W.F. Investigation of lecithin-modified insulin liposomes as carriers for oral administration. Int. J. Pharm. 2005, 294, 247–259.

4

MARINE PHOSPHOLIPID LIPOSOMES

Koretaro Takahashi[1], Masatoshi Tanouchi[2], and Tomoko Okada[3]
[1]Faculty of Fisheries Sciences Hokkaido University, 3-1-1 Minato-cho, Hakodate 041-8611, Japan; [2]Health Foods & AM Laboratory, Ltd., 302-go, Renai-Uenosakuragi, 1-10-19 Uenosakuragi, Taito-ku Tokyo 110-0002, Japan; and [3]Cosmo Foods Co., Ltd. Azuma Building 12-1, Kodenma-cho Nihonbashi, Chuo-ku Tokyo 103-0001, Japan

INTRODUCTION

Marine phospholipids can be defined as phospholipids containing docosahexaenoic acid (DHA) or eicosapentaenoic acid (EPA). Typical DHA-rich phospholipid sources are squid skin, squid muscle, and fish roe phospholipids. A typical source for EPA exclusively rich phospholipid is starfish (*Asterias amurensis*). On the other hand, typical sources of both DHA- and EPA-abundant phospholipids are those of giant scallop internal organs and krill oils. Squid muscle debris, connective tissues, and testis of marine animals are also good sources for DHA-abundant phospholipids. We may say that marine phospholipids are rich in fishery by-products or poorly used marine bioresources.

Marine phospholipids rich in DHA and EPA are beneficial for the treatment of cancer. DHA-containing phospholipids have been recognized as beneficial lipids that promote cancer cell differentiation induced by cell differentiators for both nonfixative cancer cells (Tochisawa et al., 1997; Hosokawa et al., 2001) and fixative cancer cells (Hossain et al., 2006). By applying marine phospholipids as liposomal bilayers, they would boost the effect of the encapsulated hydrophilic anticancer compounds (Murakawa et al., 2007). Also, by inserting hydrophobic antiobesity compounds into the liposomal bilayers, marine phospholipids boost the effect of the inserted hydrophobic antiobesity compounds (Okada et al., 2011). When marine phospholipids are served as liposomal drinks, they would be more effective than adding into solid foods or feeds.

FUNCTIONAL BENEFITS OF MARINE PHOSPHOLIPIDS

In a recent review (Lu et al., 2011), it is discussed in detail that marine phospholipids must be more efficient than marine triacylglycerol (TG) (i.e., fish oil) in delivering n-3 polyunsaturated fatty acids to desired tissues. Fish oils are well known to decrease the levels of total cholesterol, TG content, and low density lipoprotein (LDL) while also increasing the high density lipoprotein (HDL) in blood. How-

Table 4-A. Lipid Content and Thiobarbituric Acid Reactive Substances (TBARS) Value in the Serum of Rats Fed Experimental Diets

	Control	EPA-PC	EPA-TG	DHA-TG
Total cholesterol (mg/dL)	66.0 ± 7.8[a]	55.1 ± 3.2[b]	61.9 ± 5.3[a]	64.4 ± 6.1[a]
HDL cholesterol (mg/dL)	57.4 ± 6.0[b]	64.9 ± 5.5[a]	54.1 ± 10.1[b]	53.6 ± 9.2[b]
LDL cholesterol (mg/dL)	5.1 ± 0.5[a]	4.5 ± 0.4[b]	5.2 ± 0.6[a]	5.1 ± 0.6[a]
Free cholesterol (mg/dL)	12.6 ± 1.3	11.1 ± 1.7	13.0 ± 1.0	11.9 ± 3.0
Cholesterol ester (mg/dL)	53.6 ± 7.3[a]	43.9 ± 2.3[b]	48.9 ± 5.9[b]	52.4 ± 5.3[a]
Phospholipid (mg/dL)	124.7 ± 13.3[b]	139.9 ± 18.1[a]	102.4 ± 13.7[c]	108.0 ± 16.3[c]
Triacylglycerol (mg/dL)	41.3 ± 4.7[a]	32.4 ± 3.4[b]	28.1 ± 3.9[c]	30.1 ± 2.9[c]
Total lipid (mg/dL)	216.4 ± 27.6[a]	196.9 ± 21.9[a]	172.6 ± 27.7[b]	174.0 ± 23.4[b]
TBARS (mg/dL)	0.64 ± 0.03[a]	0.48 ± 0.006[c]	1.56 ± 0.06[b]	0.5 ± 0.09[c]

Values are means ± SD. Values with different superscript letters are significantly different, $p<0.05$, ANOVA

ever, in practical cases, decreasing the TG content and increasing the HDL content in blood cannot be achieved by moderate taking of fish oil. A large amount of fish oil administration is necessary for this purpose compared to marine phospholipids. In fact, we recently observed that among marine lipids—the phosphatidylcholine (PC) form EPA (EPA-PC), TG form DHA (DHA-TG), and TG form EPA (EPA-TG)—the EPA-PC was the only lipid molecular species effective in increasing the level of HDL with moderate amount of intake, as shown in Table 4-A (Shina, 2009).

The effect of krill oil on serum lipids and markers of oxidative stress and inflammation has recently been studied (Ulven et al., 2011). Krill oil contains approximately 35% phospholipids. Molecular species of krill phospholipids are also well studied (Winther et al., 2011). It has been observed that the metabolic effects of krill oil rich in marine phospholipids are comparable to fish oil even if the EPA+DHA dose in the krill oil is only 63% of the dose in the fish oil. A typical health benefit of fish oil is the prevention of over-production of arachidonic acid-derived eicosanoids, which often increase risk of cancer, thrombosis, allergy, and other diseases. In this sense, marine phospholipids would be more effective when they are served in the same amount. Moreover, Di Marzo and colleagues (2010) showed that DHA in the brain increased significantly after krill oil administration when compared to control animals. It is interesting to note that n-3 fatty acids linked to phospholipids may be differently distributed in the body compared to n-3 fatty acids in other lipid molecular forms. It was pointed out that in the early days the phospholipid forming polyunsaturated fatty acid dosing should result in a higher polyunsaturated fatty acid content in plasma than TG forming polyunsaturated fatty acid dosing (Galli et al., 1992). For these reasons, the phospholipids should be more effective than the TG to exert health benefits.

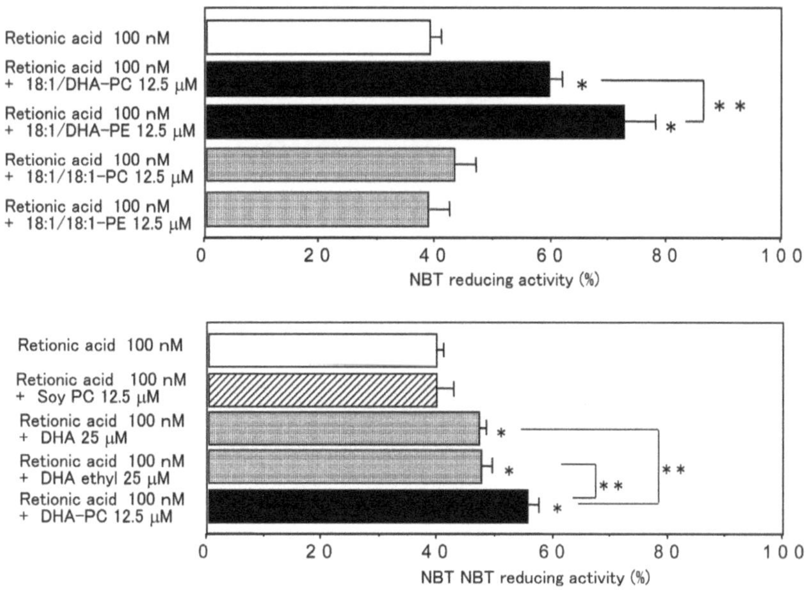

Fig. 4.1. Effect of retinoic acid (RA) and various lipids on cell differentiation of HL-60 cells (Tochisawa et al., 1997). HL-60 cells (5×10^4) were incubated with the individual lipids for 24 hours, then RA was added followed by incubation for additional 72 hours. Nitroblue tetrazolium (NBT) reducing activity in HL-60 cells was assayed. Data shown as mean value $\pm$ SD. **p*<0.01 vs. control. ***p*<0.01 vs. RA.

PROMOTIONAL EFFECT OF MARINE PHOSPHOLIPIDS ON BIOACTIVE COMPOUNDS

Since the 1990s, phospholipids containing DHA have been recognized as beneficial lipids that promote cancer cell differentiation induced by retinoic acid (Tochisawa et al., 1997)(Fig. 4.1) and dibutyryl cyclic adenosine monophosphate (Hosokawa et al., 2001) for non-fixative cancer cells such as leukemia, or butylate for fixative cancer cells such as colon cancer (Hossain et al., 2006). This functionality may be very useful in cancer therapy because phospholipids containing DHA should decrease side effects of the anticancer drugs that seriously impair quality of life. For example, a tremendous amount of retinoic acid is usually prescribed for the lives of patients as an anticancer drug, and with too much retinoic acid, side effects seriously impair the quality of life of the patients. The prescription of anticancer drugs together with marine phospholipids should facilitate a lower dosage of anticancer drugs because marine phospholipids promote the effect of the drugs. We considered that susceptibility of cancer cells should increase by the incorporation of marine phospholipid. In fact, classical data (Burns & Spector, 1987) shows that uptake of doxorubicin: (2*S*,4*S*)-4-(3-Amino-2,3,6-trideoxy-α-L-*lyxo*hexopyranosyloxy)-

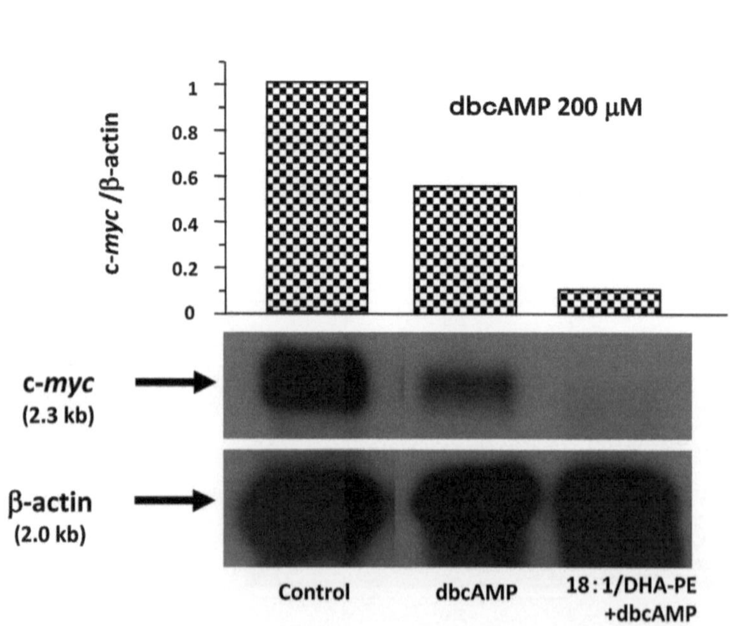

Fig. 4.2. Down regulation of *c-myc* mRNA in HL-60 cells treated with dibutyryl cAMP (dbcAMP) and PE form DHA (Hosokawa, 2004). HL-60 cells (5×10^4) were incubated with (dbcAMP) after preincubation with 50 μM 1-oleoyl, 2-docosahexaenoyl-phosphatidylethanolamine for 24 hours.

2,5,12-trihydroxy-2-hydroxyacetyl-7-methoxy-1,2,3,4-tetrahydrotetracene-6,11-dione (Adriamycin) was greater in DHA-enriched L1210 cells compared to other unsaturated fatty acid-enriched cells. This was also true for mitoxantrone (Burns et al., 1988). In these experiments, free fatty acid was used, which is no doubt less bioavailable than the phospholipids. There have been several reports that explain at least in part the mechanisms whereby DHA-inserted phospholipids exert promotional effects on functional compounds. Stillwell and colleagues (1993) considered that increase in permeability of the organelle biomembranes should be the reason why the cells become susceptible to DHA when it is incorporated. Larsson and colleagues (2004) considered that DHA may also influence gene expression, transcription factor activity, and signal transduction pathways. For example, 1-Oleoyl-2-DHA phosphatidylethanolamine (PE) was shown to facilitate up-regulating expressions of *c-jun* mRNA and c-Jun protein, while it was effective in down-regulating *c-myc* mRNA (Hosokawa, 2004) (Fig. 4.2). In part, c-Jun protein is responsible in the cell differentiation process, and c-Myc protein is responsible in cell cycle progression. Therefore, it is considered that phospholipids containing DHA may, in part, suppress leukemia by modulating the oncogene expressions. On the other hand, polyunsaturated fatty acids are known to regulate peroxisome proliferator-activated receptors (PPARs), which are involved in cell differentiation, cell prolif-

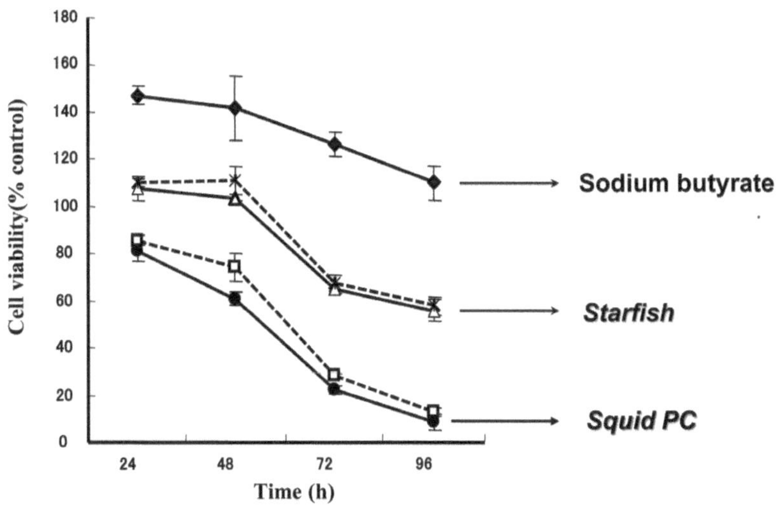

Fig. 4.3. Effect of PC from squid meal and starfish, as well as BA on cell viability of human colon carcinoma Caco-2 cells. Caco-2 cells were seeded on 96-well plate with 100 µL growth medium to give 6.2×10^3 cells. After 24 h, 1.0 mM sodium butyrate (BA) and PC from squid meal or starfish were added. Cell viability was determined using WST-1 every 24 h. WST-1 was added to each well, and the plate was incubated at 37℃ for 3h. Absorbance at 450/650 nm was measured for each well on a Micro plate reader. Data are shown as means ±S.D. (n=4). ◆, 1.0 mM BA; □, 75 µM squid meal PC; Δ, 75 µM starfish PC; ●, 75 µM squid meal PC + 1.0 mM BA; ×, 75 µM starfish PC + 1.0 mM BA.

eration, and inflammatory responses (Grimaldi, 2001: Vamecq & Latruffe, 1999). DHA was found to induce apoptosis in vascular smooth muscle cells by activation of PPAR-α and p38 mitogen-activated protein kinases (Diep et al., 2000).

GROWTH SUPPRESSION OF HUMAN COLON CARCINOMA CACO-2 CELLS

We indicated that PC from squid and starfish suppresses colon carcinoma Caco-2 cells more effectively than sodium butyrate, as shown in Fig. 4.3 (experiment done by Kota Zama). Suppression was more prominent when squid PC was used compared to that of starfish. Starfish PC mainly contains EPA as fatty acid moiety in *sn*-2 position while that of squid mainly contains DHA as fatty acid moiety in that position (The total fatty acid compositions are shown in Table 4-B.).

Therefore, we may say that PC containing DHA is more useful in suppressing colon cancer than PC containing EPA. We compared the promotional performance among squid phospholipid classes with the same fatty acid moieties on Caco-2 cell

Table 4-B. Fatty Acid Composition of Squid and Starfish PC

Fatty acid	Squid-PC	Starfish-PC
C14:0	1.5	0.9
C16:0	30.9	3.8
C18:0	1.5	6.9
C18:1	3.3	3.3
C18:2	0.2	tr
C18:3	0.1	1.0
C20:0	2.9	tr
C20:1	0.3	8.0
C20:4	0.2	4.2
C20:5	7.6	41.4
C20:6	48.5	10.3

differentiation (experiment done by Megumi Konishi). With the same fatty acid combination, phosphatidylserine (PS) exerted more effectively than PC on the promotional effect. Hossain and colleagues (2006) showed that the suppression of Caco-2 cells by the marine phospholipids is not only caused by the promotional effect of cancer cell differentiation but also the induction of apoptosis with other bioactive compounds. As shown in Fig. 4.4, enrichment factor, which is an index of DNA fragmentation relative to control (=1.0), was significantly high in squid PC and the squid PC bioconverted PS when combined with sodium butylate. Unfortunately, PS is poorly found in natural sources except brain and nerve tissues. For this reason, transphosphatidylation of marine PC to PS should become critical when obtaining DHA or EPA bounded PS in a reasonable price.

Recently, we observed that as substrates for DHA accretion in fetus rat brain, lysophosphatidylserine (LPS) form DHA seemed to be the most effective DHA carrier among PS, PC, LPS, and Lyso PC (experiment done by Kazuki Onoyama and Takeshi Ohkubo). The DHA % in fatty acid composition of the rat fetus brain is depicted in Table 4-C.

We analyzed the DHA content in the blood serum of the expectant mother rats on all of the DHA bounded phospholipid classes (PC, LPC, PS, and LPS) served to individual groups. We observed that the expectant mother rats served with DHA-bounded LPS showed the highest DHA content in their blood serum, and also the transport of DHA-bounded LPS was the highest among DHA-PC, DHA-LPC, DHA-PS, and DHA-LPS in small intestinal epithelial cell model (data not shown). By taking into consideration that fetus blood-brain barrier is not well developed to discriminate individual lipid classes, the limiting pass must be the small intestinal epithelial cell monolayer. By carrying out two steps of bioconversions, which are transphosphatidylation via phospholipase D and by partial hydrolysis via *sn*-1,-

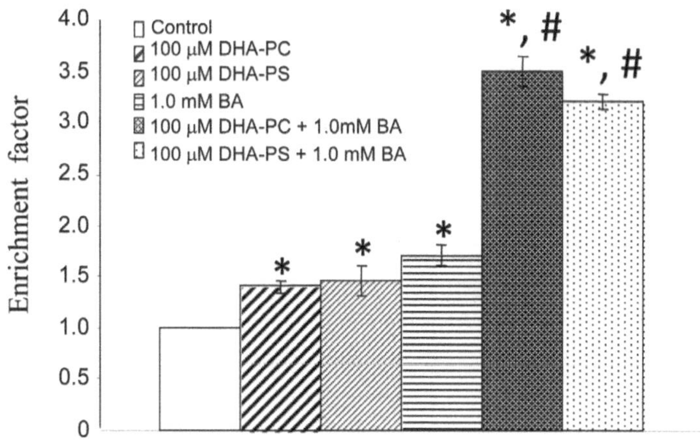

Fig. 4.4. Enrichment factors induced by sodium butyrate (BA) in Caco-2 cells with DHA-PC (from squid meal) and PS (derived from squid meal PC). Cells were pre-incubated without any treatment for 24 h followed by treatment with 1.0 mM BA, 100 M DHA-PC and DHA-PS, and 100 M DHA-PC + 1.0 mM BA and 100 M DHA-PS +1.0 mM BA for 48 h. Apoptosis was quantified using an ELISA assay. Results are represented as relative fold increase in apoptosis as compared with control. The bars represent ± SD (n=3). *p<0.01 vs control, #P <0.01 vs 1.0 mM NaBt.

Table 4-C. Fatty Acid Composition of the Fetus Brain Lipid in Pregnancy; Rat Administered Phospholipid Samples by Probe (%)

Fatty acid	Control	DHA-PC served	DHA-PS served	DHA-LPC served	DHA-LPS served
C14:0	2.2±0.1	2.7±0.7	2.3±0.2	2.2±0.1	2.1±0.2
C16:0	38.8±0.4	38.0±1.5	38.7±0.9	38.4±0.6	38.2±0.7
C18:0	16.9±1.0	16.9±1.9	16.9±1.6	18.5±1.6	18.4±1.5
C18:1	17.7±1.0	19.0±2.5	18.4±1.6	16.7±1.9	16.2±1.8
C18:2	1.0±0.1	1.2±0.4	1.2±0.1	1.1±0.2	1.4±0.9
C20:4	12.9±0.2	12.5±0.7	12.9±0.4	13.0±0.5	12.2±0.4
C22:6	9.5±0.8	9.8±1.1	9.5±1.3	10.1±1.2	11.5±1.2[a]

3 specific lipases or phospholipase A_1 on PC obtained from squid skin or squid muscle debris, exclusively DHA-rich LPS (Table 4-D) can be obtained without any difficulty.

It has been recognized that phospholipid must be more bioavailable than free fatty acid form and more dispersible, as well as bioavailable, than TG form. Therefore, the phospholipids are considered to be one of the most beneficial chemical

Table 4-D. Fatty Acid Composition of Bioconverted Marine Phospholipids

Fatty Acid	Squid PL	DHA-PC	DHA-PS	DHA-LPC	DHA-LPS
C14:0	1.9	2.3	2.2	0.15	
C15:0	1.1	1.5	1.3		
C16:0	35.1	42.4	42.9	1.2	1.8
C16:3	1.3				
C18:0	6.1	1.9	1.6	1	0.5
C18:1	3.8	3.3	3		
C20:0	3.2	1.7	1.7	0.1	0.2
C20:3	2.3				
C20:5	11.4	7.2	9.5	13.2	9
C21:0					0.8
C22:6	33.9	39.8	36.6	85.3	87.6

structures to incorporate DHA and EPA into cells. Marine phospholipids, which are rich in DHA and EPA, exactly meet this condition. Anticancer is only a part among numbers of beneficial functionality of the marine phospholipids.

Not only the functionalities of polyunsaturated fatty acids themselves but also the phospholipid chemical structures often show notable health benefits other than the functions of polyunsaturated fatty acids. A typical one for this is the preventive

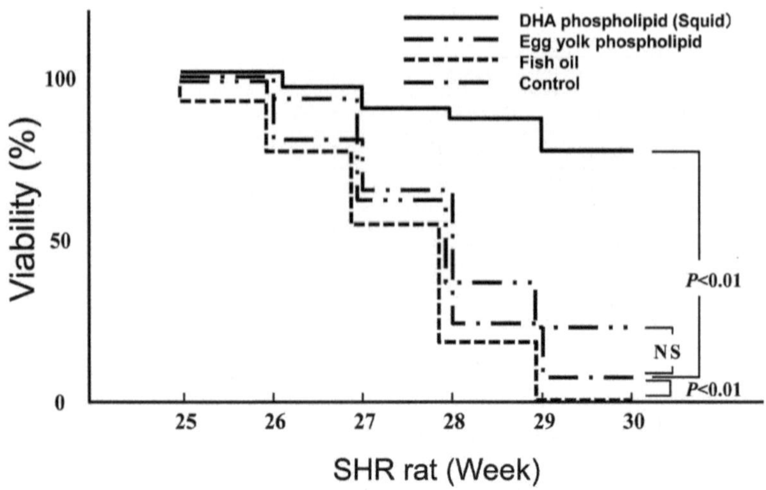

Fig. 4.5. Survival of SHR rats (easily die by apoplexy) served with phospholipid form DHA (squid PC), fish oil, and egg yolk PC (Inoue et al., 2001).

effect of phospholipids containing DHA on apoplexy. Among fish oil TG, egg yolk phospholipid, and squid phospholipid rich in *sn*-2 DHA, the squid phospholipid is the only available lipid to prevent apoplexy (Fig. 4.5).

Phospholipids containing DHA may be useful to facilitate conformational change in the extra cellular domain of alpha-chain of class I major histocompatibility complex molecules and, thereby, should modulate its function (Jenski et al., 2000).

MARINE PHOSPHOLIPIDS AND SUPPRESSION OF CANCER INITIATION

The effect of marine phospholipids on suppression of cancer initiation induced by carcinogen was studied in vivo on rats. Experimental diets for the rats contained 70 g lipids/kg. Corn oil used as basal oil was in part replaced by 1.375 g of starfish phospholipid or 1.375 g of squid phospholipid. The same molar amounts of EPA ethyl ester and DHA ethyl ester were used as positive controls. Rats were fed for 3 weeks, and 48 hours before sacrificing, then dimethylhydrazine was injected as a carcinogen to induce the initiation of cancer. After sacrificing, DNA in the crypt cells on the large intestine was dyed and the apoptotic cells were counted through a microscope. Results, as displayed in Table 4-E (experiment done by Kota Zama and Kenji Fukunaga), show that the number of the apoptotic cells were comparable among all the n-3 polyunsaturated fatty acid-related lipid served groups treated with dimethylhydrazine, regardless of the lipid classes administrated. They all showed avoidance of developing into true cancer by inducing the damaged cells to undergo apoptosis. In contrast, the rats without n-3 polyunsaturated fatty acid served showed only 1/3 to 1/4 the number of apoptotic cells. From these results, we can say that as long as n-3 polyunsaturated fatty acids are structured in the lipid molecules, there seem to be no difference in performance on avoiding cancer initiation. Nevertheless, among the

Table 4-E. Number of Apoptotic and Mitotic Cells in Crypts

Diet	Injection	*n*	Number of apoptotic cells	Number of mitotic cells
EPA-ethyl ester	DMH	5	4.4±0.8*	0.8±0.4
	Saline	6	0.5±0.5	1.3±0.7
Starfish PC	DMH	5	3.6±0.5*	0.4±0.5
	Saline	6	0.5±0.5	1.3±0.5
DHA-ethyl ester	DMH	5	4.4±0.5*	0.6±0.8
	Saline	6	0.5±0.5	1.5±0.5
Squid PC	DMH	5	5.2±0.7*	0.4±0.5
	Saline	6	0.5±0.5	0.8±0.7
Basal diet	DMH	5	1.2±0.7	1.4±0.5
	Saline	7	0.0±0.0	1.8±0.7

DMH = dimethylhidrazine; PC = phosphatidylcholine
*Significantly different from basal diet group, $p < 0.05$.

three widely used lipids such as PC, ethyl ester, and TG, the PC is the most beneficial. The ethyl esters are not allowed for foods, and TG usually requires emulsifier to disperse into foods. Another benefit of using phospholipid is the oxidative stability. By adding astaxanthin in combination with tocopherol, marine phospholipids, the n-3 polyunsaturated fatty acids containing phospholipids become extremely stable against oxidation (Takeuchi et al., 1997).

FUNCTIONALITY OF MARINE PHOSPHOLIPID LIPOSOMES

SUPPRESSION OF CANCER USING MARINE PHOSPHOLIPIDS IN COMBINATION WITH MUSHROOM IMMUNE ACTIVATOR

Certain mushroom water extracts contain anticancer β-glucans. We considered that encapsulating anticancer β-glucans in n-3 polyunsaturated fatty acid containing phospholipid liposomes may promote the anticancer effect of β-glucans. Phospholipids with n-3 polyunsaturated fatty acids (i.e., marine phospholipids) themselves exert anticancer effect, as aforementioned. To bear out this hypothesis, we carried out an in vivo study on mice to see promotional effects on cancer tumor suppression when *Agaricus blazei* Murrill water extract encapsulated marine phospholipid is orally taken as a liposomal drink. Myeloma sp2 cells were implanted (3.0×10^5 cells/BALB/c male mouse). After 20 days, mice with comparable tumor sizes were selected and divided into the following four groups by changing the drinks: (1) control (water), (2) 0.4 mg/ml squid phospholipid liposome [PC/PS/Cholesterol(Chol) = 9/1/10 molar ratio] alone, (3) 0.5 mg/ml *Agaricus blazei* Murrill water extract alone, and (4) 0.4 mg/ml squid phospholipid liposome (PC/PS/Chol=9/1/10 molar ratio) with 0.5 mg/ml *Agaricus blazei* Murrill water extract partially encapsulated. Liposomal drinks with or without *Agaricus blazei* Murrill water extract were orally administrated for 3 weeks. The orally administrated liposomal drink amount per mouse was approximately 5 ml a day. The results showed that by serving the mice with only low concentration of squid phospholipid liposome or *Agaricus blazei* Murrill water extract individually, no appreciable suppression against cancer growth was observed. However, when both the squid phospholipid and *Agaricus blazei* Murrill water extract were combined, suppression was observed, as shown in Fig. 4.6. To see the reproducibility of this promotional effect on mushroom immune activator, another in vivo study with some modifications was done. Half the number of myeloma sp2 cells (1.5×10^5 cells/BALB/c male mouse) compared to the prior experiment was implanted. After 20 days, mice with comparable size were selected and divided into four groups. Liposomal drinks with *Agaricus blazei* Murrill water extract with double concentration in the form of simple mixture or in the form of partially encapsulated liposomes were orally administrated for 3 weeks. The experimental groups were designed as follows: (1) control (water), (2) 1.0 mg/ml squid phospholipid liposome (PC/PS/Chol=9/1/10 molar ratio) alone, (3) 1.0 mg/ml *Agaricus blazei* Murrill water extract alone, (4) 1.0 mg/ml squid phospholipid liposome (PC/PS/

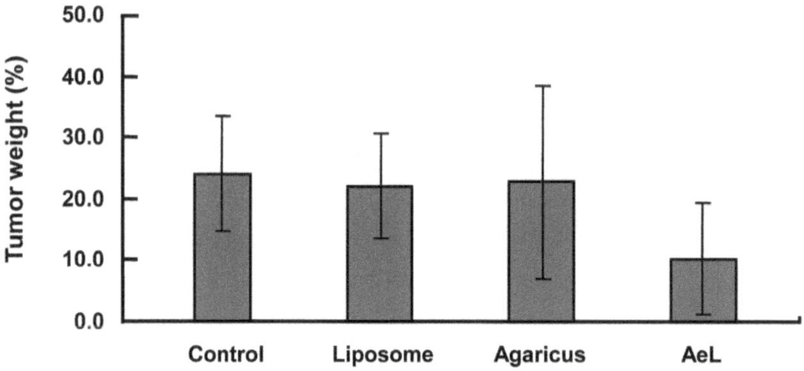

Fig. 4.6. Tumor weight % against body weight of Myeloma sp2 bearing BALB/c mice. Samples were administered 7 days after Myeloma sp2 tumor (3.0×10^5 cells) implantation. Marine phospholipid based liposomes (0.4 mg/ml) were orally administered from day 0 to 13, and Agaricus (0.5 mg/ml) from day 0 to 28. AeL; Agaricus (0.5 mg/mL) encapsulated liposomes(0.4 mg/ml) were also administered from day 0 to 28. (n=7).

Chol=9/1/10 molar ratio) with 1.0 mg/ml *Agaricus blazei* Murrill water extract in the form of simple mixture, and (5) 1.0 mg/ml squid phospholipid liposome (PC/PS/Chol=9/1/10 molar ratio) with 1.0 mg/ml *Agaricus blazei* Murrill water extract partially encapsulated. The orally administrated liposomal drink amount per mouse was approximately 5 ml per day. As seen in Fig. 4.7 (Murakawa et. al., 2007), *Agaricus blazei* Murrill water extract alone and squid phospholipid alone groups show moderate tumor suppression with total administrations of approximately 105 mg/mouse throughout the experimental term. When both *Agaricus blazei* Murrill water extract and squid phospholipid were administrated simultaneously in a simple mixture form with the same amount of the individual components, promotional effect of squid phospholipid liposome on sp2 myeloma cancer tumor suppression was observed. Similarly, when *Agaricus blazei* Murrill water extract was partially encapsulated with squid phospholipid liposome with total administrations being the same amount with the above simple mixture form (105 mg/mouse), promotional effect on cancer tumor suppression was more enhanced. Though there was no significant difference in tumor sizes between the simple mixture form administrated group (i.e., Group 4) and the β-glucan partially encapsulated form administrated group (i.e., Group 5), the promotional effect seemed more prominent in the encapsulated form. In fact, although it was only one, the mouse that had a tumor disappear was in the partially encapsulated form administrated group. If we succeed in encapsulating more amount of *Agaricus blazei* Murrill water extract in the marine phospholipid liposomes, there should be a significant difference in tumor sizes between Group 4

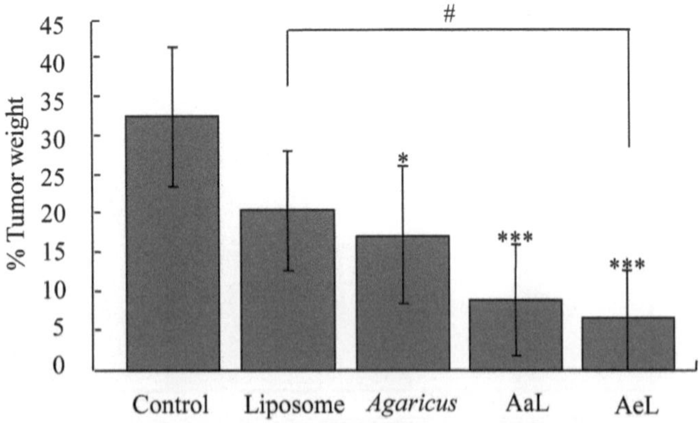

Fig. 4.7. Effect of marine phospholipid based liposomes and *Agaricus* on Tumor weight (%) against Body Weight of Myeloma sp2 Bearing BALB/c Mice (Murakawa et al., 2007). Myeloma sp2 tumor cells (1.5 $\times 10^5$ cells/mice) were implanted into mice. Twenty days after implantation of the cells, liposomes (1.0 mg/mL), *Agaricus* (0.5 mg/mL), AaL; *Agaricus* (0.5 mg/mL)-liposomes (1.0 mg/mL) mixture and AeL; *Agaricus* (0.5 mg/mL) encapsulated liposomes (1.0 mg/mL) were administered for 21 days. The results are the means SD (n=7). Asterisks indicate significant difference as compared to control (*P<0.05, ***P<0.001).

and 5. Thus it was concluded that *Agaricus blazei* Murrill water extract encapsulated marine phospholipid may be useful in myeloma sp2 therapy.

SUPPRESSION OF CANCER USING MARINE PHOSPHOLIPIDS IN COMBINATION WITH ADHESIVE CHITOSAN

Chitosan has been applied in the food, agricultural, and cosmetic industries due to its high availability, high biocompatibility, biodegradability and ease of chemical modification. Besides these industrial applications, scientists recently identified its potential application in the field of pharmaceutical sciences (Singla & Chawla, 2000; Aspden et al., 1995; Schipper et al., 1996). It is applied in several drug delivery systems due to its lack of toxicity and transmucosal penetration enhancer properties (Paul & Sharma, 2000). It may induce very little cellular toxicity (Berscht & Nies, 1995; Dornish et al., 1997) and is naturally biodegradable because it is metabolized by lysozymes both in vitro and in vivo (Onishi & Machida, 1999). Currently, it is attracting much attention as a possible mediator of transepithelial drug delivery due to its ability to mediate increased transepithelial drug permeation both in vitro (Dodane et al., 1999; Smith et al., 2004) and in vivo (Illum et al, 2002; Degim et al., 2003). Furthermore, the residence time of the drug carrier systems in the gastrointestinal tract is an important factor controlling the bioavailability of drugs of high molecular weight. In the early 1990s, the mucoadhesive function of chitosan received much attention for prolonging the residence time of dosage forms at the absorption site (Gupta et al.,

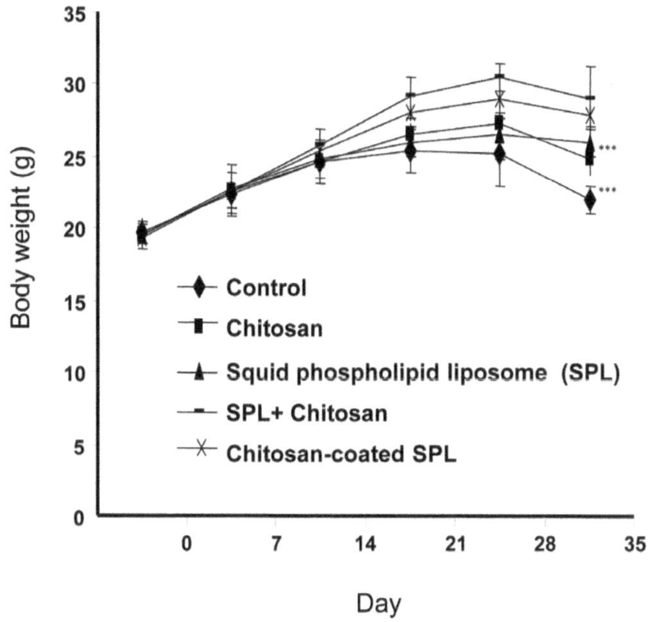

Fig. 4.8. Effect of squid meal liposome and chitosan samples on the body weight of myeloma sp2 bearing BALB/c mice. Twenty days after implantation of the cells, liposomes (1.0 mg/ml), chitosan (5.0 mg/ml), chitosan (5.0 mg/ml)-liposomes (1.0 mg/ml) mixture and chitosan-coated (5.0 mg/ml) liposomes (1.0 mg/ml) were administered for 35 days. The results are the means difference as compared to control and liposome (***$P<0.001$). SD (n=12). Asterisks indicate significant

1990). It has been reported that the stability of chitosan-coated liposomes in gastric fluid was significantly higher as compared to uncoated liposomes. It has been proposed that chitosan-coated liposomes might be utilized as a bioadhesive intestinal delivery system (Filipović-Grčić et al., 2001). Thus, we considered that chitosan-coated marine phospholipid may be more beneficial in promoting antitumor activity than marine phospholipid or chitosan itself. It is anticipated that chitosan-coated marine phospholipid liposomes may also promote the anticancer effects by prolonging the retention on intestinal wall. Therefore, we investigated the tumor suppression effects of marine phospholipids with chitosan on myeloma sp2-induced mice.

Weanling BALB/c mice were divided into five groups. The groups were (1) Control (water), (2) 1.0 mg/ml squid phospholipids liposome (PC/PS/Chol=9/1/10 molar ratio) alone, (3) 5.0 mg/ml chitosan alone, (4) chitosan (5 mg/ml) + squid phospholipid (1mg/ml) liposome mixture, and (5) chitosan-coated (5 mg/ml) squid phospholipid liposome (1 mg/ml). The orally administered liposomal drink amount per mouse was approximately 5 ml per day. As shown in Figs. 4.8 and 4.9, chitosan

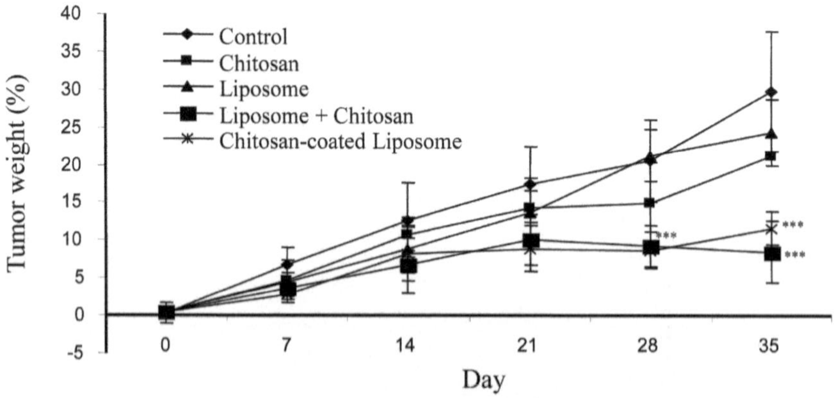

Fig. 4.9. Effect of squid meal liposome and chitosan samples on the tumor weight of myeloma sp2 bearing BALB/c mice. Twenty days after implantation of the cells, liposomes (1.0 mg/ml), chitosan (5.0 mg/ml), chitosan (5.0 mg/ml)-liposomes (1.0 mg/ml) mixture and chitosan-coated (5.0 mg/ml) liposomes (1.0 mg/ml) were administered for 35 days. The results are the means difference as compared to control and liposome (***$P<0.001$). SD (n=12). Asterisks indicate significant

alone and squid phospholipid alone groups showed slight tumor associated weight loss and tumor suppressions throughout the experiment term. There was a significant cancer tumor-associated weight loss suppression as well as tumor suppression effect when both chitosan and squid phospholipid were administrated simultaneously in a simple mixture form and chitosan-coated liposome form.

ANTIOBESITY PROPERTIES OF *UNDARIA* LIPID CONTAINING FUCOXANTHIN ENCAPSULATED WITH SCALLOP PHOSPHOLIPID LIPOSOMES

Brown seaweed (*Undaria pinnatifida*) lipid containing fucoxanthin (UL) encapsulated into scallop midgut gland phospholipids (PL) liposomes were prepared to see the promotional effect of marine phospholipid liposome on antiobesity (Okada et al., 2011). The animal models used in that study were 3-week-old male diabetic-obese mice model KK-A^y mice. Each group received different combinations of lipid drinks [i.e., 0.2%UL, 0.3%PL (scallop midgut gland phospholipid), 0.2%UL encapsulated 0.3%PL liposomes]. Animals were sacrificed after a 4-week experimental feeding period, and adipose tissues and organs were dissected and weighed. The mesenteric white adipose tissue (WAT) weight was reduced in all experimental groups as compared to the control group. This reduction was significantly different ($P < 0.05$) from the control group for mice receiving the UL encapsulated PL liposomal drink. Total WAT is composed of perirenal WAT, retroperitoneal WAT,

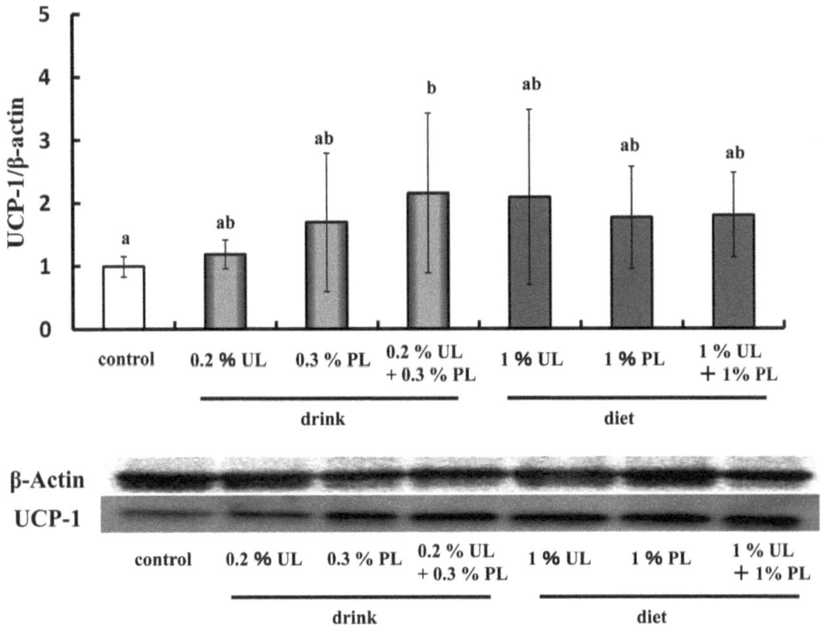

Fig. 4.10. Western blotting analysis of uncoupling protein 1 (UCP1) in epididymal WAT and relative expression level of UCP1 protein compared to β-actin (Okada et al., 2011). Columns with different superscript letters are significantly different ($p < 0.05$).

mesenteric WAT, epididymal WAT, and gluteal WAT. Only the UL encapsulated PL liposome (9.02 ±1.23 g/100 g BW) exhibited a significant weight reduction compared to the control group (10.1 ± 0.41 g/100 g BW). Uncoupling protein 1 (UCP1) mRNA expression levels were determined by real-time polymerase chain reaction analysis, and UCP 1 expression was determined by western blotting analysis. UCB 1 is known to decrease white adipose tissue by being up regulated. Both UCP 1 (Fig. 4.10) and UCP 1 mRNA expression (Fig. 4.11) in epididymal fat from mice fed with the UL encapsulated PL liposomes were significantly higher than in the control group. Although both the group receiving 0.2%UL and the group receiving 0.2%UL encapsulated 0.3%PL liposomes exhibited higher UCP 1 mRNA expression levels than the control, these levels were significantly different only in the case of the liposome served group (4.23 times higher than that of control group). These results suggest that UL and PL work in an additive manner to reduce obesity, resulting in a greater antiobesity effect than that which would be expected at these reduced concentrations.

Further, these results suggest that even if the administration amount of fucoxanthin is too small to exert its antiobesity property; it should turn out to be possible

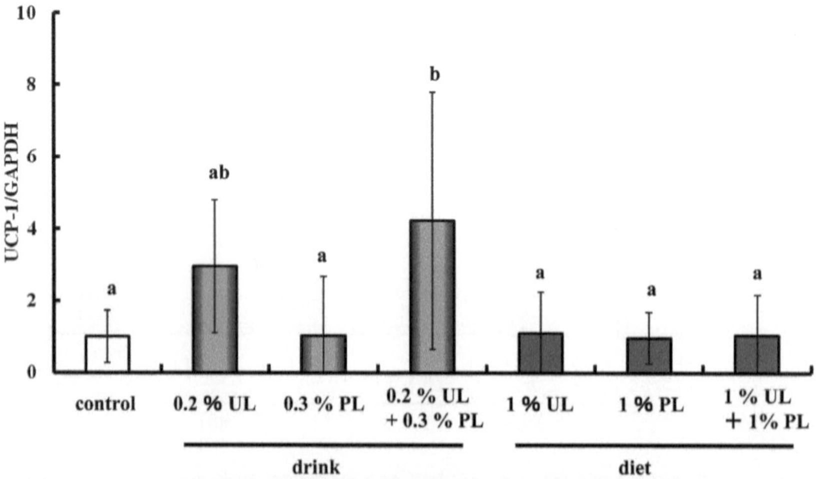

Fig. 4.11. UCP1 mRNA expression levels in epididymal WAT (Okada et al., 2011). Expression of UCP1 mRNA was estimated by quantitative real-time polymerase chain reaction. Relative values were presented as the ratio of UCP1 mRNA to GAPDH mRNA.

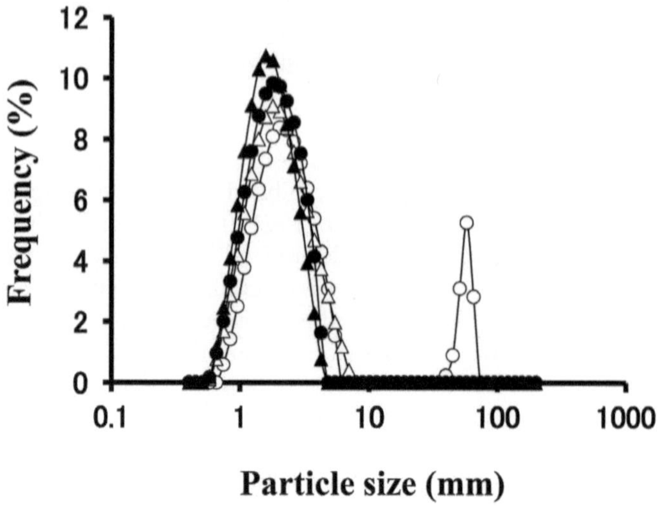

Fig. 4.12. Particle size distribution of 20 times filtered (0.8 mm) marine phospholipid based liposomes after 14 days. Liposome (PC + PS = 1 mg/ml) was diluted to about 10 times volume with PBS. Size distribution of liposomes was analyzed with a laser diffraction particle size analyzer. (Wave length 670 nm, SALD-200V, SHIMADHU). ○,PS 0%; Δ,PS 5%; ●,PS 10%; ▲,PS 20%.

to exert its function by encapsulating fucoxanthin containing lipids into marine phospholipid liposomes and serve as drinks. Though fucoxanthin is an excellent functional compound to reduce WAT, limited supply has been a bottle neck. The aforementioned β-glucan is also precious. We conclude that marine phospholipids, especially in the form of liposomes, should help to boost the effect of precious functional compounds so as to bring them to the marketplace.

STABILITY OF MARINE PHOSPHOLIPID LIPOSOMES

The most abundant phospholipid class in marine phospholipid is PC, as in most natural sources. However, uniquely with PC, liposome particles tend to be fused when they bump together. PS is known to have a negative charge, and for this reason we considered that marine PC liposomes containing PS inserted in the liposome bilayer should make the liposome more stable by repulsing each other, owing to the PS inserted liposome negative charge. We examined the minimum amount of PS essential for stabilization. As shown in Fig. 4.12 (experiment done by Shinichiro Manabe), it appeared that 10% or less is enough to avoid fusion.

Microbes can be eliminated by using membrane filtration. Autoclaving is also available for avoiding microbe harm. However, autoclaving causes phospholipid oxidation. With astaxanthin (0.03 mM) and α-tocopherol (0.03 mM), marine phospholipid liposomes become 4 times more stable against oxidation (unpublished

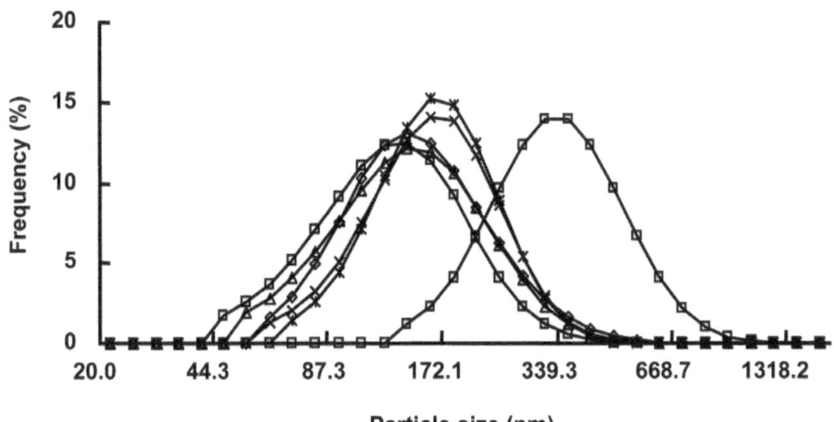

Fig. 4.13. Particle size distribution of freeze-thawed marine phospholipid based liposomes. Lipid composition of liposomes was PC:PS:Chol=1:1:2. The size distribution of liposomes was analyzed in the same manner with Fig. 4.12. Number of freeze-thawing: ◆, 0 time; ▲, 4 times; □, 5 times.

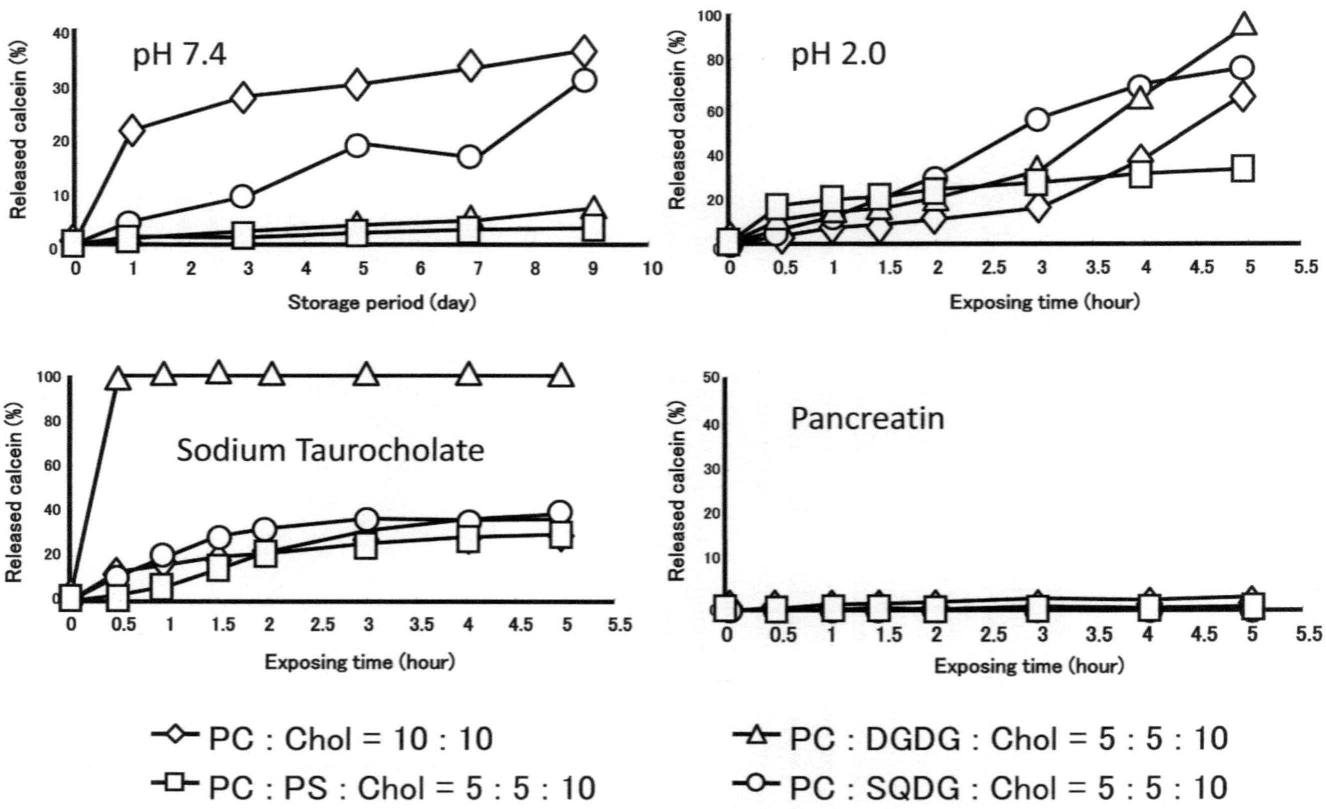

Fig. 4.14. Stability of marine phospholipid based liposomes under various digestive juice conditions.

data). Oxidative stability of marine phospholipids in the liposomal form was recently reviewed in detail (Lu et al., 2011).

MARINE PHOSPHOLIPID LIPOSOMES SIZE DISTRIBUTION AFTER FREEZING AND THAWING

In practical use of marine phospholipid liposomes, freezing is necessary for long periods of storage. Cholesterol (Chol) is essential to keep marine phospholipid-based liposomes stable in all aspects. However, if cholesterol is coupled together with negative impression, phytosterol esters can be substituted for half of it (Iwakura, 2008). The mean particle sizes of freeze-thawed liposomes (PC/PS/Chol=1:1:2) ranged from 46.2 to 560, 65.8 to 504, and 129 to 1102 nm for 0, 4, and 5 times freeze–thawing, respectively, as shown in Fig. 4.13 (experiments done by Kentaro Murakawa). It is obvious that the size of the liposomes increases by repeated freezing and thawing.

STABILITY OF MARINE PHOSPHOLIPID-BASED LIPOSOMES UNDER GASTROINTESTINAL CONDITIONS

In the early 2000s, the stability of marine phospholipid-based liposomes under acidic conditions was studied (Cansel et al., 2001; Nacka et al., 2001b). In addition to acidic condition raging from pH 1.5 to neutral, influence of bile salts, phospholipase A_2, and temperature under 37°C on marine phospholipid-based liposomes was also investigated (Nacka et al., 2001a). The studies showed that acidification induces instantaneous vesicle aggregation in marine phospholipid-based liposomes, which was partially reversed when the external medium was neutralized. It is also pointed out that acidification causes a complex morphological bilayer rearrangement and lead to the formation of small aggregates. Nevertheless, these studies conclude that marine phospholipids-based liposomes could be used as an oral administration vector. Perfect encapsulation or freedom from leakage is not required for food use. We also examined the stability of marine phospholipid liposomes with marine glycolipids incorporated in the liposome bilayer under refrigeration conditions and gastrointestinal conditions (experiments done by Yuji Hirouchi). As for gastric juice, bile acid, and digestive enzymes, we employed pH 2 Tris-hydrochloric acid (HCl) buffer, 10 mM sodium taurocholate and 150 mM sodium chloride (NaCl) containing 10 mM Tris-HCl buffer (pH 7.4), and 0.02% pancreatin containing 5.0 mM calcium chloride, and 160 mM NaCl containing 10 mM Tris-HCl buffer (pH 7.4) solution, respectively. As illustrated in Fig. 4.14, the addition of PS was effective in stabilizing marine phospholipids-based liposomes under all the conditions examined. Digalactosyldiacylglycerol (DGDG) and sulfoquinovosyldiacylglycerol (SQDG) obtained from algae were expected to enhance the transport of the liposomes in the small intestine but showed negative effect on the stability under bile salt exposure and acidic condition. It is not clear why the addition of these glycolipids weakens the liposome bilayer. A difference in fatty acid composition in relation to the combination of the fatty acid might, at least in part, affect the stability against digestive juice. It was reported (Hossain, 2009) that marine phospholipid liposome is relatively stable as long as the concentration of phospholipase A_2 is less than 5 μg/ml (10 mM Tris-HCl, 145 mM NaCl, pH 7.4).

TRANSPORT OF MARINE PHOSPHOLIPID-BASED LIPOSOMES THROUGH SMALL INTESTINAL EPITHERIAL M-CELL MODEL

It has long been believed that all the diacylglycerophospholipid molecules are partially hydrolyzed to lysophospholipids and free fatty acids in the digestive track. In the early 1990s, it was observed that diacylphospholipids should be absorbed at least in part by keeping their molecular structure (Galli et al., 1992). Besides this, marine phospholipid-based liposomes are expected to be incorporated by small intestinal epitherial M-cells, maybe by phagocytosis (Fig. 4.15). We induced M-cells in the small intestinal epitherial cell monolayer derived from Caco-2 cell monolayers by coculturing Raji cells (experiments done by Yuji Hirouchi). A typical view of M-cells prepared with obvious lack of micro villi is shown in Fig. 4.16. Into this M-cell, we inserted a small intestinal epitherial cell monolayer. To the simple small intestinal epitherial cell monolayer, we served the following hybrid marine phospholipids-based liposomes: (1) PC : Chol = 10 : 10 (mol/mol), (2) PC : PS : Chol = 5 : 5 : 10 (mol/mol/mol), (3) PC : DGDG : Chol = 5:5:10 (mol/mol/mol), (4) PC : SQDG : Chol = 5:5:10 (mol/mol/mol), and (5) PC:PS : DGDG:Chol = 5 : 3 : 2 : 10 (mol/mol/mol/mol). As illustrated in Fig. 4.17, (2), (3), and (4), namely the liposomes which have in part substituted PS or DGDG or SQDG for PC, showed higher transport through M-cells. Phagocytotic cells are known to engulf charged particles. The PS and SQDG have electric charge, and this may be one reason why the PS or SQDG inserted liposomes are well transported by the M-cells. This happens because M-cells preferably transport charged objects. On the other hand, DGDG is traditionally known as a

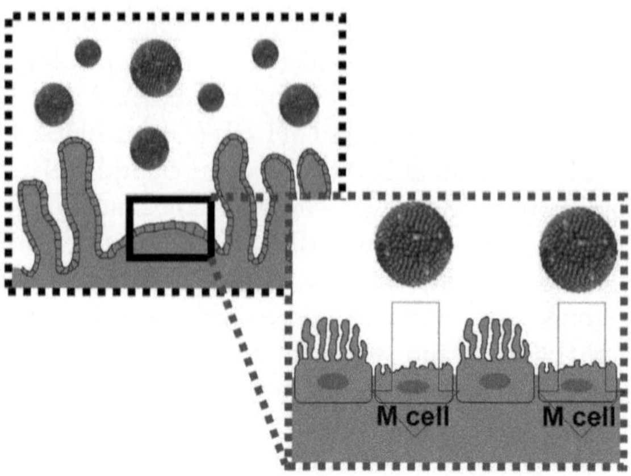

Fig. 4.15. Transport of marine phospholipid based liposomes through M-cells in the small intestine.

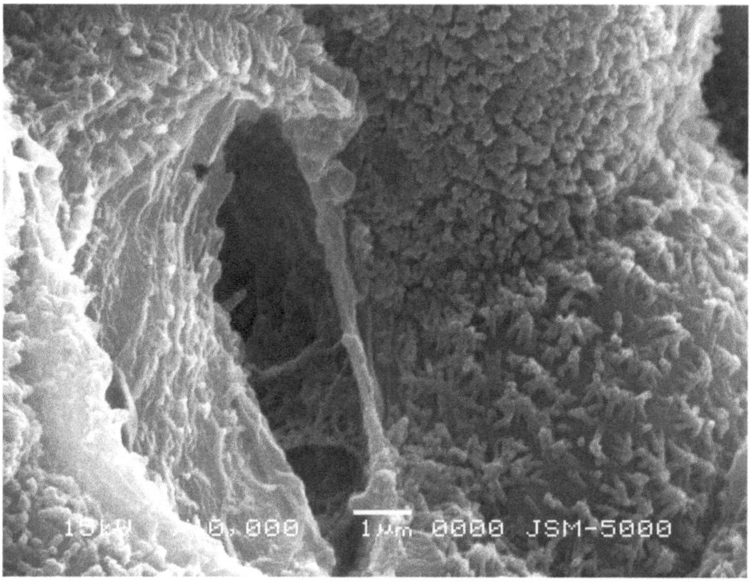

Fig. 4.16. Magnified view of M-cells prepared.

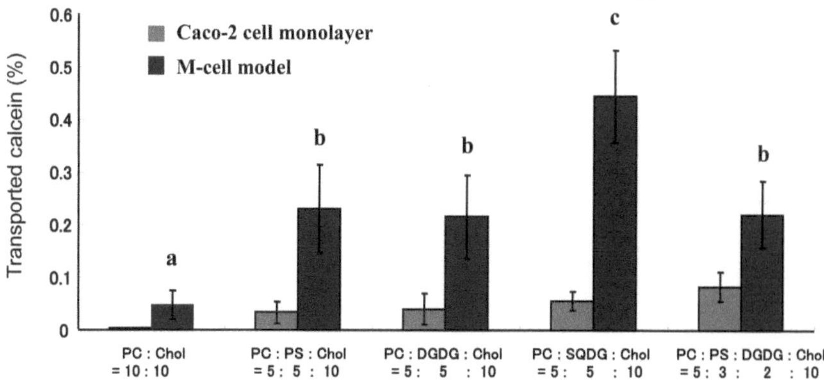

Fig. 4.17. Transport of encapsulated calcein through M-cells derived small intestinal epithelial cell model under 37 celcius. Total lipid concentration was 1.0 mM and liposomes with PC:Chol=1:1 was considered as control. Error bars represent standard deviation of mean. Group b is significantly different from group a ($p<0.01$) and group c is significantly different from group a and b ($p<0.01$).

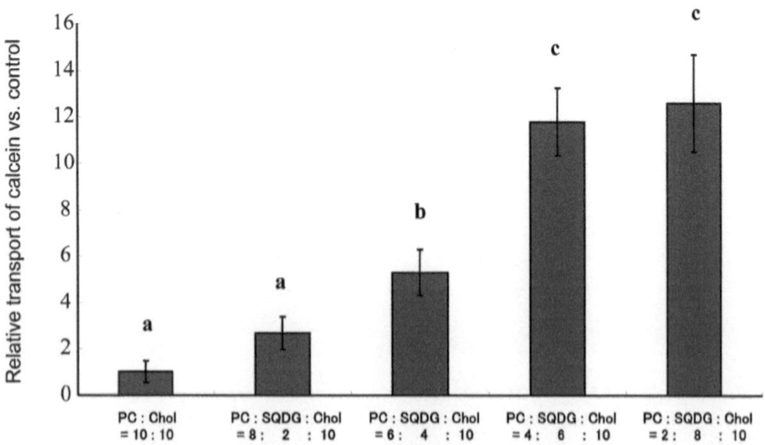

Fig. 4.18. Transport of calcein encapsulated marine phospholipid liposomes through co-cultured Caco-2 cell and Raji cell (M-cell model) monolayer (n=6-8) by varing SQDG ratio in the liposomal bilayers. Total lipid concentration was 1.0 mM and liposomes with PC:Chol=1:1 was considered as control. Error bars represent standard deviation of mean. Group b is significantly different from group a ($p<0.01$) and group c is significantly different from group a and b ($p<0.01$), Kruskal-Wallis test.

promoter in transporting functional compounds, which are considered difficult to be absorbed in the small intestine, even if the compound is soluble in aqueous media. We increased the SQDG ratio to measure the increase in negative change by measuring ζ charge of the hybrid liposomes prepared. The liposomes prepared were (1) PC : Chol = 10 : 10 (mol/mol) as basal bilayer, (2) PC : SQDG : Chol = 8 : 2 : 10 (mol/mol/mol) as 10%SQDG addition, (iii) PC : SQDG : Chol = 6 : 4 : 10 (mol/mol/mol) as 20%SQDG addition, (4) PC : SQDG : Chol = 4 : 6 : 10 (mol/mol/mol) as 30%SQDG addition, and (5) PC : SQDG : Chol = 2 : 8 : 10 (mol/mol/mol) as 40%SQDG addition. As shown in Fig. 4.18, transport of the hybrid liposomes increased in an SQDG amount-dependent manner at least until the 30% SQDG addition. When the SQDG amount increased to 40%, transport reached the ceiling. ζ charge decreased in negative proportion to the SQDG amount until 20% SQDG addition. However, at 30% SQDG addition, ζ charge became almost the same as that of 20% SQDG addition. And when the SQDG amount increased to 40%, ζ charge moderately decreased. This result shows that ζ charge alone is not responsible for the increase in transport through M-cells. We tried to learn what is responsible for the engulfment of the SQDG containing marine phospholipids-based hybrid liposomes by the M-cells. It has been recognized that there are at least three ways to engulf particles by M-cells: pinocytosis, endocytosis, and phagocytosis. No inhibition was observed by using EIPA (a pinocytosis inhibiter) and chlorpromazine (an endocytosis inhibitor). For this reason, we inferred that phagocytosis might be the most responsible for the transport of SQDG containing hybrid marine phospholipid liposomes.

CONCLUSION

By designing marine phospholipid-based liposomes with various lipid classes inserted into the liposome bilayer, many kinds of desired specifications liposomes would be possible. If the desired hydrophilic functional compound has bitter or other kinds of unacceptable taste, the liposome encapsulation would be useful even though the liposomes collapse inside the stomach. When the functional compound is intended to be delivered into M-cells in the small intestine, the liposomes must be resistant to digestive juice and recognized by M-cells. For this purpose, sterols would be available for the physical reinforcement of marine phospholipid-based liposomes. And for promoting M-cell transportation, addition of PS and SQDG to the liposome bilayer would be useful. As long as the liposome bilayer is basically composed of marine phospholipids, they would promote the encapsulated functional compounds, and this is the principal advantage of choosing marine phospholipids as liposomal carriers.

REFERENCES

Aspden, T.J.; Adler, J.; Davis, S.S.; Skaugrud, Q.; Illum, L. Chitosan as a nasal delivery system: evaluation of the effect of chitosan on mucocilliary clearance rate in the frog palate model. Int. J. Pharm. 1995, 122, 69–78.

Berscht, P.; Nies, B.; Leibendorfer, A.; Kreuter, J. In vitro evaluation of biocompatibility of different wound dressing materials. J. Mat. Sci. Mater. Med. 1995, 6, 201–205.

Burns, C.P.; Haugstad, B.N.; Mossman, C.J.; North, J.A.; Ingraham, L.M. Membrane lipid alteration: effect on cellular uptake of metoxantron. Lipids 1988, 23, 393–397.

Burns, C.P.; Spector, A.A. Membrane fatty acid modification in tumor cells: a potential therapeutic adjunct. Lipids 1987, 22, 178–184.

Cansell, M.; Moussaoui, N.; Lefrancois, C. Stability of marine lipid based-liposomes under acid conditions. Influence of xanthan gum. J. Liposome Res. 2001, 11, 229–242.

Degim, I.; Acrturk, F.; Erdogan, D.; Lortlar, N. Transdermal administration of bromocriptine. Biol. Pharm. Bull. 2003, 26, 501–505.

Di Marzo, V.; Griinari, M.; Carta, G.; Murru, E.; Ligresti, A.; Cordeddu, B.K.; Giordano, E.; Bisogno, T.; Collu, M.; Batetta, B.; et al. Dietary krill oil increases docosahexaenoic acid and reduces 2-arachidonoylglycerol but not N-acylethanolamine levels in the brain of obese Zucker rats. Int. Dairy J. 2010, 20, 231–235.

Diep, Q.N.; Touyz, R.M.; Schiffrin, E.L. Docosahexaenoic acid, a peroxisome proliferator-activated receptors-α ligand induces apoptosis in vascular smooth muscle cells by stimulation of p38 mitogen-activated protein kinase. Hypertension 2000, 36, 851–855.

Dodane, V.; Amin, K.M.; Merwin, J. Effect of Caco-2 on epithelial permeability and structure. Int. J. Pharm. 1999, 182, 21–32.

Dornish, M.; Hagen, A.; Hansson, E.; Pecheur, C.; Verdier, F.; Skaugrad, O. Safety of potasan: ultrapure chitosan salts for biomedical and pharmaceutical use. Domard, A., Roberts, G., and Varum, K., Eds.; Advances Chitin Sci., Jaques Andre: Lyon, 1997; pp. 694–670.

Filipović-Grčić, J.; Škalko-Basnet, N.; Jalšenjak, I. Mucoadhesive chitosan-coated liposomes: characteristics and stability. J. Microencap. 2001, 18, 3–12.

Galli, C.; Sirtori, C.R.; Mosconi, C.; Medini, L.; Gianfranceschi, G.; Vaccarino, V.; Scolastico, C. Prolonged retention of doubly labeled phosphatidylcholine in human plasma and erythrocytes after oral administration. Lipids. 1992, 27, 1005–1012.

Grimaldi, P.A. Fatty acid regulation of gene expression. Curr. Opin. Clin. Nutr. Metab. Care 2001, 4, 433–437.

Gupta, P.K.; Leung, S-HS.; Robinson, J.R. Bioadhesive/mucoadhesives in drug delivery to the gastrointestinal tract. Bioadhesive Drug Delivery Systems; Lenaerts, V. and Gurny, R., Eds.; CRC Press: Boca Raton, FL, 1990; pp. 65–92.

Hosokawa, M. Improvement of biological response of cells by marine lipids. Marine Functional Lipids: Sources, Functionalities, and Applications, Edition Number 142; Takahashi, K., Ed.; Kouseisha Kouseikaku: Tokyo, 2004; pp. 151–152 (In Japanese).

Hosokawa, M.; Sato, A.; Ishigamori, H.; Kohno, H.; Tanaka, T.; Takahashi, K. Synergistic effects of highly unsaturated fatty acid-containing phosphatidylethanolamine on differentiation of human leukemia HL-60 cells by dibutyryl cyclic adenosine monophosphate. Japanese J.Cancer Res. 2001, 92, 666–672.

Hossain, Z.; Fukunaga, K.; Tanouchi, M.; Takahashi, K. Chitosan and marine phospholipids reduce matrix metalloproteinase activity in myeloma SP2 tumor-bearing mice. Eur. J. Lipid Sci. Technol. 2009, 111, 877–889.

Hossain, Z.; Konishi, M.; Hosokawa, M.; Takahashi, K. Effect of polyunsaturated fatty acid-enriched phosphatidylcholine and phosphatidylserine on butyrate-induced growth inhibition, differentiation and apoptosis in Caco-2 cells. Cell Biochem. Function 2006, 24, 159–165.

Illum, R.; Watts, P.; Fisher, A.; Hinchcliffe, M.; Norbury, H.; Jabbal-Gill, I.; Nankervis, R.; Davis, S. Intranasal delivery of morphine. J. Pharm. Exp. Ther. 2002, 301, 391–400.

Inoue, Y. Recent developments of polyunsaturated fatty acids and their derivatives related research. New Food Industry 2001, 43, 22–26 (In Japanese).

Iwakura, K., Master's thesis, Hokkaido University, Hakodate, 2008 (In Japanese).

Jenski, L.J.; Nanda, P.K.; Jiricko, P.; Stillwell, W. Docosahexaenoic acid-containing phosphatidylcholine affects the binding of monoclonal antibodies to purified Kb reconstituted into liposomes, Biochem. Biophys. Acta 2000, 1467, 293–306.

Larsson, S.C.; Kumlin, M.; Ingelman-Sundberg, M.; Wolk, A. Dietary long-chain n-3 fatty acids for the prevention of cancer: a review of potential mechanisms, Am. J. Clin. Nutr. 2004, 79, 935–945.

Lu, F.S.H.; Nielsen, N.S.; Timm-Heinrich, M.; Jacobsen, C. Oxidation stability of marine phospholipids in the liposomal form and their applications. Lipids 2011, 46, 3–23.

Murakawa, K.; Fukunaga, K.; Tanouchi, M.; Hosokawa, M.; Hossain, Z.; Takahashi, K. Therapy of myeloma in vivo using marine phospholipid in combination with Agaricus blazei Muril as an immune respond activator. J. Oleo Sci. 2007, 56, 179–188.

Nacka, F.; Cansell, M.; Entressangles, B. In vitro behavior of marine lipid-based liposomes, influence of pH, temperature, bile salts, and phospholipase A^2. Lipids 2001a, 36, 35–42.

Nacka, F.; Cansell, M.; Gouygou, J.P.; Gerbeaud, C.; Meleard, P.; Entressangles, B. Physical and chemical stability of marine lipid-based liposomes under acid conditions. Colloids Surf. B 2001b, 20, 257–266.

Okada, T.; Mizuno, Y.; Sibayama, S.; Hosokawa, M.; and Miyashita, K. Antiobesity effects of undaria lipid capsules prepared with scallop phospholipids. J. Food Sci. 2011, 76, 1, H2–H6.

Onishi, H.; Machida, Y. Biodegradation and distribution of water-soluble chitosan in mice. Biomaterials 1999, 20, 175–182.

Paul, W.; Sharma, C.P. Chitosan, a drug carrier for the 21st century. S.T.P. Pharm. Sci. 2000, 10, 5–22.

Schipper, N.G.; Varum, K.M.; Artursson, P. Chitosans as absorption enhancers for poorly absorbable drugs. 1. Influence of molecular weight and degree of acetylation on drug transport across human intestinal epithelial (Caco-2) cells. Pharm. Res. 1996, 13, 1686–1692.

Shina, T., Master's thesis, Hokkaido University. Hakodate, 2009 (In Japanese).

Singla, A.; Chawla, M. Chitosan: Some pharmaceutical and biological aspects—an update. J. Pharm. Pharmacol. 2000, 53, 1047–1067.

Smith, J.; Dornish, M.; Wood, E.J. Effect of chitosan on epithelial cell tight junctions. Pharm. Res. 2004, 21, 43–49.

Stillwell, W.; Ehringer, W.; Jenski, L.J. Docosahexaenoic acid increases permeability of lipid vesicles and tumor cells. Lipids 1993, 28, 103–108.

Takeuchi, M.; Hara, S.; Totani, Y.; Hibino, H.; Tanaka, Y. Autoxidative behavior of polyunsaturated phospholipids. I. Oxidative stability of marine oil containing polyunsaturated phospholipids. J. Jpn. Oil Chem. Soc. 1997, 46, 175–181.

Tochisawa, K.; Hosokawa, M.; Kohno, H.; Odashima, S.; Takahashi, K. Effect of phospholipids containing docosahexaenoic acid on differentiation and growth of HL-60 human promyelocytic leukemia cells. J. Jpn.Oil Chem.Soc. 1997, 46, 383–390.

Ulven, S.M.; Kirkhus, B.; Lamglait, A.; Basu, S.; Elind, E.; Haider, T.; Berge, K.; Vik, H.; Pedersen, J.I. Metabolic effects of krill oil are essentially similar to those of fish oil but at lower dose of EPA and DHA, in healthy volunteers. Lipids 2011, 46, 37–46.

Vamecq, J.; Latruffe, N. Medical significance of peroxisome proliferator-activated receptors. Lancet. 1999, 354, 141–148.

Winther, B.; Hoem, N.; Berge, K.; Reubsaet, L. Elucidation of phosphatidylcholine composition in krill oil extracted from Euphausia superba. Lipids 2011, 46, 25–36.

5

FORMULATION OF LIPID MICRO/NANODISPERSION SYSTEMS

Isao Kobayashi[1], Sosaku Ichikawa[2], Marcos A. Neves[3], Takashi Kuroiwa[4], and Mitsutoshi Nakajima[5]

[1]Food Engineering Division, National Food Research Institute, National Agriculture and Food Research Organization (NARO), 2-1-12 Kannondai, Tsukuba, Ibaraki 305-8642, Japan; [2]Graduate School of Life and Environmental Sciences, University of Tsukuba, 1-1-1 Tennoudai, Tsukuba, Ibaraki, 305-8572, Japan; [3]Graduate School of Life and Environmental Sciences, The Alliance for Research on North Africa (ARENA), University of Tsukuba, 1-1-1 Tennoudai, Tsukuba, Ibaraki, 305-8572, Japan; [4]Faculty of Engineering, Tokyo City University, 1-28-1 Tamazutsumi, Setagaya-ku, Tokyo, 158-8557, Japan; and [5]Graduate School of Life and Environmental Sciences, The Alliance for Research on North Africa (ARENA), University of Tsukuba, 1-1-1 Tennoudai, Tsukuba, Ibaraki, 305-8572, Japan.

INTRODUCTION

Micro/nanodispersions are multiphase systems consisting of a continuous liquid phase and micro/nanomaterials, such as emulsion droplets, micro/nanoparticles, micro/nanocapsules, (giant) vesicles and liposomes, and micro/nanobubbles. Lipid micro/nanodispersions contain at least one lipid in the systems. The major formulations include oil-in-water (O/W) emulsions, solid lipid micro/nanoparticles, (giant) vesicles, and liposomes. These formulations have useful applications for foods, pharmaceuticals, and cosmetics. In the case of O/W emulsions, lipid droplets are dispersed in the continuous phase, and bioactive lipids (e.g., carotenoids) can be dissolved in the droplets. This type of formulation has a better water dispersibility of lipids compared to those in bulk form, which may increase lipid absorption in vitro and in vivo (Neves et al., 2008b). Solid lipid micro/nanoparticles dispersed in the continuous phase generally consist of triglycerides of high melting point and bioactive lipids. Since lipids are crystallized in the micro/nanoparticles, the mobility and exchange of bioactive substances are suppressed, leading to less degradation of bioactive substances than those in lipid droplets (Weiss et al., 2008). Nanodispersions of bioactive lipids can be formulated by solvent evaporation of O/W emulsion droplets (Tan and Nakajima, 2005a; Leong et al., 2011) and drops (Ribeiro et al., 2008). Surface-active lipids are also used as emulsifiers for formulating lipid micro/nanodispersions. Vesicles and liposomes are small compartments stabilized by bilayer forming lipids. The lipid bilayer isolates the internal aqueous phase from the external aqueous phase, allowing the encapsulation of biological substances and hydrophilic bioactive substances in the vesicles and liposomes dispersible in the external aqueous phase. Vesicles and liposomes with sizes exceeding 1 μm are usually called giant vesicles and giant liposomes. Besides the preceding formulations, self-organized nanoparticles dispersible in the aqueous phase can be formed using a hydrophilic lipid as one of their components (Chuah et al., 2009a).

Lipid micro/nanodispersions are formulated by a top-down approach, a bottom-up approach, or a combination of both approaches. A top-down approach usually uses emulsification devices to break down a bulk phase into smaller droplets with sizes of 0.1 to 100 μm. Traditional emulsification devices (e.g., colloid mills and high-pressure homogenizers) apply shear, extension, and impact to the systems (McClements, 2004), resulting in the production of polydisperse emulsions with possible denaturation of bioactive substances. New emulsification techniques, such as membrane emulsification and microchannel (MC) emulsification, are capable of producing monodisperse emulsions (Nakashima et al., 2000; Kobayashi & Nakajima, 2006). MC emulsification has a better droplet size uniformity (typically <5%). In addition, MC emulsification has a very mild droplet-generation process based on spontaneous transformation of an oil-water interface (Sugiura et al., 2001a), which can prevent denaturation of bioactive substances. Solid lipid micro/nanoparticles are obtained by cooling down or solvent evaporation of emulsion droplets. A bottom-up approach formulates nanomaterials, such as nanoparticles and vesicles (and liposomes), by condensing molecules, monomers or ions into a liquid phase. For instance, in the physical process of this approach, individual molecules spontaneously assemble in a predefined manner under thermodynamic control to nanostructures (Föster & Konrad, 2003). This process can formulate thermodynamically stable nanodispersions with sizes of ~5 to 100 nm by balancing the attraction and repulsion forces between a pair of molecules as building blocks to produce nanomaterials with novel functionalities (Sanguansri & Augustin, 2006).

This chapter presents an overview of the authors' recent investigations into the formulation of lipid micro/nanodispersion systems by top-down and bottom-up approaches. We describe MC emulsification, developed by the authors' research group. We focus on the fundamentals of MC emulsification, MC emulsification devices, and the formulation of lipid microdispersions using MC emulsification. This chapter will also outline the formulation of lipid nanodispersions by top-down approach (e.g., emulsification and subsequent solvent evaporation) or bottom-up approach.

MICROCHANNEL EMULSIFICATION

FUNDAMENTALS

DROPLET GENERATION PROCESS. Droplet generation for MC emulsification is a unique and very mild process, which is driven by spontaneous pinch-off of the dispersed phase that has passed through the channels (Sugiura et al., 2001a). The forced flow (cross-flow and co-flow) of the continuous phase is not needed to generate droplets by MC emulsification and usually requires an energy input of 10^3 to 10^4 J/m^3.

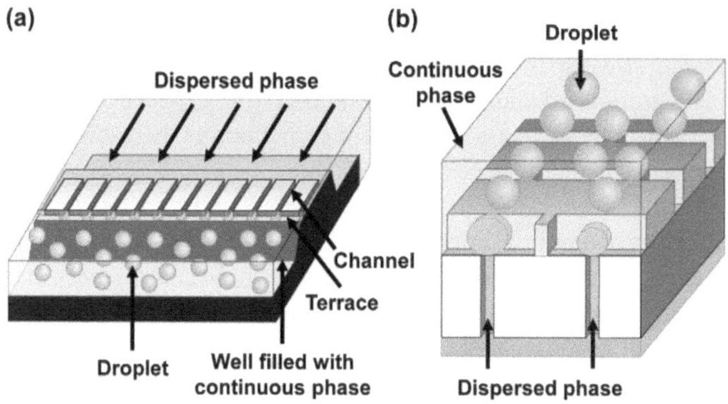

Fig. 5.1. Three-dimensional drawing of droplet generation via part of a grooved MC array (a) and part of an asymmetric straight-through MC array (b).

Typical droplet generation for MC emulsification is presented in Fig. 5.1. This droplet generation process has been investigated in detail using high-speed video microscopy (Sugiura et al., 2001a), computational fluid dynamics (CFD) (Kobayashi et al., 2004a; 2005a; van Dijke et al., 2008), and Lattice Boltzmann (LB) simulations (van der Zwan et al., 2009). Sugiura et al. (2001a) classified the droplet generation process into the expansion process and the detachment process. In the expansion process, the dispersed phase that has passed through the channels gradually expands on the terrace with a disk-like shape. Afterwards, the detachment process starts when the tip of the expanding dispersed phase passes through the terrace outlet. In the initial stage of this detachment process, the dispersed phase slowly expands in the well, since the Laplace pressure of the dispersed phase in the well is higher than on the terrace. The difference in these Laplace pressures gradually becomes smaller as the size of the expanding dispersed phase increases, and then their balance becomes opposite. The change in the Laplace-pressure balance causes rapid flow of the dispersed phase into the well, which forms a neck on the terrace. The neck instantaneously pinches off after the dispersed-phase flux at the neck exceeds that in front of the neck (van Dijke et al., 2008). This behavior leads to the generation of a droplet and completes the droplet generation process. Periodical droplet generation via the channels and a terrace yields a monodisperse emulsion containing uniform droplets. Previous MC emulsification researches have clarified some key points of the droplet generation, whereas further investigations are needed to obtain a complete understanding of the process.

EFFECT OF THE FLOW OF THE DISPERSED PHASE. The flow rate of the dispersed phase is the most important operating parameter in MC emulsification. The droplet size is not influenced by the velocity of the dispersed phase in a channel (or the pressure applied to the dispersed phase) below a critical value (Fig. 5.2). Below a critical flow rate of the dispersed phase, monodisperse emulsions were produced, and their droplet size distribution was independent of the flow rate (Kobayashi et al., 2003a). This feature is advantageous for the practical production of monodisperse emulsions. The droplet generation rate from each channel increases linearly with the increasing velocity of the dispersed phase, achieving a maximum value at a critical value (Fig. 5.2). Channel dimensions are also important parameters affecting the maximum droplet generation rate from each channel. When MC emulsification is performed using channels with a specific depth, an MC array with narrower and longer channels can produce monodisperse emulsions at higher droplet generation rates (Sugiura et al., 2002a). Thus, the channels used for MC emulsification should have a greater pressure drop of the dispersed phase, which is given by Fanning's equation:

$$\Delta P_{d,\,MC} = 4f(\rho_d U_d^2/2)\,(l_{MC}/d_{MC}) \qquad (1)$$

where $\Delta P_{d,\,MC}$ is the pressure drop of the dispersed phase in a channel, f is the Fanning's friction factor, ρ_d is the density of the dispersed phase, U_d is the velocity of the dispersed phase, l_{MC} is the channel length, and d_{MC} is the channel diameter.

In MC emulsification, it is important to appropriately control the flow of the dispersed phase during droplet generation, since the flow transition of the dispersed phase greatly influences the size uniformity of the generated droplets. Sugiura et

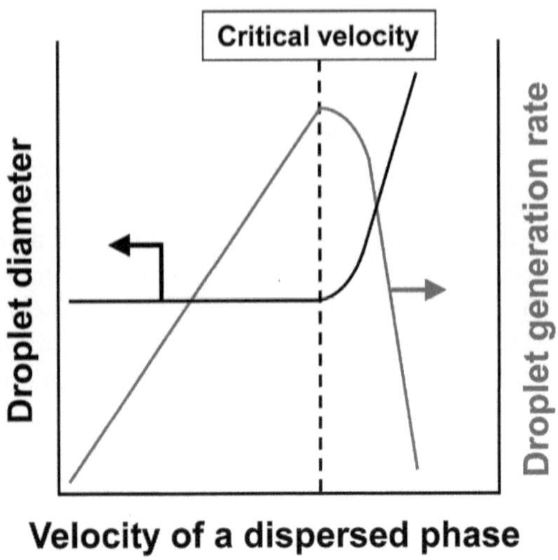

Fig. 5.2. Effect of the velocity of the dispersed phase on the droplet size and droplet generation rate for MC emulsification.

al. (2002b) characterized the flow of the dispersed phase during droplet generation using several dimensionless numbers, finding that the capillary number (*Ca*), defined as the ratio of the viscous force to the interfacial tension, can explain well the flow transition of the dispersed phase for MC emulsification. The interfacial tension dominates the successful generation of uniform droplets below a critical *Ca* (proportional to the velocity of the dispersed phase). Above the critical *Ca*, the influence of the viscous force cannot be neglected, and the resultant droplet size becomes sensitive to the velocity of the dispersed phase, resulting in the production of polydisperse emulsions. In fact, the critical *Ca* is dependent on the viscosity ratio, surface properties and geometry of the MC array, as well as the operating temperature (Butron Fujiu et al., 2011a). Further investigations are needed to completely understand the flow and its transition of the dispersed phase during droplet generation.

DROPLET SIZE. MC emulsification is capable of generating uniform droplets with sizes of 1 to 200 μm (Kobayashi et al., 2007a; 2010a). The droplet size depended primarily on the geometry of the MC array, especially the channel (and terrace) depth and the terrace length (Sugiura et al., 2002a; 2002c). The viscosity ratio also affects the droplet size to some extent at low viscosity ratio (dispersed phase/continuous phase) below a critical value (Fig. 5.3) (Kobayashi et al., 2005a; van Dijke et al., 2010). Sugiura et al. (2002c; 2004a) proposed analytical models for predicting the droplet size for MC emulsification. The predicted droplet size agreed well with the experimental results. Their prediction models consider the structure and dimensions

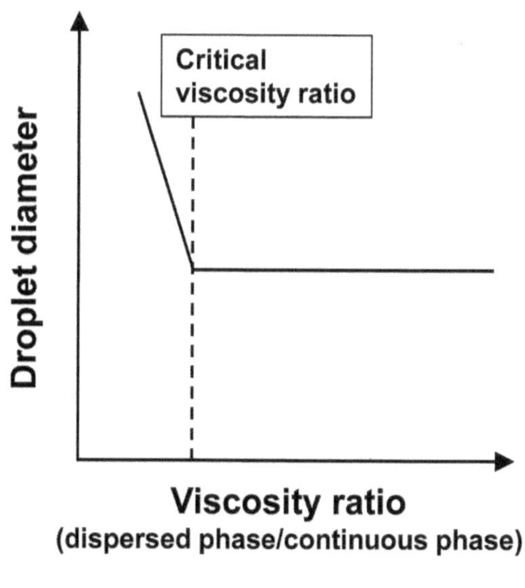

Fig. 5.3. Effect of the viscosity of the dispersed phase on the droplet diameter for MC emulsification.

of the channel and terrace but do not include the influences of the velocity of the dispersed phase and the viscosity of the two phases. van Dijke et al. (2008) derived an analytical model describing the droplet size based on the balance of the fluxes of the dispersed phase through the channel and the neck. This model includes the pressure applied to the dispersed phase as a dynamic parameter, the interfacial tension, and the contact angle besides the important dimensions of the channel and terrace. van der Zwan et al. (2009) also proposed a geometric model for describing the droplet size, which consists of an adapted *Ca* including the influences of the terrace width and height and the ratio of the terrace length to the terrace height. Although the latter two models would be useful for fast optimization of the process conditions, the viscosity of the continuous phase is not considered in these models.

INTERACTION BETWEEN THE CHANNEL SURFACE AND THE DISPERSED PHASE COATED BY SURFACTANT MOLECULES. Another key factor for successful MC emulsification is to prevent wetting of the dispersed phase to the channel and terrace surfaces. Hydrophilic and hydrophobic MC arrays are generally used for producing monodisperse O/W and W/O emulsions, respectively (Kawakatsu et al., 1997). Droplet generation by MC emulsification is critically influenced by the interaction between the channel and terrace surfaces and surfactant molecules covering the dispersed phase. A non-attractive interaction must be kept to stably produce monodisperse emulsions at high droplet generation rates (Butron Fujiu et al., 2011b). Single-crystal silicon is usually used as a material for MC array plates. The surface of the silicon plates becomes hydrophilic after a plasma-oxidation treatment. The charge of surfactant molecules greatly affects the generation process of O/W emulsion droplets by MC emulsification, since hydrophilic silicon MC ar-

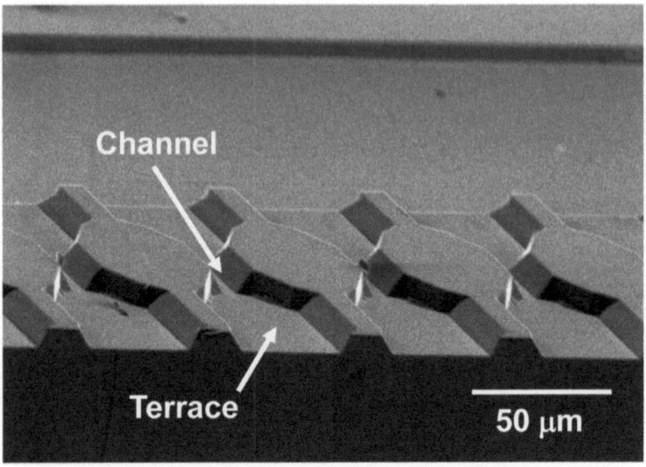

Fig. 5.4. Scanning electron micrograph of part of a silicon grooved MC array.

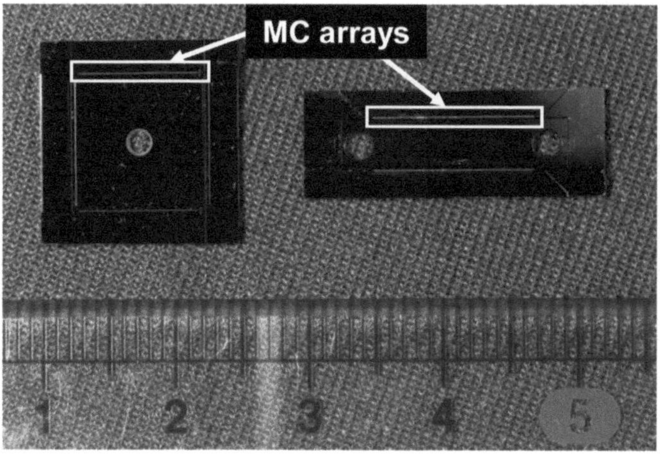

Fig. 5.5. Photographs of silicon grooved MC array plates. (Left) Dead-end type. (Right) Cross-flow type.

ray plates have a negatively charged surface. In contrast, silicon MC array plates after a hydrophobic treatment would have a non-charged surface. Tong et al. (2000) and Kobayashi et al. (2003a) demonstrated that anionic and nonionic surfactants are preferable for producing monodisperse O/W emulsions by MC emulsification. Contact angle measurements also provide useful information about the interaction between the plate surface and the dispersed phase covered by surfactant molecules (Kawakatsu et al., 2001a; Kobayashi et al., 2003a). Butron Fujiu et al. (2011a) recently proposed a novel contact-angle-measurement technique using an MC array consisting of parallel long channels. This technique allows one to acquire multiple contact angle data in a single experiment and to directly observe the interaction between the channel surface and the dispersed phase coated by surfactant molecules. The contact angle data obtained using this technique agreed well with those obtained using the conventional technique.

EMULSIFICATION DEVICES

GROOVED MICROCHANNEL ARRAYS. A grooved MC array consists of many parallel channels and terraces outside the channels (Fig. 5.4), and there are deeply etched wells outside the MC array. At least two grooved MC arrays are fabricated on each MC emulsification plate. Silicon grooved MC arrays are usually fabricated through repeated processes of photolithography and anisotropic wet etching (Kikuchi et al., 1992). Grooved MC arrays are sealed by physically attaching them to a flat glass plate in a module during experiments.

Grooved MC array plates are classified into dead-end and cross-flow types (Fig. 5.5). The first MC emulsification plate was a dead-end type and had four grooved

MC arrays (Kawakatsu et al., 1997). Dead-end type plates are useful for analyzing the droplet generation behavior via an MC array, whereas it is not straightforward to collect the produced emulsions using such plates. As a solution to this drawback, Kawakatsu et al. (1999) designed cross-flow type plates on which the generated droplets can be collected by the cross-flow of a continuous phase in a channel between two MC arrays. The cross-flow type plates enabled the long-term operation of MC emulsification and the reliable generation of uniform droplets with sizes of 1 to 100 μm, having been used in many MC emulsification researches. However, the cross-flow type plates have a low throughput of uniform droplets (e.g., <0.1 mL/h for vegetable oil-in-water emulsions). Recently, Kobayashi et al. (2010c) designed a large cross-flow type plate consisting of 14 MC arrays and demonstrated mass production of uniform vegetable oil droplets with a Sauter mean diameter of about 10 μm at 1.5 mL/h. This droplet throughput can satisfy the production of emulsion samples on a laboratory scale. We consider that parallelization of the large MC emulsification plates is preferable as a further scaling-up approach.

Single-crystal silicon has been primarily used as a material of grooved MC arrays. Grooved MC arrays have also been fabricated on polymer and stainless steel plates. Liu et al. (2004) fabricated grooved MC arrays made of poly (methyl methacrylate) (PMMA) by injection molding. PMMA is an inherently hydrophobic polymer. Monodisperse W/O emulsions were produced using cross-flow type PMMA plates consisting of uniform channels with a depth of 5 to 20 μm. Tong et al. (2001) reported grooved MC arrays made of stainless steel fabricated by mechanical dicing. Stainless steel is a practically useful material because of better chemical resistance and greater mechanical strength than the other above-mentioned materials. The fabricated parallel channels with an equivalent diameter of 10 to 80 μm were of relatively uniform. The uniform O/W emulsion droplets were generated from individual channels with a hydrophilic surface after a plasma oxidation treatment. Technological development is needed for fabricating non-silicon grooved MC arrays available for the practical production of emulsion samples.

STRAIGHT-THROUGH MICROCHANNEL ARRAYS. Kobayashi et al. (2002a) developed straight-through MC arrays for greatly improving the droplet productivity of MC emulsification. A straight-through MC array consists of numerous channels arranged perpendicular to a plate (Fig. 5.6). For instance, a straight-through MC array can compactly position 10^4 channels with an equivalent diameter of 10 μm within a 10×10 mm^2 area, which exceeds a grooved MC array plate by more than 100 times. Silicon straight-through MC arrays are usually fabricated through a process including deep-reactive-ion etching to create channels (Kobayashi et al., 2002). In MC emulsification, the channel size distribution greatly affects the size distribution of the resultant droplets. The fabricated channels have a very narrow size distribution (e.g., <2% in Fig. 5.6.A), demonstrating that straight-through MC arrays satisfy a necessary condition for producing monodisperse emulsions. Existing straight-through MC arrays were also designed to have deep channels of which the

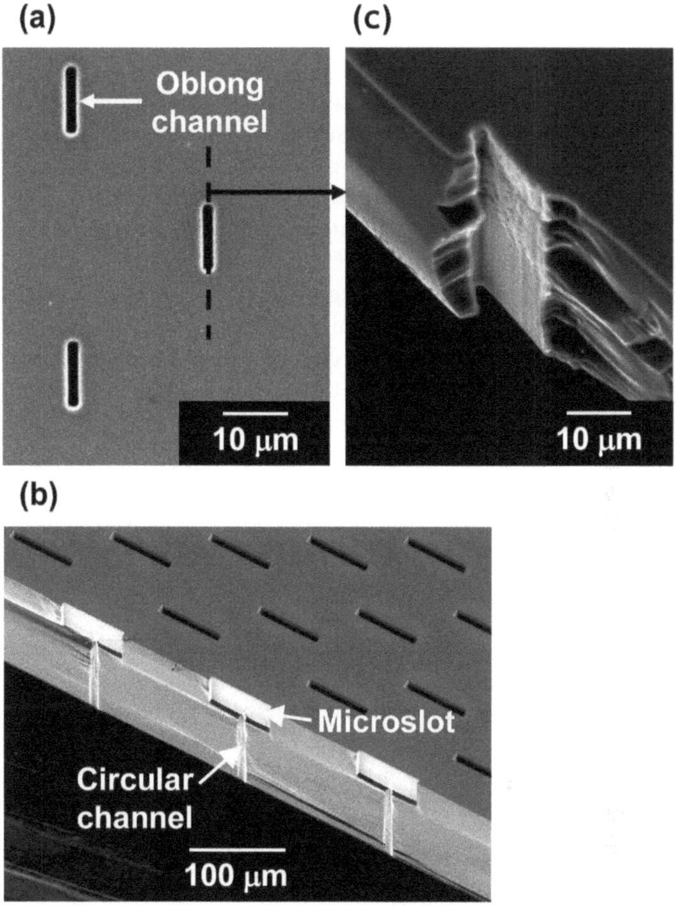

Fig. 5.6. (a, b) Scanning electron micrographs of part of a symmetric straight-through MC array and a cross section of an oblong channel. Reproduced from Kobayashi et al. (2008a) with permission from Springer. (c) Scanning electron micrograph of part of an asymmetric straight-through MC array.

channel depth/diameter ratio is more than ten times. These deep (long) channels are advantageous for producing monodisperse emulsions at high droplet generation rates, as described in the "Effect of the Flow of the Dispersed Phase" section.

The channel cross-sectional shape is one of the most important factors affecting droplet generation via a straight-through MC array. A straight-through MC array with capillary-like circular channels resulted in the generation of polydisperse large droplets with a droplet/channel size ratio of >10. The resultant droplet size was sensitive to the cross-flowing continuous phase. In contrast, straight-through MCs with oblong channels were capable of stably generating uniform droplets with a typical droplet/channel size ratio between 3 and 4, when the channel cross-sectional aspect ratio exceeded a threshold value of about 3 (Kobayashi et al., 2004b; 2009a). Droplet generation from oblong channels was simulated and analyzed using a CFD method (Kobayashi et al., 2007b). The visualized CFD results strongly suggest that a necessary condition for generating uniform droplets is to maintain a sufficient space at the channel outlet where the continuous phase can flow into the channel during droplet generation. However, there is a major drawback in emulsification using the preceding symmetric straight-through MC arrays: the difficulty in producing monodisperse emulsions when using a dispersed phase with very low viscosity. Kobayashi et al. (2005b) developed asymmetric straight-through MC arrays consisting of numerous pairs of microslots and circular channels (Fig. 5.6.B), enabling the stable production of monodisperse emulsions using two phases of low viscosity. Asymmetric straight-through MC arrays are mainly used for current MC emulsification researches.

Fig. 5.7 depicts a straight-through MC array plate of standard size. A straight-through MC array is fabricated at the center of this plate. Modules, which can be equipped with a straight-through MC array plate, are designed for continuously producing emulsions. The droplets generated in the module can be collected by the gentle cross-flow of the continuous phase in a channel formed between the plate and a flat glass plate. Monodisperse emulsions with sizes of 5 to 200 μm have been produced using straight-through MC arrays. Asymmetric straight-through MC array plates of standard size recently achieved the mass production of uniform lipid droplets with a maximum production rate of 60 mL/h (Vladisavljević et al., 2011).

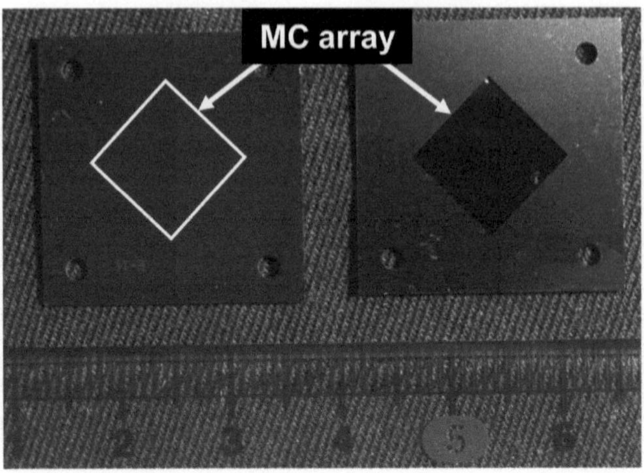

Fig. 5.7. Photographs of a silicon straight-through MC array plate. (Left) Top view. (Right) Bottom view.

Kobayashi et al. (2005c) scaled up the straight-through MC array plate. This large plate has four straight-through MC arrays where the total surface area exceeds the standard plate by about 10-fold. The large straight-through MC array plate demonstrated the mass production of uniform lipid droplets, which is about 6-fold higher than the droplet productivity for the standard counterpart. Straight-through MC array plates can be further scaled up to the size of a silicon wafer.

Single-crystal silicon wafers have been mostly used as a material of symmetric straight-through MC array plates, and silicon-on-insulator wafers have been used as a material of asymmetric straight-through MC array plates. PMMA straight-through MC array plates were also fabricated by a process of synchrotron radiation lithography and etching (Kobayashi et al., 2008b). The fabricated oblong channels of uniform size were extremely deep with a depth/diameter ratio of 18. Monodisperse W/O emulsion droplets with a number mean diameter of 25 μm were stably generated from the hydrophobic PMMA oblong channels. Further investigation is in progress to develop new non-silicon straight-through MC array plates.

FORMULATION OF LIPID-BASED MICRODISPERSIONS

SINGLE AND MULTIPLE EMULSIONS. Much of the MC emulsification literature has reported the production of O/W emulsions containing lipids, since triglyceride oils have been used as a control dispersed phase. MC emulsification has produced monodisperse O/W emulsions containing uniform droplets of vegetable oil (refined soybean oil and high oleic sunflower oil) or medium-chain triglyceride (MCT) oil (Kawakatsu et al., 1997; Kobayashi et al., 2001). Fig. 5.8 depicts the generation of uniform soybean oil droplets via a grooved MC array and an asymmetric straight-through MC array. Monodisperse O/W emulsions were also produced by MC emul-

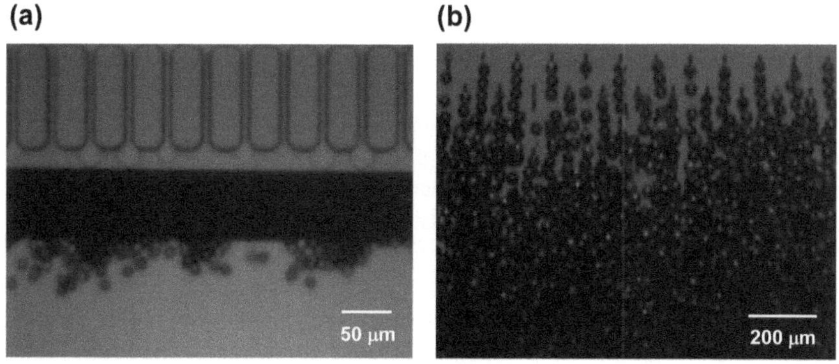

Fig. 5.8. Generation of uniform soybean oil-in-water emulsion droplets via a grooved MC array (a) and an asymmetric straight-through MC array (b). Reproduced from Kobayashi et al. (2010c) with permission from Springer.

sification when refined palm olein was used after removing monoglycerides and diglycerides (Tan et al., 2005c). This result suggests that the hydrophobicity of the dispersed oil phase is a critical parameter affecting the generation of oil droplets in MC emulsification. Triglyceride oils are generally viscous liquids at room temperature; therefore, the ratio of the dispersed-phase viscosity to the continuous phase viscosity is usually high unless the continuous water phase contains a considerable amount of thickeners. Droplet production per MC array plate tends to increase as the dispersed phase viscosity decreases, indicating that MC emulsification at an elevated temperature can increase the production of lipid droplets (Butron Fujiu et al., 2011b).

The effects of the type of surfactants and proteins on MC emulsification have been reported in the literature (Tong et al., 2000; Kobayashi & Nakajima, 2002b; Saito et al., 2005). Uniform lipid droplets can be generated via an MC array in the absence of surface active molecules, where the generated droplets were unstable, as was expected (Kobayashi et al., 2002b). Tong et al. (2002) demonstrated the production of monodisperse O/W emulsions stabilized by an anionic phospholipid, such as lysophosphatidylcholine (LPC). Interestingly, droplets were generated more stable by using lecithin in the dispersed phase and LPC in the continuous phase.

Monodisperse O/W emulsions containing lipids produced by MC emulsification have been applied to produce monodisperse microparticles and microcapsules. Monodisperse solid lipid microparticles were prepared by cooling uniform droplets of melted triglyceride oils (tripalmitin and hydrogenated fish oil) generated using MC emulsification (Sugiura et al., 2000). Tripalmitin microspheres were obtained by MC emulsification and subsequent solvent evaporation (Kobayashi et al., 2003b). The MC emulsification in this work produced monodisperse O/W emulsions consisting of tripalmitin and hexane. The hexane in the oil droplets successfully transferred to the continuous water phase during solvent evaporation at atmospheric pressure and room temperature, considerably reducing the droplet size. Monodisperse gelatin/acacia complex coacervate microcapsules were also obtained by MC emulsification and subsequent coacervation (Nakagawa et al., 2004). The single-core microcapsules were prepared using uniform triglyceride droplets stabilized by appropriate types of gelatin.

In MC emulsification, W/O emulsions have generally been produced using alkane oils with low and medium carbon numbers as the continuous phase due to their low viscosity. Only a few studies have reported the production of water-in-triglyceride oil emulsions using MC emulsification (Sugiura et al., 2001b; Liu et al., 2005; Kobayashi et al., 2009b). Uniform aqueous droplets could be generated in triglyceride oils (MCT oil, soybean oil, or triolein) as the continuous phase via MC arrays (Fig. 5.9). The viscosity ratio of the dispersed aqueous phase to the continuous triglyceride-oil phase is very low (typically between 0.02 and 0.05), resulting in droplet generation sensitive to operating conditions. In addition, MC emulsification using the viscous continuous triglyceride-oil phase causes a quite low droplet-generation rate per channel, which can be somewhat increased by high-temperature MC emulsification.

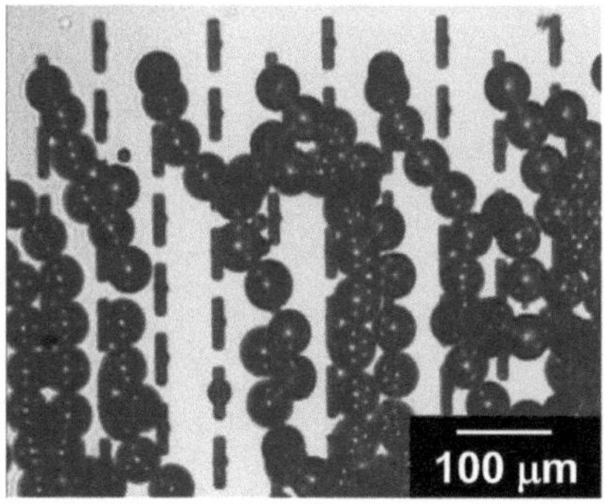

Fig. 5.9. Generation of uniform water-in-soybean oil emulsion droplets via an asymmetric straight-through MC array. Reproduced from Kobayashi et al. (2009b) with permission from Springer.

Appropriate hydrophobic surfactants, such as sorbitan fatty acid esters and polyglycerol condensed ricinoleic acid esters, were successfully used to produce monodisperse W/O emulsions (Kawakatsu et al., 2001a; Suguira et al., 2001b). In contrast, W/O systems containing crude soybean lecithin or crude egg-yolk lecithin resulted in unstable generation of water droplets and their immediate coalescence. Monodisperse W/O emulsions stabilized by a mixture containing highly purified phosphatidylcholine (>95%) were produced using a grooved MC array plate and an asymmetric straight-through MC array plate (Kuroiwa et al., 2009a).

There are only a few studies reporting on production of W/O/W emulsions using MC emulsification. The W/O/W emulsions were produced by two-step emulsification processes. W/O emulsions were first prepared by homogenization (Kawakatsu et al., 2001b; Sugiura et al., 2004b) or microfluidization (Kobayashi et al., 2005d). W/O emulsions with droplet sizes on the order of several micrometers to several tens of micrometers were prepared by homogenization. In contrast, fine W/O emulsion droplets with a Sauter mean diameter as small as 150 nm were produced by microfluidization. Afterwards, oil droplets containing smaller water droplets were generated by injecting a W/O emulsion stabilized by a hydrophobic surfactant into a continuous external aqueous phase through channels. Water-in-triglyceride oil (soybean oil, triolein, or MCT oil) emulsions have been successfully used to produce monodisperse W/O/W emulsions by MC emulsification (Kawakatsu et al., 2001b; Sugiura et al., 2004b; Kobayashi et al., 2005d). The resultant oil droplets, each containing aqueous droplets, were stabilized by a hydrophilic surfactant. The volume fraction of fine aqueous droplets dispersed in uniform oil droplets can be precisely controlled (Fig. 5.10) (Kobayashi et al., 2005d). Optical microscopy demonstrated that the internal aqueous droplets did not leak from the medium oil phase during MC emulsification.

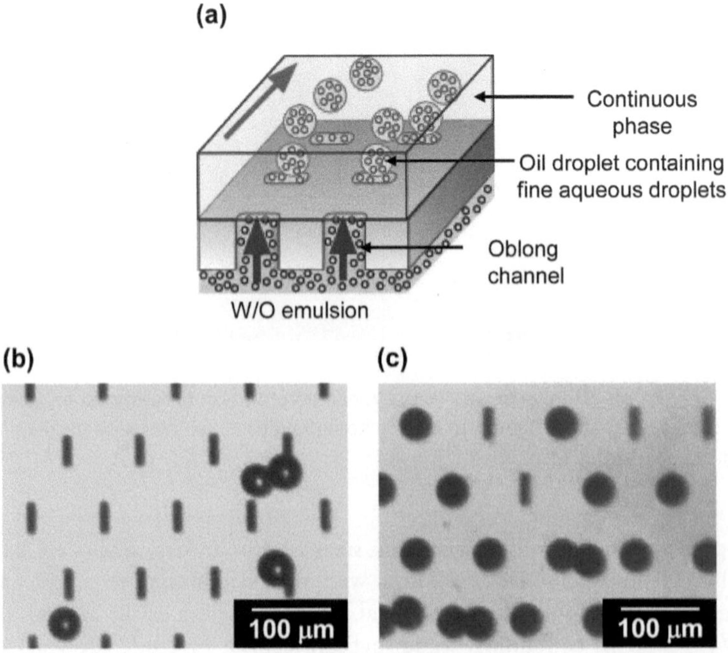

Fig. 5.10. (a) Three-dimensional drawing of the production of a W/O/W emulsion via a symmetric straight-through MC array. (b, c) Generation of uniform soybean oil droplets containing fine water droplets from oblong channels. The volume fraction of the internal water phase in the generated oil droplets was 10 vol% in (b) and 30 vol% in (c). Reproduced from Kobayashi et al. (2005d) with permission from American Oil Chemists' Society.

As described above, monodisperse single and multiple emulsions containing lipids have been produced by MC emulsification. Monodisperse lipid-based emulsions are useful templates or products in the fields of foods and pharmaceuticals. Some of their interesting applications will be described in later sections.

EMULSIONS CONTAINING FUNCTIONAL LIPIDS. The health benefits associated with diets high in foods containing antioxidants have been studied for many decades. Bioavailability of those active substances plays a critical role in development of their delivery systems (Ribeiro et al., 2011).

Carotenoids are very sensitive to heat, oxidation, and light due to their unsaturated chemical structures. They are almost insoluble in water and slightly soluble in oil at room temperature (about 0.2 g/L oil), but their solubility in oil increases greatly with increasing temperature (Ax et al., 2001). It has been found that only a minor part of the carotenoids in raw fruits or vegetables is absorbed in the intestine, probably due to the fact that carotenoids in nature exist as crystals or bound in

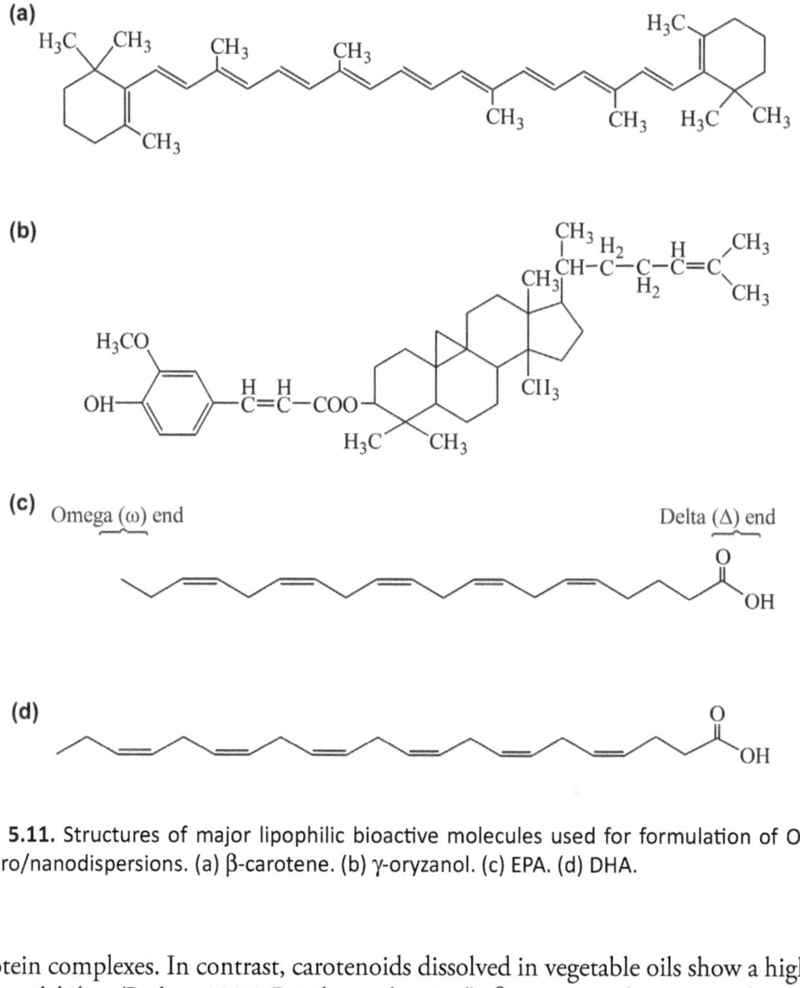

Fig. 5.11. Structures of major lipophilic bioactive molecules used for formulation of O/W micro/nanodispersions. (a) β-carotene. (b) γ-oryzanol. (c) EPA. (d) DHA.

protein complexes. In contrast, carotenoids dissolved in vegetable oils show a higher bioavailability (Parker, 1997; Briviba et al., 2004). β-carotene, the most widespread of all carotenoids, can be found in fruits, vegetables, and palm oil, among other sources. The chemical structure of β-carotene containing 11 conjugated double bonds (Fig. 5.11.A) is accountable for its high antioxidant and provitamin-A activity (Unlu et al., 2007).

γ-Oryzanol mainly occurs in the unsaponifiable fraction of rice bran oil, corn, and barley. Its structure has a combination of sterols and ferulic acid esters (Fig. 5.11.B), and it is presently investigated, mainly in Japan, because of its high antioxidant activity (Ardiansyah et al., 2006).

Another group of active compounds, the omega-3 polyunsaturated fatty acids (ω-3 PUFAs) reportedly have a beneficial role on early brain development and function (Let et al., 2004). The conversion of dietary α-linolenic acid [ALA, 18:3(ω-3)] to long-chain PUFAs, such as eicosapentaenoic acid [EPA, 20:5(ω-3)] and docosahexaenoic acid [DHA, 22:6(ω-3)] (Fig. 5.11.C and 5.11.D, respectively), is lower than 1%, thus requiring the fortification of the latter essential dietary nutrients (Innis, 2007). By enriching foods with fish or algae oils, the general consumption of ω-3 PUFAs in the population may be considerably increased (Ruxton et al., 2004). Fig. 5.11.C shows a schematic view of EPA, depicting the ω-end (methyl group, which is oil soluble) and the Δ-end (carboxylic acid group, which is water soluble).

Most recently, various reports on emulsions containing lipophilic substances dissolved in finely dispersed oil droplets have mentioned increased absorption of these substances either in vitro (Ribeiro et al., 2006; Yonekura et al., 2007), or in vivo (Garaiova et al., 2007). For preparing such O/W emulsions, the bioactive substance is dissolved in a vegetable oil and subsequently emulsified with an aqueous phase containing an emulsifier to stabilize the oil droplets. Such substances in the form of fine droplets have a better water dispersibility compared to those in bulk form.

Neves et al. (2008a) investigated the formulation of O/W emulsions loaded with bioactive molecules, such as β-carotene (3.2 g β-carotene/L oil) or γ-oryzanol (25 g γ-oryzanol/L oil), prepared using MC arrays. Refined soybean oil containing the dissolved lipophilic molecule was dispersed in a continuous water phase containing 1 wt% of either sucrose monolaurate or gelatin. The emulsification process conducted using the asymmetric straight-through MC array enabled the production of monodispersed O/W emulsions, resulting in β-carotene-loaded O/W emulsions with a number mean diameter of 27.6 μm and coefficient of variation (CV) of 2.3% (Fig. 5.12), and γ-oryzanol-loaded droplets with a number mean diameter of 28.8 μm and CV of 3.8% (Fig. 5.13). The highly monodisperse β-carotene-loaded droplets were physically stable throughout the storage period observed, resulting in

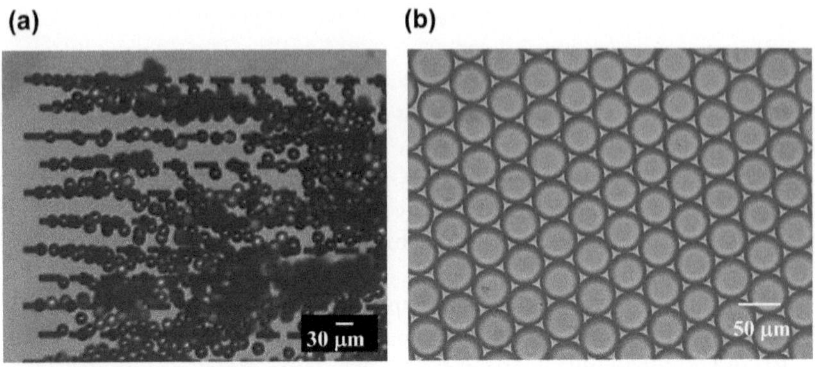

Fig. 5.12. Photomicrographs of β-carotene-loaded O/W emulsion formulated using straight-through MC array. (a) Droplet generation. (b) Monodisperse oil droplets. Reproduced from Neves et al. (2008a) with permission from Springer.

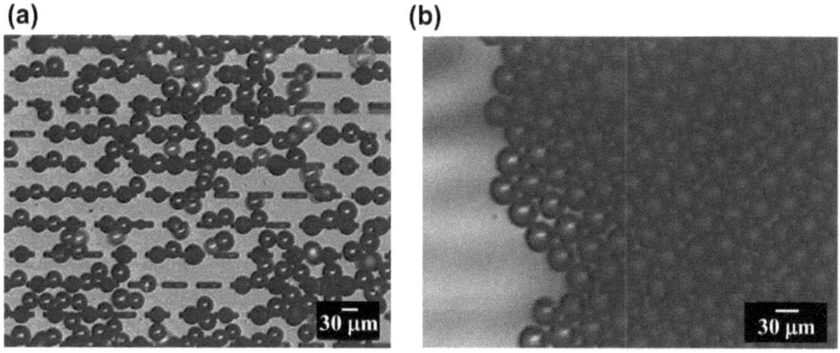

Fig. 5.13. Photomicrographs of γ-oryzanol-loaded O/W emulsion formulated using straight-through MC array. (a) Droplet generation. (b) Monodisperse oil droplets. Reproduced from Neves et al. (2008a) with permission from Springer.

droplets with a number mean diameter of 28.2 μm and CV of 2.9% after 4 months storage in darkness at 5°C. Moreover, single micrometer-sized monodisperse emulsions loaded with β-carotene were successfully formulated using the grooved MC emulsification, resulting in droplets with a number mean diameter of 9.1 μm and CV of 6.2%.

The authors evaluated the formulation of size-controlled O/W emulsions loaded with long-chain ω-3 PUFAs using straight-trough MC arrays (Neves et al., 2008b). The lipid phase consisting of palm oil rich in β-carotene (Red Palm Superolein) loaded with a commercial mixture of PUFAs (45 g/L) was dispersed in a continuous water phase containing sucrose monolaurate and β-lactoglobulin (1 wt% of each emulsifier). This combination of emulsifiers was recently reported to increase carotenoid absorption in colon carcinoma cells (Ribeiro et al., 2006). The fine carotenoid O/W emulsions loaded with PUFAs were prepared using an MC emulsification device and recovered by controlling the continuous phase flow rate. The effect of various levels of the dispersed phase flux (J_d) and the average continuous phase flow velocity along the MC plate surface (V_c) on the behavior of droplet formation, droplet size, and droplet size distribution were evaluated. As for J_d values lower than 40 L/(m²·h), which corresponds to the droplet formation volume rate of 4 mL/h, highly uniform-sized O/W emulsions with most droplets smaller than 28.5 μm and CV lower than 3.3% were successfully formulated. Further increase of J_d led to the formation of polydisperse emulsions. On the other hand, the effect of V_c on the droplet formation could be neglected for V_c ranging from 0.23 to 2.33 μm/s. The number mean diameter of the droplets formed (between 27.9 and 28.8 μm) as well as the resulting CV values (ranging from 3.5% to 4.3%, respectively) changed little over the range of V_c examined, indicating that highly monodisperse droplets with narrow size distribution were obtained, regardless of V_c.

Neves et al. (2009a) have applied straight-through MC emulsification for the formulation of olive oil based O/W emulsions. Monodisperse olive oil droplets stabilized by an aqueous solution of polyglycerol monooleate (1 wt%) were obtained, and the resultant droplet diameter was independent of the continuous phase flow rates. These results agree well with those reported by Kobayashi et al. (2002a). Those authors found that number mean diameter and CV were independent of V_c when soybean oil and 0.3 wt% sodium dodecyl sulphate aqueous solution were emulsified using an oblong straight-through MC.

As demonstrated in the results above, the challenge for manufacturers is to formulate easy-to-disperse emulsions, which are suitable for various applications. Optimization of processing conditions, choice of emulsifier and other ingredients are the most important variables to achieve a certain droplet size and optimal size distribution for each application.

STABILIZATION OF EMULSION BY THE SURFACE MODIFICATION (SURFACE COATING) OF DROPLETS. Since emulsions are thermodynamically unstable systems, stability of emulsion is one of the most important factors governing the preservation stability of products. In the case of O/W emulsion, it is known that the emulsion is stabilized by the electrostatic or steric interactions between the charged surfaces of liquid droplets coated with polyelectrolytes. In this section, we introduce our study on the stabilization of the O/W emulsion using the ability to form a self-assembled layer between a charged polysaccharide and a bio-based emulsifier (Chuah et al., 2009b). Monodisperse O/W emulsions produced by MC emulsification were used to evaluate the emulsion stability. As a result, the stabilization effect of the emulsion by the polysaccharide addition could become obvious because the state of the oil droplets was able to be examined by both instrumental analysis and direct microscopic observation.

Modified lecithin (ML) was used as a hydrophilic bio-based emulsifier. It behaves as anionic emulsifier in the neutral or weak acidic pH conditions, and the surface of oil droplets in O/W emulsions stabilized by the ML should be negatively charged. Chitosan (CHI), a cationic polysaccharide, is expected to interact with the negatively charged modified lecithin through electrostatic interaction to form a film at the surface of the oil droplets, thus providing further stability to the emulsions.

ML-stabilized O/W emulsion was successfully generated by MC emulsification using refined soybean oil as the dispersed phase and 0.25 wt% ML solution (pH 3) as the continuous phase. The number mean diameter and standard deviation of droplets measured from captured photomicrographs were 24.9 μm and 1.1 μm, respectively. ζ-potential of the droplets was around -84 mV, confirming the presence of negative ML on the droplet surface. Fig. 5.14 shows the effect of CHI concentration on the Sauter mean diameter and ζ-potential of emulsion oil droplets. At the range of the CHI concentration between 0.01 and 0.1wt%, their Sauter mean diameter and standard deviation increased significantly. The instability

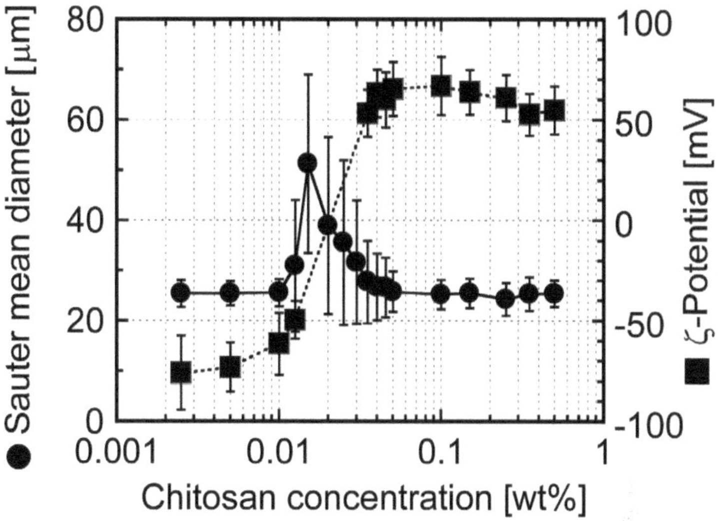

Fig. 5.14. Effect of CHI concentration on the Sauter mean diameter (circles) and ζ-potential (squares) of soybean oil-in-water monodisperse emulsions. Error bars represent the magnitude of standard deviation. Reproduced from Chuah et al. (2009b) with permission from Elsevier.

of droplets would be caused by decrease of electrostatic repulsion between droplets due to charge neutralization of droplet surface by adsorbed CHI. For concentrations exceeding that range, changes in droplet diameter were not noticeable and droplets had better size uniformity. The absolute values of ζ-potential at lower or higher CHI concentrations were high, and they should be enough to avoid the droplet aggregation.

Stability against heating-cooling treatment and the long-term preservation stability of ML-stabilized emulsion with and without the addition of CHI were examined. The emulsions were subjected to thermal treatment at 30–90°C for 30 min, after which they were cooled down to room temperature prior to analysis. As for CHI-free emulsions, the droplet coalescence was gradually observed after treatments at 70–90°C. However, emulsion in the presence of 0.25wt% CHI remained relatively stable even upon heating up to 90°C. Furthermore, as shown in Fig. 5.15, upon prolonged storage at room temperature for 3 months, coalescence of the oil droplets was observed in the ML-stabilized emulsion without CHI addition, whereas in the emulsion containing 0.25wt% CHI, droplet coalescence was not observed. The droplet size distributions of the emulsions with CHI measured by a laser diffraction particle size analyzer were almost identical before and after storage. In the presence of CHI, the emulsion tended to be more stable to aggregation due to the relatively strong electrostatic and steric repulsion associated with the relative-

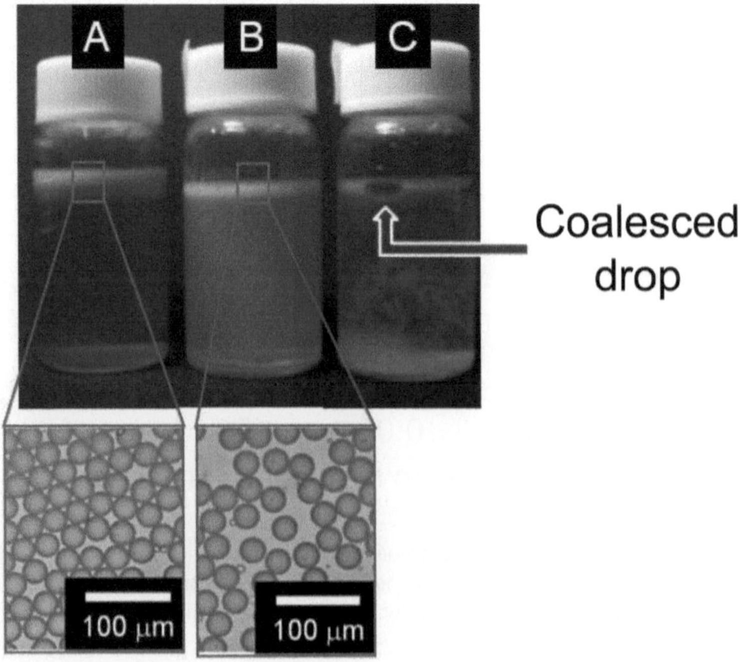

Fig. 5.15. Appearance of a freshly prepared soybean oil-in-water monodisperse emulsion stabilized by ML (a), and emulsions stored at room temperature for 3 months with the addition of 0.25 wt% CHI (b) and without CHI (c). Reproduced from Chuah et al. (2009b) with permission from Elsevier. Bottom pictures are photomicrographs of oil droplets taken from samples (a) and (b). In sample (c), a drop formed by coalescence can be found.

ly thick and electrically charged double interfacial layer. These results show that the stability of the oil droplet improved by adding CHI at appropriate concentrations.

In addition, the authors examined the effect of addition of carboxymethylcellulose (CMC), an anionic polysaccharide, on the stability of O/W emulsion stabilized by β-lactoglobulin (β-Lg). Under weak acidic conditions of pH 4–6, β-Lg is positively charged and CMC is negatively charged. MC emulsification was used for preparing the O/W emulsion as well as the above-mentioned study. Similar behavior to the ML-CHI system was observed as a result of examining the change in the droplet diameter of the emulsion in the presence of CMC at various concentrations. The stability against heat treatment and the preservation stability were improved by adding CMC under suitable conditions (Chuah, 2009c).

Therefore, the authors suggested that it is possible to achieve the stabilization of the emulsion by the appropriate combination of charged polysaccharides with emulsifiers, which forms the molecular assembly on the surface of the emulsion droplets.

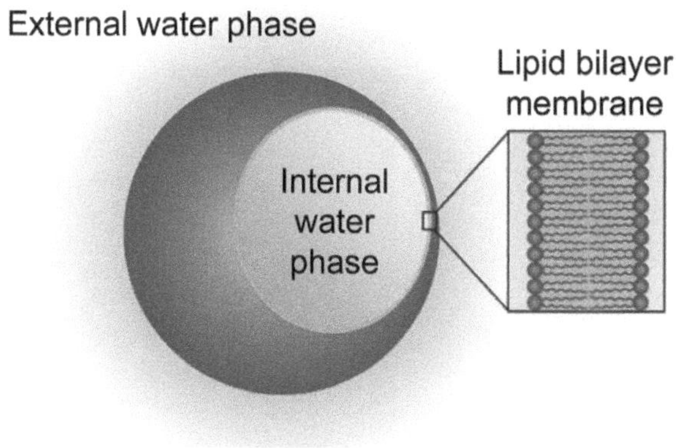

Fig. 5.16. Schematic illustration of a lipid vesicle.

LIPID VESICLES. Lipid vesicles (also called liposomes) are compartments that are formed in a water phase by the molecular assembly of bilayer-forming lipids. A vesicle has a closed structure that consists of a lipid bilayer membrane separating internal and external water phases (Fig. 5.16). Lipid vesicles can be utilized as carriers of functional materials in many practical and potential application fields. However, the controlled preparation of lipid vesicles containing hydrophilic materials with high entrapment efficiency to entrap hydrophobic materials is rather difficult. In this section, the authors describe two preparation methods of lipid vesicles, developed with the intention to improve entrapment efficiency of hydrophilic materials and ability to control size. To achieve this goal, an MC emulsification technique was utilized for preparing lipid vesicles because of its ability to generate monodisperse droplets under mild emulsification conditions.

Vesicles larger than 1 μm are called giant vesicles (GVs). GVs have the potential to be utilized as artificial biochemical microcompartments, allowing the integration of various cell-mimicking functions, including encapsulated enzyme-catalyzed transformations (Menger & Keiper, 1998; Hotani et al., 1999; Walde & Ichikawa, 2001; Walde et al., 2009; Walde et al., 2010). However, the controlled preparation of GVs containing entrapped materials is difficult because there are no methods for efficient, quantitative production of GVs that entrap water-soluble biomacromolecules (i.e., enzymes). Sugiura et al. (2008) developed a novel method, "lipid-coated ice droplet hydration method", for producing GVs containing water-soluble molecules such as proteins and polysaccharides in their internal water phases. The lipid-coated ice droplet hydration method consists of two main processes: (1) preparation of dispersed water droplets and (2) conversion of the water droplet system into giant vesicles. In the first step, a monodisperse W/O emulsion containing water soluble molecules in the dispersed water droplets is produced by means of MC emulsification. The water droplets obtained in the W/O emulsions are immediately frozen, fol-

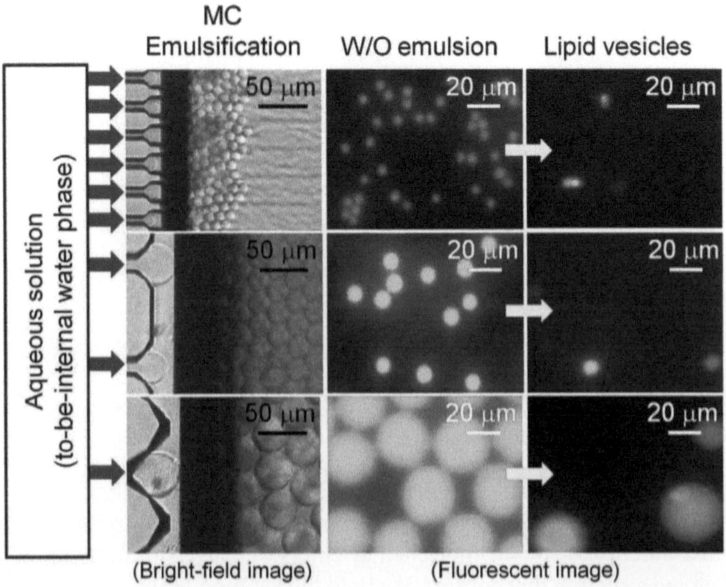

Fig. 5.17. Photomicrographs of each step in the "lipid-coated ice droplet hydration method": MC emulsification behaviors using MC arrays with different geometry (left pictures); starting W/O emulsions used as template for GV preparation (middle pictures); the resultant GVs (right pictures). The starting W/O emulsions contained a water-soluble fluorescent dye, calcein, in the dispersed water phase. Reproduced from Kuroiwa et al. (2007; 2008; 2009b) with permission from the Membrane Society of Japan, the American Chemical Society, and Wiley-VCH.

lowed by the replacement of the emulsifier used in the MC emulsification by bilayer-forming lipids. During this step, the water droplets are kept frozen. The organic solvent in the continuous phase is then removed by evaporation at a temperature below 0°C. After this procedure, the ice droplets are covered by lipid layers that will form the vesicles' membranes at the very end of the procedure. Finally, an external water phase is added to hydrate the lipid layers and to form the vesicles. The procedure described above successfully formulated spherical GVs having diameters similar to those of the water droplets in the starting W/O emulsions (7–40 μm), as shown in Fig. 5.17 (Kuroiwa et al., 2007; Kuroiwa et al., 2008; Kuroiwa et al., 2009b). It is confirmed by fluorescence-microscopy that the vesicles entrapped a water-soluble fluorescent marker (calcein) in their interior. By refining the preparation processes, we can accomplish high entrapment efficiencies between 20% and 50% for hydrophilic materials. A modified procedure without the surfactant replacement step was also presented (Kuroiwa et al., 2009a). This novel method is a highly reproducible procedure for the preparation of GVs with high entrapment yields. Furthermore,

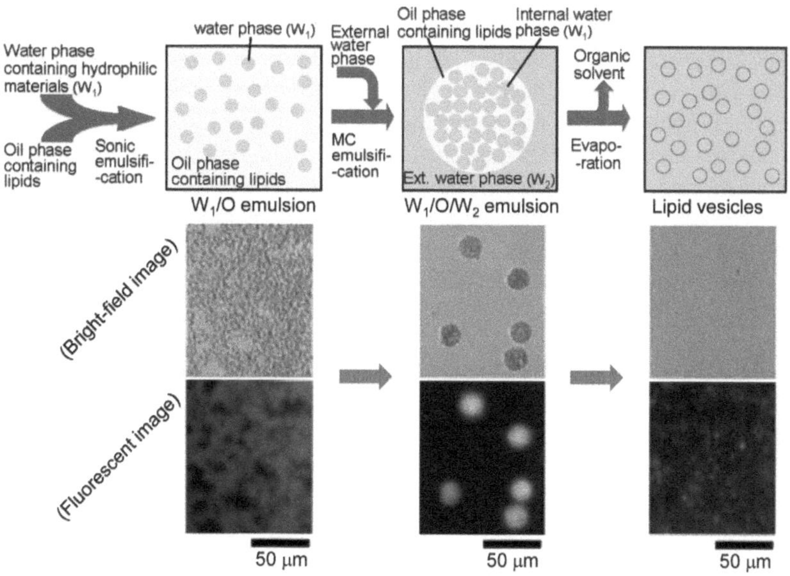

Fig. 5.18. Schematic procedure for preparation of lipid vesicles entrapping hydrophilic materials with high efficiency using W/O/W emulsion as a template. Bottom photomicrographs represent the products obtained by each step in the preparation: the primary W/O emulsion (left pictures), the secondary W/O/W emulsions (middle pictures), and the resulting lipid vesicles (right pictures). Upper and lower pictures are taken under blight-field mode and fluorescent mode, respectively.

enzymatic reactions inside GVs could be performed for α-chymotrypsin or carboxylesterase systems by the permeation of externally added substrates through the lipid bilayer.

Recently, the authors developed a method for preparing lipid vesicles with a multilamellar bilayer structure that entraps water-soluble materials using W/O/W multiple emulsions (Kuroiwa et al., 2009c). The preparation procedure is depicted in Fig. 5.18. First, emulsification by ultrasonic treatment was applied to produce a primary water-in-oil (W_1/O) emulsion consisting of aqueous droplets that contain to-be-entrapped molecules and a volatile organic solution containing a mixture of bilayer-forming lipids. Secondary emulsification for obtaining W_1/O/W_2 emulsions was conducted by means of MC emulsification. The primary emulsion was pressed into an aqueous phase through channels, then oil droplets containing the primary water droplets were obtained. From the obtained W_1/O/W_2 emulsion, the solvent of the oil phase was removed via gentle evaporation at ambient conditions. After the above processes, lipid vesicles dispersed in the aqueous solution were successfully obtained. The vesicles had a diameter similar to that of the primary W/O emulsion droplets (~1 μm), as determined by means of a laser diffraction particle size analyzer. The lipid layer had a lamellar structure, as confirmed by transmission electron microscopy. The entrapment yield of a hydrophilic fluorescent marker (calcein) reached ca. 80%. This value is considerably high as an entrapment yield of water soluble molecules, compared to the values obtained by conventional liposome preparation methods. This method enables the formulation of lipid vesicles with high entrapment efficiency and controlled size, potentially useful for application of vesicles containing hydrophilic functional materials in the pharmaceutics, food, cosmetic, and other related industrial fields.

FORMULATION OF LIPID NANODISPERSIONS

FORMULATION OF LIPID NANODISPERSIONS USING EMULSIFICATION AND SOLVENT EVAPORATION TECHNIQUES

Functional lipids, such as carotenoids, ω-3 PUFAs, and numerous other compounds, are widely used as bioactive ingredients in various food products. However, the poor water solubility of these molecules in food formulations is a major consideration for the food industry. Moreover, active ingredients with low water solubility might have a reduced bioaccessibility (Yin et al., 2009). On the other hand, in many cases it is advantageous to deliver bioactive lipids in an aqueous system, so that increasing their palatability, desirability, and bioactivity. For example, a bioactive lipid might be incorporated into a beverage or food that could be consumed by drinking or eating (McClements et al., 2007). Nanotechnology provides a good opportunity to improve the solubility of such active ingredients and to increase bioavailability. Today, nanotechnology is one of the new frontiers in the field of food science, and it offers one of the most exciting prospects for technological innovation. With a particle size in the nanometer range, nanodispersions are physically more stable compared to the conventional dispersions which contain particles in micrometer sizes (Moraru et al., 2003).

Preparation of an emulsion generally requires energy input via a homogenization process. The energy must overcome the Laplace pressure, which increases with decreasing emulsion droplet size (McClements et al., 2007). High-pressure homogenization is extensively used in the food, pharmaceutical, and biotechnology industries to mix, disperse, emulsify, and process many products (Floury et al., 2006). Microfluidization is a high-pressure homogenization technique that is efficient for preparing O/W emulsions with fine droplet size. The interaction chamber of the microfluidizer has a fixed geometry and is present in the form of a confined capillary tube where disruption of the droplets occurs due to liquid-liquid and, above all, liquid-solid shear forces (Perrier-Cornet et al., 2005). High-pressure homogenization is a top-down approach, by which a bulk phase is broken down into smaller droplets, usually with sizes of 0.1 to 100 μm (Kanafusa et al., 2007).

Yin et al. (2008) investigated the effects of six different polyglycerol esters (PGEs) of fatty acids and two different particle sizes produced using various pro-

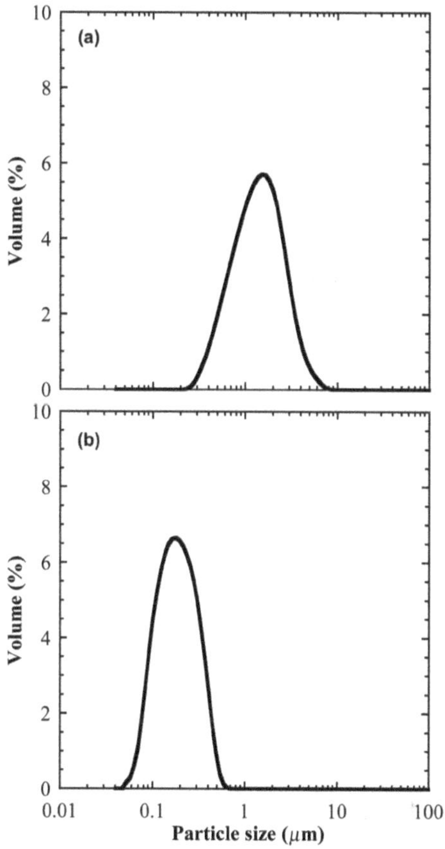

Fig. 5.19. Particle size distribution of β-carotene O/W micro/nanodispersions stabilized by decaglycerol monolaurate (ML750). (a) Coarse emulsion (10 MPa), (b) Fine emulsion (100 MPa). Reproduced from Yin et al. (2008) with permission from Springer.

cessing parameters on the physicochemical properties and stability of the β-carotene emulsions during digestion in simulated gastric fluid. Different operating pressure levels were utilized to obtain a coarse emulsion (10 MPa) and a fine emulsion (100 MPa). β-carotene (0.1 wt%) was initially dissolved in soybean oil, prior to dispersion in a continuous water phase containing 1 wt% PGE. The Sauter mean diameter values obtained were 0.17 to 0.27 μm for fine emulsions (size distribution shown in Fig. 5.19.B) and 1.16 to 1.59 μm for coarse emulsions (size distribution depicted in Fig. 5.19.A). The results showed that the particle size decreased with increasing polymerization of the glycerol in PGEs, and the higher polymerization of the glycerol also increased the stability of emulsions during digestion in simulated gastric fluid.

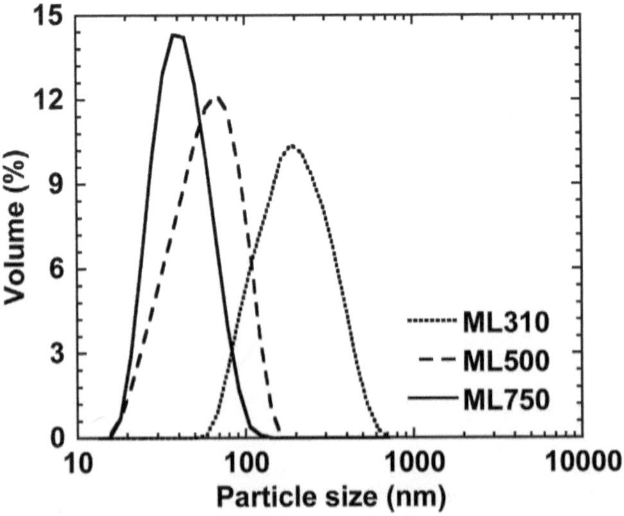

Fig. 5.20. Particle size distribution of β-carotene nanodispersions stabilized by polyglycerol monolaurate with different polymerization degrees. Reproduced from Yin et al. (2009) with permission from Elsevier.

The previous results indicate that a number of variables affect the droplet size and probably the properties of the resulting emulsions, including phase ratio, type and concentration of emulsifier, mixing technique, processing temperature, and other conditions. Yin et al. (2008) reported that the type of PGE and the particle size resulting from various homogenization pressures could affect the physicochemical properties and the stability of β-carotene emulsions during digestion in simulated gastric fluid digestion.

The effect of using PGE with different degrees of polymerization on the average particle size, size distribution, and on the stability of β-carotene nanodispersions was investigated (Yin et al., 2009). As a result, a higher polymerization degree of glycerol in the continuous phase resulted in considerable smaller droplet size (Fig. 5.20), as well as in higher stability of β-carotene emulsions.

The size of droplets obtained can be further reduced by combining high-pressure homogenization with solvent evaporation (Fig. 5.21). This technique has been reportedly used in the formulation of nanodispersions containing bioactive lipids by solvent evaporation of O/W emulsion droplets (Tan & Nakajima, 2005a; Neves et al., 2009b; Wang et al., 2010; Leong et al., 2011). Nanodispersions containing β-carotene were formulated by an emulsification-evaporation technique (Tan & Nakajima, 2005b). The preparation method consisted of emulsifying an organic solution of β-carotene in an aqueous solution containing emulsifier using two different homogenizers (a conventional homogenizer and a microfluidizer), followed

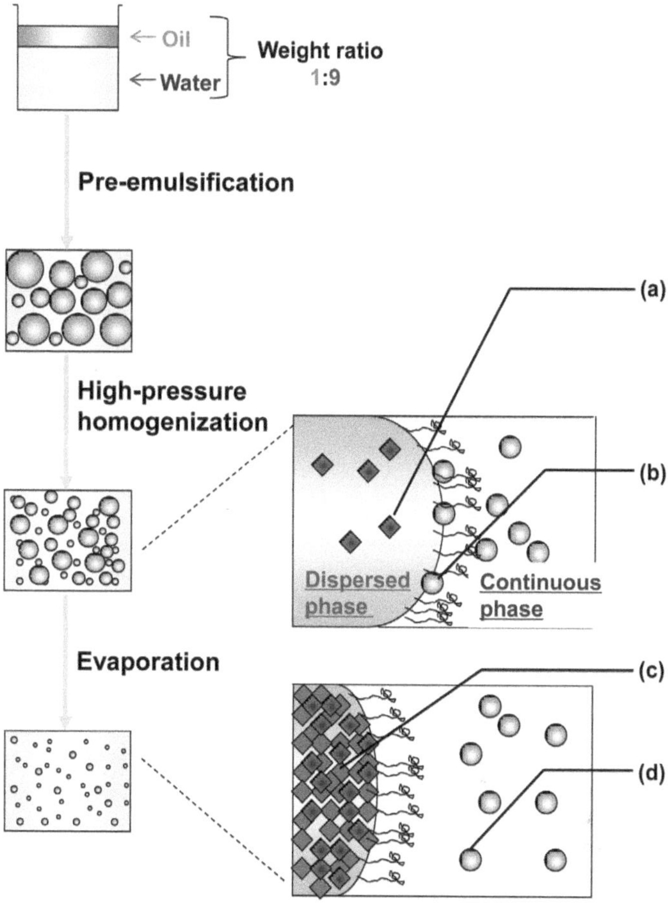

Fig. 5.21. Schematic diagram of O/W emulsion formulation using a top-down approach: high-pressure homogenization and solvent evaporation. (a) Concentration of carotene increases and precipitates when the solubility limit is crossed, (b) Hexane evaporates at the droplet interface, (c) More carotene precipitates, (d) Almost all hexane is evaporated and the particle size decreases.

by direct solvent evaporation under reduced pressure. The influence of different homogenizing conditions and organic/aqueous phase ratios on particle size parameters and content of β-carotene was investigated. In addition, the stability of β-carotene nanodispersions was carried out at a storage temperature of 4°C. In general, homogenization pressure and cycle had significant ($P < 0.05$) effects on various particle size parameters. A volume-weighted mean diameter ($d_{4,3}$) of β-carotene nanoparticles ranged from 60 to 140 nm.

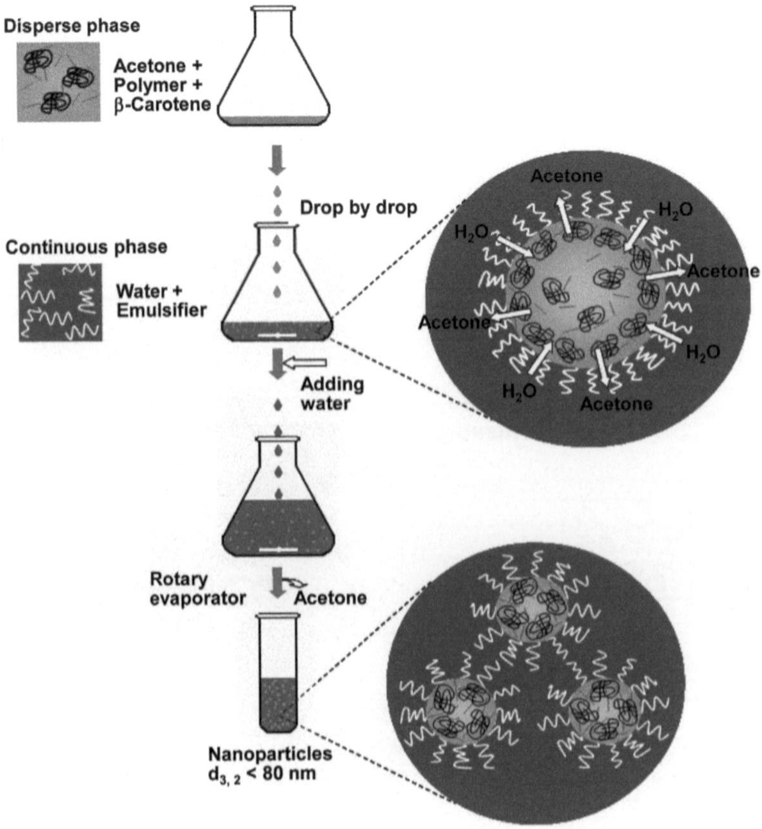

Fig. 5.22. Schematic representation of the production of β-carotene nanodispersion by solvent displacement technique. Reproduced from Ribeiro et al. (2008) with permission from Elsevier.

Chu et al. (2007a) investigated the preparation of protein-stabilized β-carotene nanodispersions by an emulsification-evaporation method. Sodium caseinate was depicted as the most effective emulsifier among selected proteins in preparing the nanodispersions, with a monomodal β-carotene particle size distribution and a Sauter mean diameter of 17 nm.

The mechanism for lipid oxidation in O/W emulsions differs from bulk lipids because emulsions have an aqueous phase that contains both prooxidants and antioxidants and oil-water interface interactions between oil and water components (Waraho et al., 2011). Aiming to suppress lipid oxidation in fish-oil-based emulsions, Neves et al. (2009c) have evaluated the effect of lipid structuring on the oxidative stability of fish O/W nanoemulsions. An oil-triglyceride-water system was used to evaluate the effect of lipid physical state on the oxidative stability of PUFA-rich O/W nanoemulsions. Structuring of Menhaden (*Brevoortia tyrannus*) fish oil was accomplished by its partial replacement (1:1) with tripalmitin, a long-chain high melting point (66°C) saturated triglyceride. The emulsification was conducted at 70°C, except for fish oil emulsions. The lipid phase was dispersed in a continuous water phase (1:9) containing 1 wt% decaglycerol monolaurate, pre-mixed by rotor-stator (5,000 rpm/5 min) and homogenized using high-pressure (Nanomizer, 150 MPa, 1 cycle). The resulting fish oil emulsions had monomodal size distributions with Sauter mean diameter of 130 nm. Tripalmitin emulsions were observed using polarized light microscopy, revealing lipid crystals enclosed in the oil phase. After 1 month storage at 5°C, the lipid hydroperoxides (LOOH) content in fish oil emulsions (190 μmol-LOOH/kg-oil) was higher compared to that of fish oil emulsions containing tripalmitin (55 μmol/kg), depicting nearly 70% reduction on LOOH by structuring fish oil with tripalmitin.

FORMULATION OF LIPID NANODISPERSIONS BY DROP-BY-DROP AND SOLVENT DISPLACEMENT TECHNIQUES

Nanodispersions of β-carotene were prepared by modifying the method by Fessi et al. (1989), based on interfacial nanodeposition of β-carotene at the interface of acetone and aqueous phase after drop-by-drop and solvent displacement (Chu et al., 2007b). A typical preparation procedure is shown in Fig. 5.22 (Ribeiro et al., 2008). Firstly, the crystalline β-carotene was dissolved in acetone, as a water-miscible organic solvent. Solubilization of biodegradable polymers in this organic phase is optional in order to prepare polymer-based nanoparticles. After this step, the organic phase was slowly introduced, drop by drop, to an aqueous phase containing emulsifier, such as sodium caseinate, gelatin, and Tween 20, under moderate magnetic stirring. Acetone in the colloidal suspension system was evaporated off using a rotary evaporator under a reduced pressure.

The β-carotene nanodispersions prepared with sodium caseinate exhibited a bimodal particle size distribution: large particles (stabilized by casein micelles) with a mean particle size of 171 nm and small particles (stabilized by casein submicelles) of 13 nm (Chu et al., 2007b). This was confirmed with centrifugal analysis and transmission electron microscopic observation shown in Fig. 5.23. Most of the β-carotene (87.6%) in the nanodispersions was stabilized in the small particles.

Stability of protein-stabilized β-carotene nanodispersions against heating, salts, and pH was evaluated (Chu et al., 2008). Sodium caseinate-stabilized nanodispersions were generally more stable compared to those stabilized by whey proteins. Types of emulsifiers and their combinations affected the size distribution of β-carotene particles and their stability (Yin et al., 2009).

When the biodegradable polymers poly (D, L-lactic) and poly (D, L-lactic-co-glycolic) acids were dissolved in the organic phase, β-carotene-enriched polymer-based nanoparticles were formulated according to this drop-by-drop and solvent displacement technique (Ribeiro et al., 2008). The size of particles, stabilized with gela-

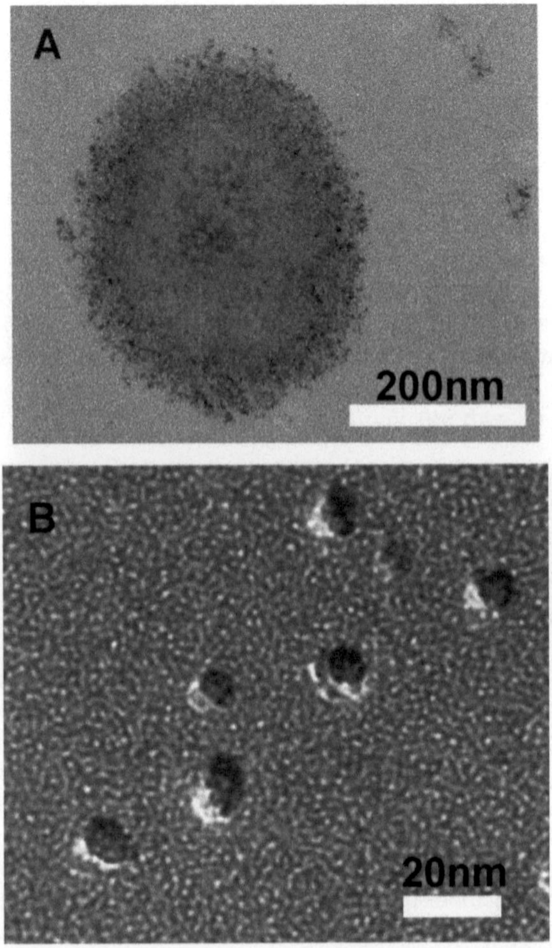

Fig. 5.23. Transmission electron microscopic images of β-carotene nanodispersions prepared by drop-by-drop and solvent displacement technique with sodium caseinate. Large (A) and small (B) particles correspond to the casein-micelle and -submicelle stabilized β-carotene particles, respectively. Reproduced from Chu et al. (2007b) with permission from the American Chemical Society.

tin or Tween 20 was around 74 and 77 nm, respectively. They were found to hardly vary after preparation and over 5 months of storage. The nanoparticles have more rigid and thicker outer layers than those prepared without any polymer, providing better stability against ripening. In this kind of formulation, polymers assumed the function of protective colloids and possibly also provided chemical stabilization of the nanodispersions.

FORMULATION OF NANODISPERSIONS BY SELF-ASSEMBLY OF MODIFIED LECITHIN AND CHITOSAN

Chuah et al. (2009a) reported that nanoparticles of 120–350 nm in hydraulic mean diameter could be prepared only by mixing solutions of ML and CHI under appropriate conditions. Nanoparticle formation procedure is quite simple: only mix aqueous solution of the above two components. No organic solvents or cross-linking agents are required during the formation of nanoparticles. Therefore, this is an ideal system for application in pharmaceutics, foods, cosmetics, and related industries.

The nanoparticles were formed within the wide range of the mixing molar ratio of the amino group of CHI and the phosphate group of ML. As mentioned in the "Stabilization of Emulsion by the Surface Modification (Surface Coating) of Droplets" section, ML is anionic surfactant under neutral or weak acidic pH condition, and CHI is a cationic natural polysaccharide, and they could form a complex via electrostatic interaction with each other. The primary aggregates of ML might be trapped and cross-linked by oppositely charged CHI chains (Fig. 5.24). Nanoparticles were obtained when the cationic CHI formed complexes with the anionic ML. As more CHI is available for interaction with ML, the uncomplexed segment of the positively charged CHI is expected to constitute the exterior of the nanoparticles, giving the particles a highly positively charged property. When the molar mixing ratio was low, ML aggregates were present in excess, rendering the particles negatively charged. This nanoparticle formation mechanism is supported by ζ-potential measurement (Fig. 5.24, bottom graph) and electron microscopic observation. The ζ-potential of nanoparticles changed from a negative into a positive value as the molar mixing ratio increased. The value did not depend on the total concentration of each component. Stable nanoparticles could be formed over a wide range of molar mixing ratios and total polyelectrolyte concentrations, except when the surface charge of the nanoparticles was close to neutrality (large precipitation was generated at the condition that ζ-potential was near 0).

Bovine serum albumin (BSA) and coomassie brilliant blue (CBB), as models of positively and negatively charged compounds, could be entrapped within the nanoparticles with relatively good efficiency (32.7 and 62.7% for BSA and CBB, respectively). The results showed that these nanoparticles have the potential to act as biocompatible carriers for encapsulation of water-soluble bioactive food components.

To further extend their application and to facilitate ease of handling, the nanoparticle suspensions were condensed by evaporation and were lyophilized with the aid of various cryoprotectants such as sucrose. The hydraulic mean diameter of nanoparticles in 5-times concentrated suspension was still in submicron order (below 500 nm). In addition, interestingly, the nanoparticles after lyophilizaton could be suspended again in buffer solution without modifying their original particle size in the presence of as little as 5 wt% of sucrose.

As described above, nanoparticles could be formed spontaneously via the self-assembly of ML and CHI, as driven by electrostatic interactions. These nanoparticles have the potential to be used as biocompatible carriers for water-soluble bioactive food components.

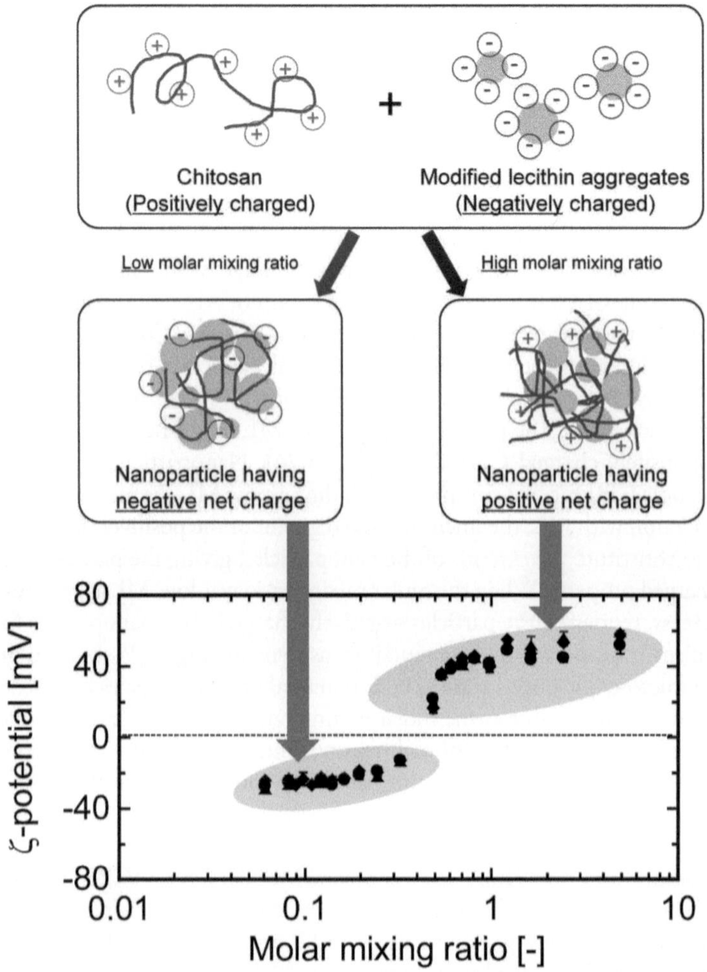

Fig. 5.24. Possible mechanism of nanoparticle formation via interaction between ML and CHI (upper illustration, not drawn to scale) and ζ-potential of nanoparticles prepared at different molar mixing ratios of amino groups of CHI to phosphate groups of ML (lower graph). Total concentration of ML and CHI: 0.1wt% (circles); 0.5wt% (diamonds), 1.0wt% (triangles). Reproduced from Chuah et al. (2009a) with permission from Institute of Food Technology.

CONCLUSION

This chapter has demonstrated that lipid micro/nanodispersions with precisely controlled droplet/particle sizes and/or unique interfacial properties can be formulated by a top-down approach, a bottom-up approach, or a combination of both.

MC emulsification, a top-down approach, is a promising technique to produce lipid monodisperse single and multiple emulsions. This technique hardly causes temperature elevation during emulsification due to a very mild droplet-generation process, differing from conventional emulsification devices. This feature might have a great potential for increasing the shelf life of food emulsions, as well as stabilizing O/W emulsions loaded with bioactive lipophilic molecules, which generally are very sensitive to oxidation and heat. Practically, MC emulsification is also a robust process, since the droplet size and its distribution of the resultant emulsions are basically insensitive to the flow of the dispersed and continuous phases. Although current MC emulsification devices are capable of generating uniform droplets with sizes over 1 μm, the authors are developing nanochannel (NC) emulsification for producing monodisperse submicron emulsions. Droplet productivity for MC emulsification mostly satisfies laboratory-scale production, whereas further scale-up of the MC emulsification system is desired to produce monodisperse emulsions at droplet production rates over 1 kg/h necessary for the industrial production of premium lipid micro/nanodispersions.

Monodisperse emulsions obtained by MC emulsification are useful templates for formulating monodisperse microparticles, microcapsules, and giant vesicles. Stability of the O/W emulsion droplets against temperature stresses (heating, freezing, and thawing) can be improved by the formation of a self-assembled layer between two types of bio-based molecules with different charges at the droplet surface. A key point in formulating the oil droplets with higher stability is the presence of the appropriate amount of molecules with a charge opposite to the emulsifier molecules adsorbing at the oil-water interface. Interaction between the surface-coated oil droplets with a size of ca. 25 μm was easily observed by optical microscopy. In addition, homogeneous lipid vesicles with high encapsulation efficiencies of hydrophilic biomacromolecules were formulated using W/O or W/O/W emulsions obtained by MC emulsification as templates of internal water phases of lipid vesicles. This new vesicle formulation technique demonstrated the ability to control the size of the resultant lipid vesicles. The use of MC emulsification plays a critical role in formulating homogeneous vesicles.

This chapter also demonstrates the controlled formulation of lipid-based nanodispersions by top-down and bottom-up approaches. Considering the emulsification-evaporation technique, a top-down approach, the mean diameters and particle-size distributions of the β-carotene nanodispersions depend on the experimental parameters, especially the PGE and β-carotene concentrations and the ratio of organic phase to aqueous phase. Microfluidization parameters also affect the particle-size distribution considerably. Detailed understanding of the properties of the PGE as emulsifiers and their performance as functions of emulsification parameters might play a major role while preparing stable nanodispersions. The drop-by-drop and solvent displacement technique, considered a kind of bottom-up approach, enables one to prepare stable β-carotene nanodispersions spontaneously. The procedure simply consists of adding drop-by-drop β-carotene, dissolved in organic phase, to the aqueous phase containing emulsifier, followed by evaporation of organic solvent. Size-controlled nanoparticles consisting of ML and CHI were formulated using a bottom-up approach by mixing two aqueous solutions containing a charged component. The particle size, surface charge, and dispersion stability were significantly affected by the mixing ratio of both components. The nanoparticles could entrap various water-soluble compounds with good efficiency. The ability of the nanoparticle dispersions to be converted to lyophilized powder or concentrated suspension was also demonstrated.

Finally, the lipid micro/nanodispersions discussed in this chapter have the potential for food, pharmaceutical, and cosmetic applications. Scale-up of their formulation processes including MC emulsification devices could lead to a range of novel food, pharmaceutical, and cosmetic products.

REFERENCES

Ardiansyah; Shirakawa, H.; Koseki, T.; Ohinata, K.; Hashizume, K.; Komai, M. Rice bran fractions improve blood pressure, lipid profile, and glucose metabolism in stroke-prone spontaneously hypertensive rats. J.Agric. Food Chem. 2006, 54, 1914–1920.

Ax, K.; Schubert, H.; Briviba, K.; Rechkemmer, G.; Tevini, M. Oil-in-water emulsions as carriers of bioavailable carotenoids. In: *PARTEC* Proceedings, 2001, 1, 1.

Briviba, K.; Kulling, S.E.; Möseneder, J.; Watzl, B.; Rechkemmer, G.; Bub, A. Effects of supplementing a low-carotenoid diet with a tomato extract for two weeks on endogenous levels of DNA single strand breaks and immune functions in healthy non-smokers and smokers. Carcinogenesis. 2004, 25, 2373–2378.

Butron Fujiu, K.; Kobayashi, I.; Uemura, K.; Nakajima, M. Temperature effect on microchannel oil-in-water emulsification. Microfluid. Nanofluid. 2011a, 10, 773–783.

Butron Fujiu, K.; Kobayashi, I.; Neves, M.A.; Uemura, K.; Nakajima, M. Effect of temperature on production of soybean oil-in-water emulsions by microchannel emulsification using different emulsifiers. Food Sci. Technol. Res. 2011b, 10, 773–783.

Chu, B.S.; Ichikawa, S.; Kanafusa, S.; Nakajima, M. Preparation of protein-stabilized β-carotene nanodispersions by emulsification–evaporation method. J. Am. Oil Chem. Soc. 2007a, 84, 1053–1062.

Chu, B.S.; Ichikawa, S.; Kanafusa, S.; Nakajima, M. Preparation and characterization of β-carotene nanodispersions prepared by solvent displacement method. J. Agric. Food Chem. 2007b, 55, 6754–6760.

Chu, B.S.; Ichikawa, S.; Kanafusa, S.; Nakajima, M. Stability of protein-stabilized β-carotene nanodispersions against heating, salts and pH. J. Sci. Food Agric. 2008, 88, 1764–1769.

Chuah, A.M.; Kuroiwa, T.; Ichikawa, S.; Kobayashi, I.; Nakajima, M. Formation of biocompatible nanoparticles via the self-assembly of chitosan and modified lecithin. J. Food Sci. 2009a, 74, N1–N8.

Chuah, A.M.; Kuroiwa, T.; Kobayashi, I.; Nakajima, M. Effect of chitosan on the stability and properties of modified lecithin stabilized oil-in-water monodisperse emulsion prepared by microchannel emulsification. Food Hydrocoll. 2009b, 23, 600–610.

Chuah, A.M. Preparation and stabilization of micro/nanodispersions by biopolymers. Ph.D. Thesis, University of Tsukuba, 2009c, 101–125.

Fessi, H.; Puisieux, F.; Devissaguet, J.P.; Ammoury, N.; Benita, S. Nanocapsule formation by interfacial polymer deposition following solvent displacement. Int. J. Pharm. 1989, 55, R1–R4.

Floury, J.; Desrumaux, A.; Axelos, M.A.V.; Legrand, J. Degradation of methylcellulose during ultra-high pressure homogenisation. Food Hydrocoll. 2002, 16, 47–53.

Föster, S.; Konrad, M. From self-organising polymers to nano- and biomaterials. J. Mater. Chem. 2003, 13, 2671–2688.

Garaiova, I.; Guschina, I.A.; Plummer, S.F.; Tang, J.; Wang, D.; Plummer, N.T. A randomised cross-over trial in healthy adults indicating improved absorption of omega-3 fatty acids by pre-emulsification. Nutr. J. 2007, 6, 1.

Hotani, H.; Nomura, F.; Suzuki, Y. Giant liposomes: from membrane dynamics to cell morphogenesis. Cur. Opn. Coll. Interf. Sci. 1999, 4, 358–368.

Innis, S.M. Dietary (n-3) fatty acids and brain development. J. Nutr. 2007, 137, 855.

Kanafusa, S.; Chu, B.S.; Nakajima, M.; Factors affecting droplet size of sodium caseinate stabilized O/W emulsions containing β-carotene. Eur. J. Lipid Sci. Technol. 2007, 109, 1038–1041.

Kawakatsu, T.; Kikuchi, Y.; Nakajima, M. Regular-sized cell creation in microchannel emulsification by visual micro-processing method. J. Am. Oil Chem. Soc. 1997, 74, 317–321.

Kawakatsu, T.; Komori, H.; Nakajima, M., Kikuchi, Y.; Yonemoto, T. Production of monodispersed oil-in-water emulsion using crossflow-type silicon microchannel plate. J. Chem. Eng. Japan 1999, 32, 241–244.

Kawakatsu, T.; Trägårdh, G.; Trägårdh, Ch.; Nakajima, M., Oda, N.; Yonemoto, T. The effect of hydrophobicity of microchannels and components in water and oil phase on droplet formation in microchannel water-in-oil emulsification. Colloids Surf. A: Physicochem. Eng. Aspects 2001a, 179, 29–38.

Kawakatsu, T.; Trägårdh, G.; Trägårdh, Ch.; Production of W/O/W emulsions and S/O/W pectin microcapsules by microchannel emulsification. Colloids Surf. A: Physicochem. Eng. Aspects 2001b, 189, 257–264.

Kikuchi, Y.; Sato, K.; Kaneko, T. Modified cell-flow microchannels formed in a single-crystal silicon substrate and flow behavior of blood cells. Microvasc. Res. 1992, 47, 126–139.

Kobayashi, I.; Nakajima, M.; Nabetani, K.; Kikuchi, Y.; Shohno, A.; Satoh, K. Preparation of micron-scale monodisperse oil-in-water microspheres by microchannel emulsification. J. Am. Oil Chem. Soc. 2001, 78, 797–802.

Kobayashi, I.; Nakajima, M.; Chun, K.; Kikuchi, Y.; Fujita, F. Silicon array of elongated through-holes for monodisperse emulsions. AIChE J. 2002a, 48, 1639–1644.

Kobayashi, I.; Nakajima, M. Effect of emulsifiers on the preparation of food-grade oil-in-water emulsions using a straight-through extrusion filter. Eur. J. Lipid Sci. Technol. 2002b, 104, 720–727.

Kobayashi, I.; Nakajima, M.; Mukataka, S. Preparation characteristics of oil-in-water emulsions using differently charged surfactants in straight-through microchannel emulsification. Colloids Surf. A: Physicochem. Eng. Aspects 2003a, 229, 33–41.

Kobayashi, I.; Iitaka, Y.; Iwamoto, S.; Kimura, S.; Nakajima, M. Preparation characteristics of lipid microspheres using microchannel emulsification and solvent evaporation methods. J. Chem. Eng. Japan 2003b, 36, 996–1000.

Kobayashi, I.; Mukataka, S.; Nakajima, M. CFD simulation and analysis of emulsion droplet formation from straight-through microchannels. Langmuir 2004a, 20, 9868–9877.

Kobayashi, I.; Mukataka, S.; Nakajima, M. Effect of slot aspect ratio on droplet formation from silicon straight-through microchannels. J. Colloid Interface Sci. 2004b, 279, 277–280.

Kobayashi, I.; Mukataka, S.; Nakajima, M. Effects of type and physical properties of oil phase on oil-in-water emulsion droplet formation in straight-through microchannel emulsification: experimental and CFD studies. Langmuir 2005a, 21, 5722–5730.

Kobayashi, I.; Mukataka, S.; Nakajima, M. Novel asymmetric through-hole array microfabricated on a silicon plate for formulating monodisperse emulsions. Langmuir 2005b, 21, 7629–7632.

Kobayashi, I.; Mukataka, S.; Nakajima, M. Production of monodisperse oil-in-water emulsions using a large silicon straight-through microchannel plate. Ind. Eng. Chem. Res. 2005c, 44, 5852–5856.

Kobayashi, I.; Lou, X.; Mukataka, S.; Nakajima, M. Preparation of monodisperse water-in-oil-in-water emulsions using microfluidization and straight-through microchannel emulsification. J. Am. Oil Chem. Soc. 2005d, 82, 65–71.

Kobayashi, I.; Nakajima, M. Generation and multiphase flow of emulsions in microchannels. Micro Process Engineering; Kockmann, N. Ed.; Advanced Micro and Nanosystems Volume 5; Wiley-VCH Verlag GmbH & Co. KGaA: Weinheim, Germany, 2006, 149–171.

Kobayashi, I.; Uemura, K.; Nakajima, M. Formulation of monodisperse emulsions using submicron-channel arrays. Colloids Surf. A: Physicochem. Eng. Aspects 2007a, 296, 285–289.

Kobayashi, I.; Uemura, K.; Nakajima, M. CFD analysis of generation of soybean oil-in-water emulsion droplets from rectangular straight-through microchannels. Food Sci. Technol. Res. 2007b, 13, 187–192.

Kobayashi, I.; Takano, T.; Maeda, R.; Wada, Y.; Uemura, K.; Nakajima, M. Straight-through microchannel devices for generating monodisperse emulsion droplets several microns in size. Microfluid. Nanofluid. 2008a, 4, 167–177.

Kobayashi, I.; Hirose, S.; Katoh, T.; Zhang, Y.; Uemura, K.; Nakajima, M. High-aspect-ratio through-hole array microfabricated in a PMMA plate for monodisperse emulsion production. Microsyst. Technol. 2008b, 14, 1349–1357.

Kobayashi, I.; Uemura, K.; Nakajima, M. Effect of channel and operation parameters on emulsion production using oblong straight-through microchannel arrays. Japan J. Food Eng. 2009a, 10, 69–75.

Kobayashi, I.; Murayama, Y.; Kuroiwa, T.; Uemura, K.; Nakajima, M. Production of monodisperse water-in-oil emulsions consisting of highly uniform droplets using asymmetric straight-through microchannel arrays. Microfluid. Nanofluid. 2009b, 7, 107–119.

Kobayashi, I.; Hori, Y.; Uemura, K.; Nakajima, M. Production characteristics of large soybean oil droplets by microchannel emulsification using asymmetric through holes. Japan J. Food Eng. 2010a, 11, 34–48.

Kobayashi, I.; Vladisavljević, G.T.; Uemura, K.; Nakajima, M. High-performance microchannel emulsification device with microfabricated asymmetric through-holes. 11th International Conference on Microreaction Technology. 2010b, OB-11.

Kobayashi, I.; Wada, Y.; Uemura, K.; Nakajima, M. Microchannel emulsification for mass production of uniform fine droplets: integration of microchannel arrays on a chip. Microfluid. Nanofluid. 2010c, 8, 255–262.

Kuroiwa, T.; Nakajima, M.; Sato, S.; Mukataka, S.; Ichikawa, S. Preparation of giant vesicles larger than 30 μm that entrap a model hydrophilic substance using a size-controlled water-in-oil emulsion. Membrane 2007, 32, 229–233.

Kuroiwa, T.; Ichikawa, S. Efficient preparation of material-encapsulating giant vesicles and their utilization for developing intravesicular microbioreaction systems. Membrane 2008, 33, 294-299 (in Japanese).

Kuroiwa, T.; Kiuchi, H.; Noda, K.; Kobayashi, I.; Nakajima, M.; Uemura, K.; Sato, S.; Mukataka, S.; Ichikawa, S. Controlled preparation of giant vesicles from uniform water droplets obtained by microchannel emulsification with bilayer-forming lipids as emulsifiers. Microfluid. Nanofluid. 2009a, 6, 811–821.

Kuroiwa, T; Nakajima, M.; Uemura, K.; Sato, S; Mukataka, S.; Ichikawa, S. Preparation characteristics of giant vesicles with controlled size and high entrapment efficiency using monodisperse water-in-oil emulsions. Emulsion Science and Technology; Tadros, T.F. Ed.; Wiley-VCH Verlag GmbH & Co. KGaA: Weinheim, Germany, 2009b, 229–242.

Kuroiwa, T.; Kobayashi, I.; Uemura, K.; Ichikawa, S. Preparation of lipid microcapsules that entrap water-soluble materials using water-in-oil-in-water multiple emulsions via solvent evaporation. J. Biosci. Bioeng. 2009c, 108 (S1: Abstracts of Asia Pacific Biochemical Engineering Conference 2009), S145–S146.

Leong, W.F.; Man, Y.B.C.; Lai, O.M.; Long, L.K.; Nakajima, M.; Tan, C.P. Effect of sucrose fatty acid esters on the particle characteristics and flow properties of phytosterol nanodispersions. J. Food Eng. 2011, 104, 63–69.

Let, M.B.; Jacobsen, C.; Meyer, A.S. Effects of fish oil type, lipid antioxidants and presence of rapeseed oil in oxidative flavour stability of fish oil enriched milk. Eur. J. Lipid Sci. Technol. 2004, 106, 170.

Liu, H.-J.; Nakajima, M.; Nishi, T.; Kimura, T. Production of monodispesed water-in-oil emulsions using polymer microchannels. J. Am. Oil Chem. Soc. 2004, 81, 705–711.

Liu, H.-J.; Nakajima, M.; Nishi, T.; Kimura, T. Effect of channel structure on preparation of a water-in-oil emulsion by polymer microchannels. Eur. J. Lipid Sci. Technol. 2005, 107, 481–487.

McClements, D.J. Food emulsions: principles, practice and techniques, 2nd Ed.; CRC Press: Boca Raton, FL, 2004; chapter 6.

McClements, D.J., Decker, E.A., Weiss, J. Emulsions-based delivery systems for lipophilic bioactive components. J. Food Sci. 2007, 72, 8, R109–R123.

Menger, F.M.; Keiper, J.S. Chemistry and physics of giant vesicles as biomembrane models. Cur. Opn. Chem. Biol. 1998, 2, 726–732.

Moraru, C.I.; Panchapakesan, C.P.; Huang, Q.; Takjistov, P.; Liu, S.; Kokini, J.I. Nanotechnology: a new frontier in food science. Food Technol. 2003, 57, 24–29.

Nakashima, T.; Shimizu, M.; Kukizaki, M. Particle control of emulsion by membrane emulsification and its application. Adv. Drug Deliver. Rev. 2000, 45, 47–56.

Nakagawa, K.; Iwamoto, S.; Nakajima, M.; Shono, A.; Satoh, K. Microchannel emulsification using gelatin and surfactant-free coacervate microencapsulation. J. Colloid Interface Sci. 2004, 278, 198–205.

Neves, M.A.; Ribeiro, H.S.; Kobayashi, I.; Nakajima, M. Encapsulation of lipophilic bioactive compounds by microchannel emulsification. Food Biophys. 2008a, 3, 126–131.

Neves, M.A.; Ribeiro, H.S.; Fujiu, K.; Kobayashi, I.; Nakajima, M. Formulation of controlled size PUFA-loaded oil-in-water emulsions by microchannel emulsification using β-carotene rich palm oil. Ind. Eng. Chem. Res. 2008b, 47, 6405–6411.

Neves, M.A.; Kobayashi, I.; Nakajima, M. Development of microchannel emulsification technology for monodispersed soybean and olive oil-in water emulsion. J. Arid Land Studies. 2009a, 19, 97–100.

Neves, M.A.; Nakajima, M.; Kobayashi, I. Development of production technology for nanoscale antioxidant food materials; Food Nanotechnology. CMC Press: Tokyo, 2009b; pp. 95–103 (in Japanese).

Neves M.A.; Kobayashi I.; Nakajima M. Oxidation processes of O/W nanoemulsions as affected by lipid melting point. J. Biosci. Bioeng. 2009c, 108 (S1: Abstracts of Asia Pacific Biochemical Engineering Conference 2009), S137.

Parker, R.S. Bioavailability of carotenoids. Eur. J. Clin. Nutr., 1997, 51, 86–90.

Perrier-Cornet, J.M.; Marie, P.; Gervais: Comparison of emulsification efficiency of protein-stabilized oil-in-water emulsions using jet, high pressure and colloid mill homogenisation. J. Food Eng. 2005, 66, 211–217.

Ribeiro, H.S.; Guerrero, J.M.M.; Briviba, K.; Rechkemmer, G.; Schuchmann, H.P.; Schubert, H. Cellular uptake of carotenoid-loaded O/W emulsions in colon carcinoma cells in vitro. J. Agric. Food Chem. 2006, 54, 9366–9369.

Ribeiro, H.S.; Chu, B.S.; Ichikawa, S.; Nakajima, M. Preparation of nanodispersions containing β-carotene by solvent displacement method. Food Hydrocoll. 2008, 22, 12–17.

Ribeiro, H.S.; Schuchmann, H.P.; Engel, R.; Walz, E.; Briviba, K. Encapsulation of Carotenoids. Encapsulation Technologies for Active Food Ingredients and Food Processing; Zuidam, N.J., Ed.; Springer: New York, 2011; pp. 211–252.

Ruxton, C.H.S.; Reed, S.C.; Simpson, M.J.A.; Millington, K.J The health benefits of omega-3 polyunsaturated fatty acids: a review of the evidence. J. Hum. Nutr. Dietet. 2004, 17, 449.

Saito, M.; Yin, L.-J.; Kobayashi, I.; Nakajima, M. Preparation characteristics of monodispersed oil-in-water emulsions with large particles stabilized by proteins in straight-through microchannel emulsification. Food Hydrocolloids 2005, 19, 745–751.

Sanguansri, P.; Augustin, M.A. Nanoscale materials development–a food industry perspective. Trends Food Sci. Technol. 2006, 17, 547–7556.

Sugiura, S.; Nakajima, M.; Tong, J.; Nabetani, H.; Seki, M. Preparation of monodispersed solid lipid microspheres using a microchannel emulsification technique. J. Colloid Interface Sci. 2000, 227, 95–103.

Sugiura, S.; Nakajima, M.; Iwamoto, S.; Seki, M. Interfacial tension driven monodispersed droplet formation from microfabricated channel array. Langmuir 2001a, 17, 5562–5566.

Sugiura, S.; Nakajima, M.; Ushijima, H.; Yamamoto, K.; Seki, M. Preparation characteristics of monodispersed water-in-oil emulsions using microchannel emulsification. J. Chem. Eng. Japan 2001b, 34, 757–765.

Sugiura, S.; Nakajima, M.; Seki, M. Effect of channel structure on microchannel emulsification. Langmuir 2002a, 18, 5708–5712.

Sugiura, S.; Kumazawa, N.; Nakajima, M.; Seki, M. Characterization of spontaneous transformation-based droplet formation during microchannel emulsification. J. Phys. Chem. B 2002b, 106, 9405–9409.

Sugiura, S.; Nakajima, M.; Seki, M. Prediction of droplet diameter for microchannel emulsification. Langmuir 2002c, 18, 3854–3859.

Sugiura, S.; Nakajima, M.; Seki, M. Prediction of droplet diameter for microchannel emulsification: prediction model for complicated microchannel geometries. Ind. Eng. Chem. Res. 2004a, 43, 8233–8238.

Sugiura, S.; Nakajima, M.; Yamamoto, K.; Iwamoto, S.; Oda, T.; Satake, M.; Seki, M. Preparation characteristics of water-in-oil-in-water multiple emulsions using microchannel emulsification. J. Colloid Interface Sci. 2004b, 270, 221–228.

Sugiura, S.; Kuroiwa, T.; Kagota, T.; Nakajima, M.; Sato, S.; Mukataka, S.; Walde, P.; Ichikawa, S. Novel method for obtaining homogeneous giant vesicles from a monodisperse water-in-oil emulsion prepared with a microfluidic device. Langmuir 2008, 24, 4581–4588.

Tan, C.P.; Nakajima, M. Effect of polyglycerol esters of fatty acids on physicochemical properties and stability of β-carotene nanodispersions prepared by emulsification/evaporation method. J. Sci. Food Agric. 2005a, 85, 121–126.

Tan, C.P.; Nakajima, M. β-carotene nanodispersions: preparation, characterization and stability evaluation. Food Chem. 2005b, 92, 661–671.

Tan, C.-P.; Kobayashi, I.; Nakajima, M. Preparation of monodispersde refined-bleached-deodorized (RBO) palm-olein-in-water emulsions by microchannel emulsification. Proceedings of the International Chemical Congress of Pacific Basin Societies, 2005c.

Tong, J.; Nakajima, M.; Nabetani, H.; Kikuchi, Y. Surfactant effect on production of monodispersed microspheres by microchannel emulsification method. J. Surfactants Detergents 3, 2000, 285–293.

Tong, J.; Nakajima, M.; Nabetani, H.; Kikuchi, Y.; Maruta, Y. Production of oil-in-water microspheres using a stainless steel microchannel. J. Colloid Interface Sci. 2001, 237, 239–248.

Tong, J.; Nakajima, M.; Nabetani, H. Preparation of phospholipid oil-in-water microspheres by microchannel emulsification technique. Eur. J. Lipid Sci. Technol. 2002, 104, 216–221.

Unlu, N.Z.; Bohn, T.; Francis, D.; Clinton, S.K.; Schwartz, S.J. Carotenoid absorption in humans consuming tomato sauces obtained from tangerine or high-beta-carotene varieties of tomatoes. J. Agric. Food Chem. 2007, 55, 1597–1603.

van der Zwan, E.; van der Sman, R.; Schröen, K.; Boom, R. Lattice Boltzmann simulations of droplet formation during microchannel emulsification. J. Colloid Interface Sci. 2009, 335, 112–122.

van Dijke, K.C.; Schröen, K.C.P.G.H.; Boom, R.M. Microchannel emulsification: from computational fluid dynamics to predictive analytical model. Langmuir 2008, 24, 10107–10115.

van Dijke, K.; Kobayashi, I.; Schröen, K.; Uemura, K.; Nakajima, M.; Boom, R. Effect of viscosities of dispersed and continuous phases in microchannel oil-in-water emulsification Microfluid. Nanofluid. 2010, 9, 77–85.

Vladisavljević, G.T.; Kobayashi, I.; Nakajima, M. Effect of dispersed phase viscosity on maximum droplet generation frequency in microchannel emulsification using asymmetric straight-through channels. Microfluid. Nanofluid. 2011, 10, 1199–1209.

Walde, P.; Ichikawa, S. Enzymes inside lipid vesicles: preparation, reactivity and applications. Biomol. Eng. 2001, 18, 143–177.

Walde, P.; Ichikawa, S.; Yoshimoto, M. The fabrication and applications of enzyme-containing vesicles. Bottom-Up Nanofabrication, Volume 2; Ariga, K., Nalwa, H.S., Eds.; American Scientific Publishers: Los Angeles, California, 2009; pp. 199–221.

Walde, P; Cosentino, K.; Engel, H.; Stano, P. Giant vesicles: preparations and applications. ChemBioChem 2010, 11, 848–865.

Wang, Z.; Yin, L.; Neves, M.A.; Kobayashi, I.; Uemura, K.; Nakajima, M. Influence of emulsifier type on the properties of O/W submicron emulsions passing through an in vitro digestion model. Lyon, France, The 5th World Congress on Emulsions, Proceedings, 2010, 136.

Waraho, T.; McClements, D.J.; Decker, E.A. Mechanisms of lipid oxidation in food dispersions. Trends Food Sci. Technol. 2011, 22, 3–13.

Weiss, J.; Decker, E.A.; McClements, J.D.; Kristbergsson, K.; Helgason, T.; Awad, T. Solid lipid nanoparticles as delivery systems for bioactive food components. Food Biophys. 2008, 3, 146–154.

Yin, L.J.; Kobayashi, I.; Nakajima, M; Effect of polyglycerol esters of fatty acids on the physicochemical properties and stability of β-carotene emulsions during digestion in simulated gastric fluid. Food Biophys. 2008, 3, 213–218.

Yin, L.J.; Chu, B.S.; Kobayashi, I.; Nakajima, M. Performance of selected emulsifiers and their combinations in the preparation of β-carotene nanodispersions. Food Hydrocoll. 2009, 23, 1617–1622.

Yonekura, L; Nagao, A. Intestinal absorption of dietary carotenoids. Mol. Nutr. Food Res. 2007, 51, 107–115.

6

SELF-EMULSIFYING DELIVERY SYSTEMS AND LIPID TRANSPORT

Edgar Acosta[1], Levente Diosady[1], A. Venketeshwer Rao[2], Dérick Rousseau[3], and Christopher Yip[1]

[1]University of Toronto, Department of Chemical Engineering and Applied Chemistry, Canada; [2]University of Toronto, Department of Nutritional Sciences, Canada; and [3]Ryerson University, Department of Chemistry and Biology, Canada.

INTRODUCTION

The last decade has seen an emergence of nanotechnology in food and pharmaceutical products. Having searched through a journal database, the first article that discusses the idea of food and nanotechnology was a brief review by Watkins (2003) outlining the efforts of the food industry in this area and particularly the use of lipids as part of nano-structured assemblies to introduce encapsulation, delivery, and novel physical properties to food products. Further, nanotechnology has also been proposed as a method to introduce novel sensors for food processing, food safety, and food packaging (Neethirajan & Jayas, 2011; Srinivas et al., 2010). Soon after these early efforts a substantial amount of research was dedicated to investigating the safety aspects of nano-structured materials in food products (Chau et al., 2007; Zolnik & Sadrieh, 2009). One of the conclusions of these efforts is that solid nanoparticles and nanotubes (<100 nm) represent a considerable risk due to their ability to penetrate epithelial tissue, to remain in the blood, and to accumulate in tissue, leading to potentially harmful effects (Chau et al., 2007; Xu et al., 2010). Soft lipid-based nanostructured materials have emerged as alternative advanced delivery systems, having lower risks of side effects and even reducing the toxic effects of the delivered drug or nutraceutical (Musthaba et al., 2009).

Soft, lipid-based nano-structures, including self-emulsifying systems, mimic nature's mechanism of lipid adsorption and transport. The first part of this chapter reviews the mechanisms of lipid transport along with the preliminary evidence on the effect of emulsifiers and additives used in engineered self-emulsifying systems on lipid transport and metabolism. In the second part of the chapter we discuss the physical principles that explain the advantages of nano-structured self-emulsifying lipid delivery systems as a way of introducing the design criteria for these formulations. Basic aspects of the equilibrium of surfactant-oil-water systems (microemulsions) are described as a way of connecting the selection of the ingredients, their phase behavior, the physical properties of the formulations, and their performance as delivery systems for poorly soluble drugs and nutraceuticals. Examples of self-emulsifying formulations are discussed in light of the different strategies used to bridge the hydrophilic-lipophilic gap between water and the poorly soluble solute. Finally, we introduce an overview of mathematical models that could be used to engineer self-emulsifying delivery systems.

While the field of self-emulsifying delivery systems has seen rapid progress in pharmaceutical and food applications, future developments require a multidisciplinary approach to take full advantage of the technology. This integral approach should include the biology of lipid uptake, transport and metabolism (including safety considerations), the physicochemical properties of the formulation, the manufacturing aspects of the product, consumer perception, and cost-benefit considerations.

LIPID UPTAKE AND TRANSPORT

Lipid uptake (absorption through the small intestine) involves a number of processes that take place in the enterocytes (epithelial cells that line the small intestine). Fig. 6.1 presents a schematic of an enterocyte and the processes involved in lipid transport. The apical side of the enterocyte faces the gut lumen and contains microvilli (brush-border membranes) that increase the surface area for mass transfer. The microvilli are covered by a mucous (mucin) layer of glucoproteins that aid in capturing lipid suspensions. Once the lipid suspension is on the surface of the microvilli, lipid molecules or small lipid aggregates (< 100 nm) are absorbed via passive diffusion mechanisms. This process of lipid diffusion, at least for the case of triglycerides, is aided by lipolysis in the stomach and in the gut via lipases in the pancreatic juice. The free fatty acids (FFA), monoglycerides (MG), and lyso-phospholipids (LysP) produced during lipolysis not only contribute to the formation of micelles that intervene in passive diffusion, they also form the building blocks in a process of re-assembly that takes place in the enterocyte. These small molecular weight lipids along with cholesterol are recruited by the smooth endoplasmic reticulum (S-ER). In the S-ER, cholesterol esters are produced in a process catalyzed by the enzyme ACAT-2, and the FFA and MG are transformed back into triglycerides by mono- and di-acylglycerol acyltransferase (MGAT and DGAT); LysP are also transformed back into phospholipids via phosphoglyceride-synthesizing enzymes. These newly formed cholesterol esters, triglycerides, and phospholids are then transferred to the rough endoplasmic reticulum where microsomal transfer protein (MTP) guide the formation of nascent chylomicrons (CM) by assembling these lipids with the apolipoprotein apoB-48. From this point the nascent CM are then transported to the Golgi apparatus where the CM are then packaged and secreted to the basolateral membrane where they are collected by the lymphatic system via the lacteals that discharge into the thoracic tube. The thoracic tube discharges into the left subclavian vein where the chylomicrons are distributed by the bloodstream to the tissue (Patel & Brocks, 2009; Shen et al., 2001; Williams, 2008). Tissue uptake of lipids is medi-

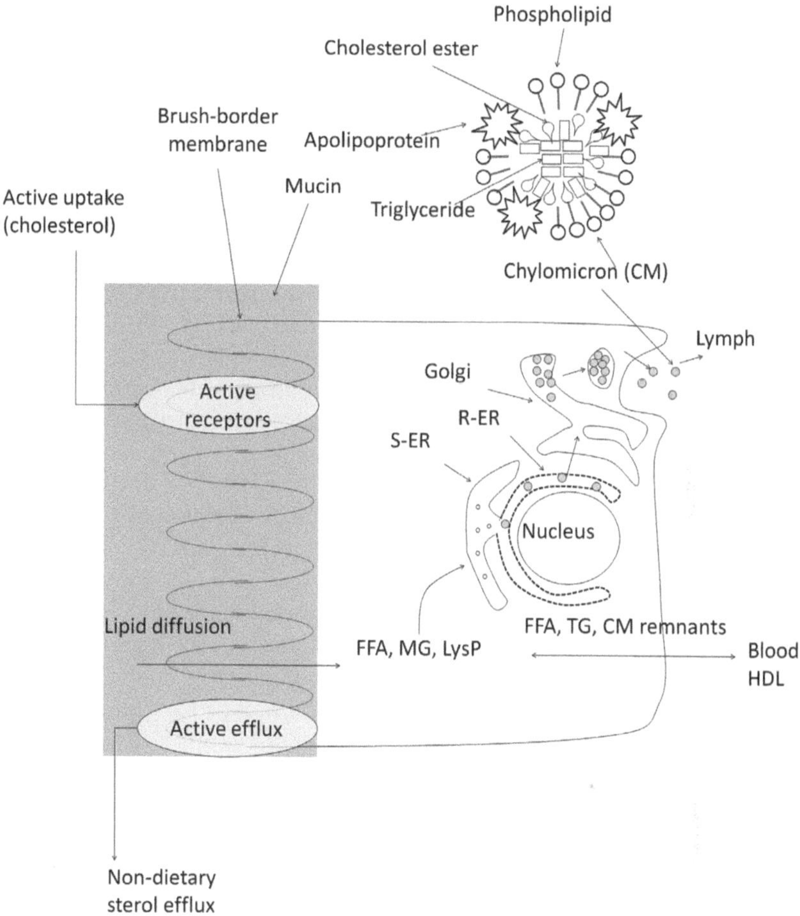

Fig. 6.1. Schematic of lipid transport processes in enterocytes. Details of these transport processes are provided in the text. FFA: free fatty acids, TG: triglycerides, CM: chylomicrons, MG: monoglyceride, LysP: lysophosphatidylcholine (lysolecithin), HDL: high density lipoprotein. Adapted from schematics of Williams (2008) and Patel and Brooks (2009).

ated via signaling pathways activated by the apolipoproteins in the chylomicrons. As the chylomicrons are depleted of triglycerides and lipids they become chylomicron remnants (De Jong et al., 2003). These remnants (mainly the lipoproteins and phospholids) are hydrolyzed in the liver and the lipids and lipoproteins are recycled into the lipid transport cycle.

The hepatic portal vein also offers the opportunity of transporting high density lipoproteins (HDLs), which are also lipoprotein assemblies that can exchange small lipids with the enterocytes. As the HDL collect small lipids from the enterocytes they transform into very low density lipoproteins (VLDL). VLDL assemblies are similar to chylomicrons, but they are smaller in size and have relatively lower lipid content. As VLDLs release the lipids in the tissue (a process mediated by apolipoproteins), they transform into low density lipoproteins (LDL), intermediate density lipoproteins (IDL), and back to high density lipoproteins (HDL). These HDL assemblies are the ones capable of carrying cholesterol from tissues to the liver where it is discharged with the bile.

The uptake of cholesterol by enterocytes, unlike other lipids, has an active absorption component that involves the Niemann-Pick C1–like 1 protein (NPC1L1) expressed on the microvilli (active receptors in Fig. 6.1). There are active efflux mechanisms that selectively reject phytosterols from the enterocyte, involving ABCG5 and ABCG8 proteins located in the microvilli. Furthermore, it has been suggested that bile salts are involved in activating the active transport of cholesterol (Watt & Symmond, 1976); however, the same signaling cascade initiated by bile salts has been linked to the activation of growth factors in enterocytes, a process associated with colon cancer (De Jong et al., 2003; Rao & Janezic, 1992).

Given that the transport of lipids through the body (mainly a hydrophilic environment) is dependent on the self-assembly of lipoproteins, it is not difficult to anticipate that the presence of surfactants (used in self-emulsifying formulations) can interfere with these mechanisms. For example, pluronic surfactant L-81 (ethylene oxide–propylene oxide block copolymer) has been found to interfere with the assembly of chylomicrons, but it does not affect the incorporation of lipids into HDL assemblies (Tso et al., 1981). Another pluronic surfactant, F-68 has been found to improve the solubilization of cholesterol in micelles but not its uptake by enterocytes, a process that seems to be facilitated by the packing of bile salts in the micelle (Watt & Simmonds, 1976). Recently it has been found that surfactants also interfere with the activity of a family of enzymes known as Cytochrome P450, or CYP, which control the metabolism (mainly oxidation, hydroxylation) of a number of drugs and other lipophilic solutes (Patel & Brocks, 2009). For example, the use of polyunsaturated fatty acids (PUFAs) has been found to inhibit the activity of CYP3 enzymes, an effect that has been associated with the increase in bioavailability of halofantrine. The role of surfactants (particularly those containing PUFAs) in inhibiting CYP activity has also been linked to the increase bioavailability of other hydrophobic drugs such as cyclosporine A and saquinavir (Patel & Brocks, 2009). While the current view is that CYP inactivation by surfactants in most cases contributes to the improvement of bioavailability of drugs (preventing undesired metabolism), there could be cases where this metabolism is necessary for some prodrugs or nutraceuticals (Christiansen et al., 2011).

Surfactants can also interfere with the activity of lipases. Hydrophilic surfactants, particularly those with polyethylene glycol groups, have been found to produce a barrier at the oil-water interface that prevents the access of lipases to the triglycerides (MacGregor et al., 1997). Solomon et al. (1996) plotted the time to achieve 25% and 50% lipolysis of a medium chain triglyceride (MCT) as a function of the hydrophilic-lipophilic balance (HLB) of nonylphenol ethoxylate surfactants used to emulsify the MCT. They found that the maximum conversion time (minimal lipase activity) was obtained at HLB ~15. More recently, the IC_{50} (surfactant concentration to inhibit 50% of the activity of pancreatic lipase) of common surfactants used to produce self-emulsifying delivery systems (SEDS) has been measured: for PEG-40 castor oil (Cremophor RH 40) IC_{50} = 0.04 mM, PEG-35 castor oil (Cremophor EL) IC_{50} = 0.19 mM, PEG-20 sorbitan monooleate (polysorbate 80) IC_{50} = 0.13 mM, and the more hydrophobic sucrose laurate IC_{50} = 0.46 mM (Christiansen et al., 2010). These authors also determined that only 6% to 14% of the surfactant itself is hydrolyzed by pancreatic lipase. Although the fact that surfactants used in SEDS inhibit lipase activity is well-established, the question then remains whether this is a positive or a negative outcome. The answer depends on the delivered active compounds and if the objective is to provide protection against lipases or not. If this protection is not necessary, then reducing lipase activity can be counter-productive to the delivery. A number of emulsion-based delivery systems use lipase activity to ensure optimal delivery, and in fact the combination of bile salts with lipases to promote the production of fatty acids and mono- and di-glycerides is nature's way of producing self-emulsifying systems (Velikov & Pelan, 2008). Recently, the mechanism of lipase inhibition induced by a pluronic surfactant (F68) was investigated via interfacial tension and interfacial rheology (Torcello-Gómez et al., 2011). Those studies confirmed that the PEG-containing pluronic surfactant hinders the adsorption of pancreatic lipase at the oil-water interface. On the other hand, the presence of lecithin not only enables the adsorption of pancreatic lipase but, at high enough concentration, it reverses the inhibitory effect of the pluronic surfactant. These results provide additional evidence that bile salts act in combination with lipases to produce nature's version of SEDS.

The discharge of pancreatic juice, containing pancreatic lipase, and the discharge of bile salts and cholesterol from the bile duct and into the upper section of the small intestine are mainly regulated by the hormones secretin and cholecystokinin, respectively. The signaling pathway that controls the release of pancreatic juices and bile is, however, complex and involves a larger number of hormones, peptide transmitters, and growth factors (Rehfeld, 1998). The cooperative action of bile salt and lipases produces a mixture of fatty acids, mono- and di-glycerides, lectihins, lysolecithins, cholesterol, and various forms of cholic acids. Fig. 6.2 presents one example of the lipid composition of a fed-state simulated intestinal fluid containing taurocholic acid, lecithin (a mixture of phosphatidyl cholines, only the dioleyl form is shown), monoglyceride, and a fatty acid salt (Jantratid et al., 2008). The phase behavior of mixtures of bile salts with fatty acids and mono-, di-, and triglyceride is an aspect that has preoccupied scientists in the field as this phase behavior dictates the fate of the emulsified lipids and their uptake by the enterocytes (Hernell et al., 1990; Koynova & Caffrey, 2002; Staggers et al., 1990). Fig. 6.3 presents a partial pseudo-ternary phase diagram of bile salts, intestinal lipids, and cholesterol as an example of these phase behavior studies. The system of Fig. 6.3 contains 99 wt% of an isotonic

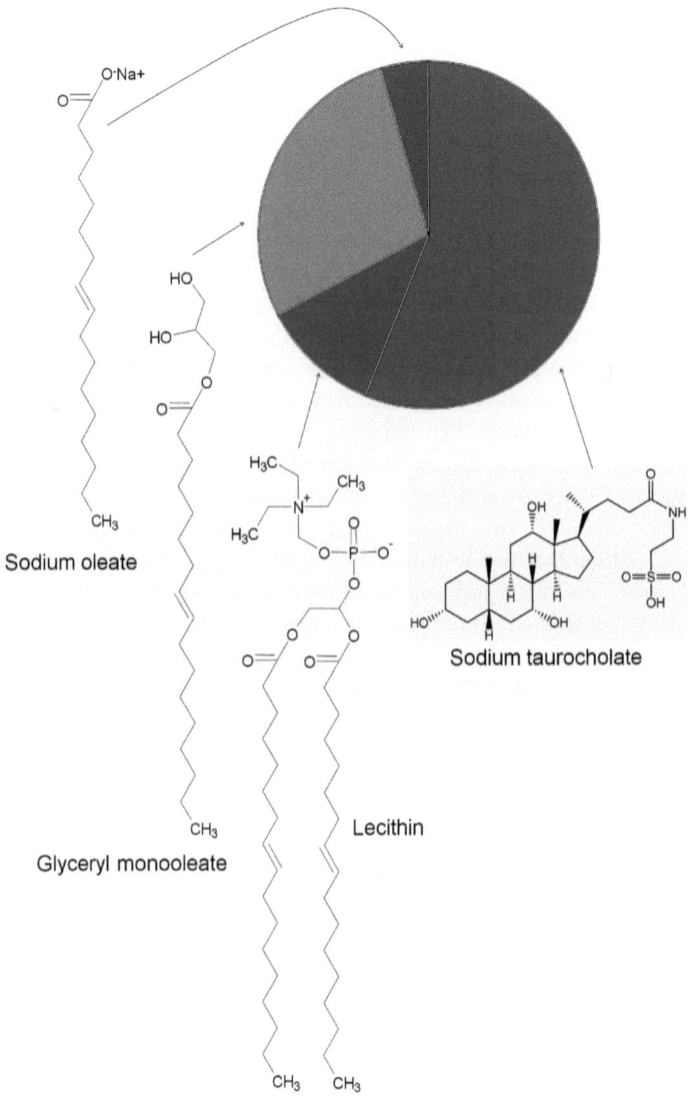

Fig. 6.2. Composition and molecular structure of lipids used in fed-state simulated intestinal fluid. Composition data from Jantratid et al. (2008).

solution (0.15M of NaCl) at 37°C and pH 6.5. The intestinal lipids in Fig. 6.3 consist of a mixture of 69.4% oleic acid, 13.9% monolein, 2.8% diolein, and 13.9% lecithin (Staggers et al., 1990). One way to interpret the phase behavior of Fig. 6.3 involves the concepts of curvature (~1/radius of curvature) and surfactant packing

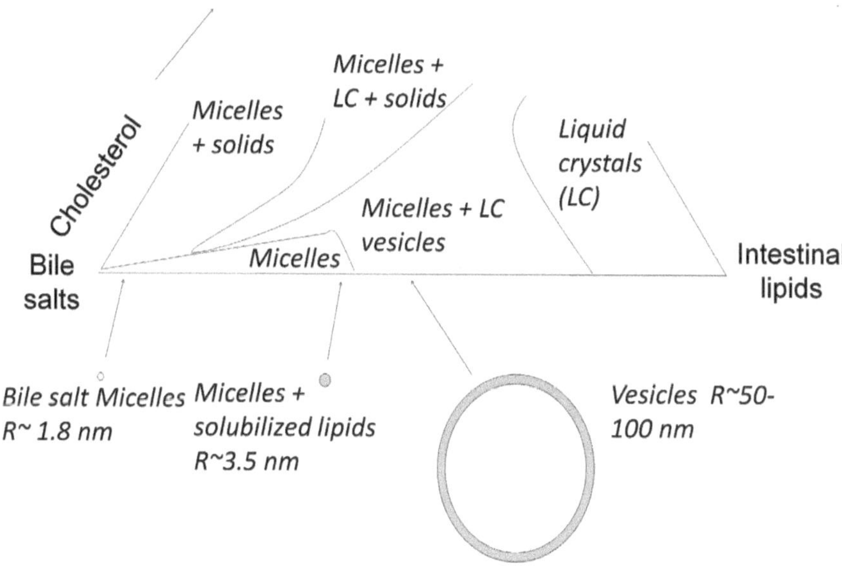

Fig. 6.3. Partial ternary (bile salt-cholesterol-intestinal lipids) phase diagram indicating the type of phases and colloidal structures produced in low-cholesterol mixtures. Figure adapted from Staggers et al. (1990) for a system containing 99% of isotonic solution, pH= 6.5 and 37°C.

at the interface. In the case of bile salts alone, they tend to form micelles (high curvature systems), evidencing the hydrophilic nature of bile salts. The addition of intestinal lipids produces an increase in the size of the micelles (i.e., reduction in curvature), which shows that these intestinal lipids are more hydrophobic than bile salts. In the absence of cholesterol, systems containing more than 40% of intestinal lipids produce near zero curvature systems that lead to the formation of vesicles and lamellar liquid crystals. Higher cholesterol content in the mixture of bile salts and lipids further induces the formation of precipitate (solid) phases and liquid crystals, likely associated with the ability of cholesterol to form crystalline phases and its association to bile salts. While micelle solutions are ideal delivery systems for lipids, three phase systems (water, emulsified lipid, and solids enriched with cholesterol) have been observed in intestinal fluid samples (Wang & Carey, 1996). Given the variability in diet and eating habits, it is not surprising that despite the best efforts of the hormonal control system to regulate the concentration of bile salts (with respect to the intake of lipids), less desirable phases are still formed. This issue also explains the variability of the bioaviability of a number of drugs and nutraceuticals that are poorly soluble and whose formulation relies solely on the body's own lipid emulsification mechanisms.

ENGINEERED LIPID DELIVERY SYSTEMS

To ensure the bioavailability of poorly soluble drugs and nutraceuticals, various delivery systems such as SEDS, SMEDS, solid lipid nanoparticles, and dispersed liquid crystal systems (e.g., cubosomes) have been developed. The idea behind all these engineered systems is that particles with sizes near 100 nm or lower can penetrate the enterocytes and avoid numerous metabolic pathways, thus ensuring high bioavailability. There are four reasons that justify the use of nanoscale delivery systems: increase in solubility in high curvature systems (predicted by the Kelvin equation), increase in surface area for mass transfer, increase of partitioning into mucin layer (as predicted by size exclusion chromatography principles), and direct uptake by type M cells in Peyer's patches.

According to the Kelvin equation, the excess free energy of highly curved systems increases the solubility of the solute (Cs) over its reported solubility (Cs_o), as follows (Jinno et al., 2006; Kipp, 2004):

$$\left(\frac{Cs}{Cs_o}\right) = \exp\left(\frac{2Mw \times \gamma}{\rho RT} \times \frac{1}{r}\right) \qquad [1]$$

where Mw is the molecular weight of the solute, γ is the interfacial tension at the oil-water interface, ρ is the density of the oil phase, R is the universal gas constant, T is the temperature, and r is the radius of curvature (assuming that the nano-particle or droplet is spherical).

The increase in area per unit of volume to improve mass transport has been considered in Amidon's transport numbers. These transport numbers represent the bases of the biopharmaceutical classification system (BCS), used to determine if the solubility/permeability of the drug is suitable for effective adsorption or if an appropriate formulation can be used to improve its absorption (Amidon et al, 1995; Oh et al., 1993). According to the transport model of Amidon, there are three dimensionless transport numbers that describe the process, one of them is the absorption number, An = radial absorption rate across the intestinal wall/axial convection rate; the second one is the dose number, Do = concentration of the solute in the dose/solubility of solute in the intestinal fluid, and finally the dissolution number, Dn = residence time/dissolution time. Within a given An, higher Dn numbers help improve the absorption; however, most of the improvement in absorption takes place when Dn increases from 1 to 10. The value of Dn is proportional to the solubility of the solute and inversely proportional to the particle size (Acosta, 2009; Amidon et al., 1995). The increase in solubility obtained when reducing particle size, as described by equation 1, reduces the value of the dose number, which is a desirable outcome to improve bioavailability. Increasing the area per unit of volume increases the rate of dissolution (i.e., reduce dissolution time), which increases Dn, also a desirable outcome. This improvement in the area per unit of volume is described by the equation

$$\left(\frac{\text{Area}}{\text{Volume}}\right) = \frac{3}{r} \qquad [2]$$

The dissolution number (Dn) has a relatively minor effect on absorption when dealing with high dose (Do) numbers, which is the case for most poorly soluble drugs and nutraceuticals (Oh et al., 1993). The only possible method to improve absorption in those cases is to increase the absorption number (An). While the reduction in particle size might accelerate passive diffusion through the enterocyte brush border membrane, thus far there is little evidence of that effect. On the other hand, retarding the axial convection rate is possible if the particle or droplet is small enough to be entrapped in the mucin layer (Lai et al., 2007). There have been several transport equations proposed for the diffusion of molecules and particles through mucin (Cu & Saltzman, 2009). Perhaps the simplest and most relevant equation to illustrate the use of nanoparticles or nanodroplets is the partition (K) of particles of hydrodynamic radius r_H into a gel network of pore radius a as follows (Giddings et al., 1968):

$$K \propto \left(1 - \frac{r_H}{a}\right)^2 \qquad [3]$$

Equation 3 is valid for systems where $r_H < 0.6a$. At conditions when $r_H \sim a$, the particle cannot penetrate into the gel. Furthermore, equation 3 is also valid in the absence of interactions between the material of the gel and the particle.

Whole particle uptake is also possible if the particle size is close to 100 nm in diameter as it can diffuse through the porous Type M cells in the Peyer's patches located in the ileum (Hussain et al., 2001). The contribution of this type of uptake is considered to be minor because the fraction of the area of the intestine covered by these Peyer's patches is small.

Fig. 6.4 presents the relative transport property—solubility (equation 1), surface area (equation 2), and partition into the mucin gel (equation 3)—calculated as the value of the property at a given particle radius divided by the value of the property for a particle of 50 nm (100 nm diameter). The shaded area in Fig. 6.4 represents the range of values of bioavailability and uptake obtained in the literature for various nanoparticle delivery systems (Acosta, 2009). As shown in Fig. 6.4, the improvement in solubility with decreasing particle size for a typical drug or nutraceutical is relatively small. The effect of area is pronounced, but it is the relative partition into the mucin gel that more closely follow the experimental observations.

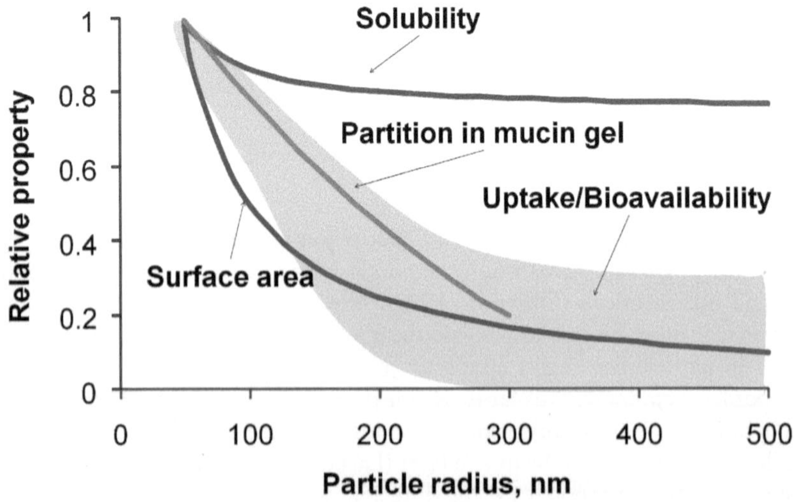

Fig. 6.4. Relative transport property (transport property at any given radius/transport property for a 50 nm particle) as a function of particle radius. Solubility curve calculated for a drug of molecular weight (Mw) 300 g/mol, interfacial tension of 50 mJ/m^2, density of 0.8 g/mL, at 37°C, using equation 1. Surface area curve calculated using equation 2. Partition curve calculated assuming gel pore radius of 500 nm and equation 3. The shaded area represents the range of relative bioavailability and uptake reported in various in vivo studies (Acosta, 2009).

The mucous layer that lines the small intestine is composed of water (~95%), glycoprotein (mucin itself) + lipids (0.5% to 5%), mineral salts (up to 1%), and free proteins (up to 1%), and its thickness is approximately 50 μm (Khanvilkar et al., 2001). The pore size of the mucin network has been reported to be 14 nm via molecular diffusion studies; however, particle diffusion studies show that 415 nm particles can penetrate through the mucin network, but 1μm particles cannot penetrate the network (Szentkuti, 1997). The predicted relative partition in Fig. 6.4 was obtained with a pore radius a = 500 nm. In addition to size, physical interactions (electrostatic, hydrophobic interactions, hydrogen bonding, etc.) also play a role in the penetration and retention in the mucin layer. A number of strategies have been implemented to impart bioadhesive properties to self-emulsifying systems and other nanoscale lipid delivery systems (Acosta 2008; 2009). Among these strategies, the use of lectins (carbohydrate-binding proteins) is worth mentioning (Ohno et al., 2006). Unfortunately, the costs and potential regulatory issues with the introduction of lectins have prevented further advances in this area. There is still much to be understood about the interaction of nanoparticles and self-emulsifying systems with the mucin layer; however, the work of Araya et al. (2006) illustrates the importance of that interaction. In that work the authors evaluated the permeation of a poorly soluble drug (ibuprofen) in SEDS and the permeation of dissolved ibuprofen and showed that when the mucin layer was disrupted (via n-acetyl-L-cysteine), the permeation of dissolved ibuprofen was not affected but the permeation from the SEDS system decreased. The authors suggested that the delivery mechanism from SEDS involves release into the mucin layer.

MICROEMULSIONS (μES) AND SELF-EMULSIFICATION

The term *microemulsion* (μE) was coined by Schulman and collaborators in 1959 after using X-ray methods and finding drop diameters in potassium oleate μEs in the order of 250Å (25 nm) (Schulman et al., 1959). However, μEs were first described in 1943 with the names oleomicelles and hydromicelles (Hoar & Schulman, 1943). Considering that at the time the prefix "micro" was the equivalent of today's perception of "nano" helps explain the misnomer micro. In today's nanotechnology definition, the term would have been nano-emulsions. However, the contemporary use of the term nano-emulsion is reserved for emulsions that are not thermodynamically stable having droplet sizes smaller than 100 nm. This introduces the second inaccuracy to the misnomer microemulsions—they are not emulsions. The fact that μEs are systems in thermodynamic equilibrium was established in the late 1960s by evaluating the solubilization of oils in surfactant systems as a function of temperature (Shinoda & Kunieda, 1973). Even at that time, the problem of the misnomer for these systems was identified, but alternative names like swollen micellar solutions failed to take hold. The term micelle solubilization is typically reserved for systems where the volume of oil solubilized is substantially smaller than the volume of surfactant (Miller, 2010). While μEs are not too different from micelles or reverse micelles, thermodynamic models encompassing both systems are yet to be developed.

Currently there are more than ten thousand articles describing various aspects of μEs, but there are a few key reviews for those interested in a general overview on the subject, particularly as it applies to food and drug delivery applications (Engström & Larsson, 1999; Garti, 2003; Garti & Yuli-Amar, 2008; Lawrence & Rees, 2000; Malmsten, 1999; Salager et al., 2005). There are several notations for different types of μEs, but the most widely accepted is Winsor's nomenclature (Winsor, 1954). The reader must be careful when using the Winsor's nomenclature because it was developed for μEs prepared with oil to water volume ratios close to 1. According to this nomenclature, there are four types of μEs. A μE where oil is the dispersed (internal) phase and water is the continuous phase is classified as a Type I (O/W) system. In Type II (W/O) μEs water is the dispersed (internal) phase and oil is the continuous phase. A Type III μE is bicontinuous in oil and water and coexists with excess water and oil phases. A Type IV μE is basically the same as a Type III system only that there is enough surfactant in the system to solubilize all the oil and water initially added to the vial (i.e., no excess phases). The problem with Type IV μEs is that a Type I or Type II system can be mistaken for a Type IV system if there is enough surfactant to solubilize all the oil and water in the system. For that reason, electrical conductivity or self-diffusion measurements are used to elucidate the type of μE (Kogan et al., 2007; Salager et al., 2005).

Fig. 6.5 illustrates vials of Type I, IV, and II μEs prepared with oil to water volume ratio = 1/1. The yellow-colored (dark grey in the figure) phases in each vial correspond to each of these μEs. The yellow color of these μEs comes from the surfactant, lecithin, used in this particular example. The ternary phase diagrams at the top of Fig. 6.5 present a schematic of the type of phase behavior obtained with each formulation when varying the surfactant-oil-water ratio. The numbers inside each phase diagram indicates the number of coexisting phases. For systems containing 2 phases, the location of the bar (upper or lower) indicates the location of the μE in the 2-phase system. Fig. 6.5 also presents schematics of the orientation of the surfactant at the oil-water interface for each μE type. This representation is valid for systems prepared at oil to water ratios close to 1. The ternary phase diagram shown at the bottom of Fig. 6.5 indicates that within the single phase region it is possible to obtain micelles, bicontinuous systems, and reverse micelle phases without observing any type of phase separation. In the formulation of self-emulsifying systems it is important to recognize that these phase transitions will occur as the surfactant system is diluted by the gastric and/or intestinal fluid. The group of Garti, in particular, has advocated for awareness of this issue and has presented examples of phase transitions in μEs that can be fully diluted (Kogan et al., 2007).

Besides changing the oil-water ratio, there are other methods to induce phase transitions in μEs. Changes in temperature, electrolyte concentration, the hydrophobicity of oil and surfactant, and the surfactant to cosurfactant ratio are used to change the type of μE (Salager et al., 2005). Thousands of phase behavior studies have been conducted to understand the effect of each variable. Most of that information was compiled in the late 1970s via empirical formulation correlations. Later, it was observed that predictions from these formulation correlations were consistent with the partition coefficient of nonionic surfactants between oil and water (Salager et al., 2000), suggesting that this empirical equation reflected the free energy cost (normalized by RT) of transferring a surfactant from the oil phase (oil continuous systems) to water (water continuous systems). At that moment, a new name for those correlations was adopted—the hydrophilic-lipophilic difference, or HLD. There are numerous references to the HLD and its parameters (Acosta et al, 2008; Salager & Antón, 1999; Salager et al., 2001; 2005). There are two basic forms of the HLD equation:

For ionic surfactants:

$$\mathrm{HLD} = \ln S - K(\mathrm{EACN}) - f(A) + Cc_i - a_T\,(\Delta T) \qquad [4]$$

For nonionic surfactants:

$$\mathrm{HLD} = Cc_n + b\,S - k\,(\mathrm{EACN}) - \phi(A) + c_T\,(\Delta T) \qquad [5]$$

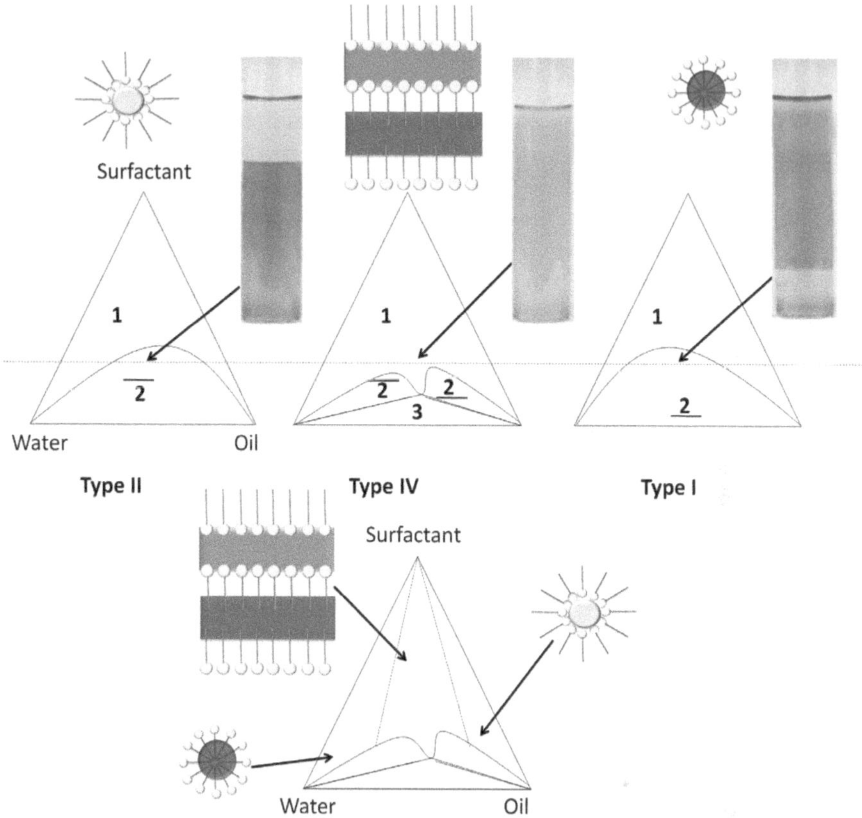

Fig. 6.5. Typical ternary phase diagrams and μE assemblies obtained with Type I (oil-in-water), Type II (water-in-oil), and Type IV (bicontinuous) systems. μE assembly drawings not to scale. The numbers inside the phase diagrams represent the number of coexisting phases. The upper and lower bars in 2-phase systems represent the location of the microemulsion phase in the test tube.

The electrolyte concentration (g/100ml of NaCl) is represented by S; the hydrophobicity of the oil is represented by EACN; the equivalent alkane carbon number (e.g., EACN of n-hexane = 6), f(A), and ϕ(A) are functions of the alcohol (cosurfactant) type and concentration; the hydrophobicity of ionic and nonionic surfactants are represented by the terms Cc_i and Cc_n, respectively. ΔT is the temperature deviation measured from the reference temperature (25°C), and k, K, a_T, and c_T are empirical constants that depend on the type of surfactant.

According to the free energy interpretation of the HLD, a negative value means that the surfactant spontaneously partitions into the aqueous phase, forming micelles

(Type I μE). Positive values of HLD means the surfactant stays in the oil phase, forming reverse micelles (Type II μE). Zero values of HLD means the surfactant has the same affinity for oil and water, thus producing bicontinuous systems (i.e., Type III/IV μE). This interpretation of HLD being related to the free energy of surfactant transfer, along with the Kelvin equation that relates surface excess free energy with curvature, was used to propose the net-average curvature (NAC) model (Acosta et al., 2003). This model, also known as HLD-NAC, makes use of the HLD to predict solubilization of oil and water, phase transitions, some elements of phase diagrams, drop size and shape, and μE viscosity (Acosta et al., 2003; Acosta, 2008; Kiran & Acosta, 2010). According to the NAC model, any μE can be represented as the average of two coexisting hypothetical states, one where water drops of radius Rw are dispersed in oil and one where oil drops of radius Ro are dispersed in water. The NAC considers two forms of curvature averages (Acosta et al., 2003):

$$\text{Net curvature: } Hn = 1/R_o - 1/R_w \quad [6]$$

$$\text{Average curvature: } Ha = (1/2)(1/R_o + 1/R_w) \quad [7]$$

The net curvature identifies the curvature of the surfactant at the oil-water interface. μEs containing the same volume of solubilized oil and water ($R_o = R_w$) are bicontinuous with net-zero curvature. The net curvature relates to the value HLD as Hn= -HLD/L where L is the extended length of the surfactant tail. The average curvature addresses the characteristic size of the μE domain. The advantage of the NAC definition over the traditional Gaussian curvature is that it allows for the mathematical treatment of the bicontinuous phase where Hn tends to zero, yet the value of Ha can be finite. At HLD = Hn = 0, the size of the μE domain can only grow to a maximum value known as the characteristic length (ξ). Thus, 1/Ha ≤ ξ. One more element of the HLD-NAC model is the prediction of the interfacial tension (γ) of μE systems, where a simple consideration of the free energy of μE drops leads to the equation (Acosta et al., 2003):

$$\gamma = Er/(4\pi R_o^2) \quad [8]$$

The parameter Er in equation 8 refers to the interfacial rigidity of the self-assembly, which is normally in the order of $1K_BT$. Higher values of Er are indicative of rigid surfactant assemblies, which lead to slower coalescence and solubilization for these systems. Although equation 8 was written specifically for oil-in-water systems, the same equation can be used for water in oil systems by simply replacing R_o for R_w. Fig. 6.6 illustrates the use of the HLD-NAC model to predict the phase volumes and drop size of μEs formulated with isopropyl myristate, lecithin as the surfactant, and lipophilic and hydrophilic additives (linkers). One important feature of Fig. 6.6 is that the drop size tends to increase, along with the solubilization of the internal phase, as HLD approaches zero. Some deviations between the predicted and the

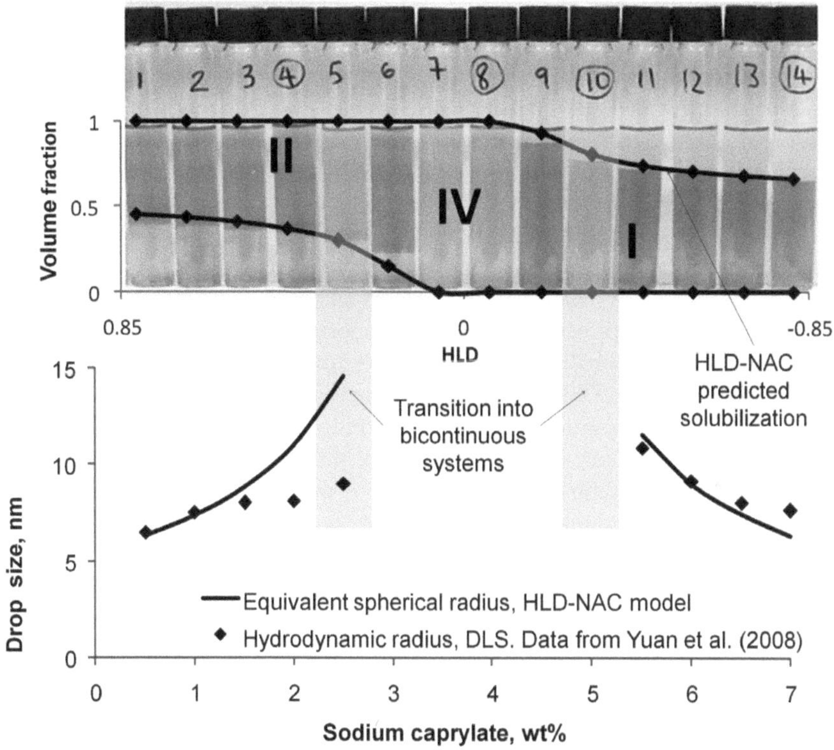

Fig. 6.6. Example of a μE phase scan for a lecithin-linker system. Increasing the concentration of the hydrophilic additive (sodium caprylate) induces a phase transition Type II-IV-I. The top part of the figure compares the phase volumes predicted by the HDL-NAC model (equations 4-7, solid lines) and the phase volumes obtained experimentally. The bottom part of the figure compares the drop sizes predicted by the HLD-NAC model and the drop size measured via dynamic light scattering (DLS). Phase behavior and particle size data obtained from Yuan et al. (2008).

experimental values are observes at the Type II–IV transition region, likely due to drop-drop interactions. Even when considering the deviations, the simple HLD-NAC model can predict the range of μE drop size, which is important in designing delivery systems. Yuan et al. (2010) used the HLD-NAC to design a dilution path of constant drop size for μEs used in transdermal delivery. As indicated in Fig. 6.4, drop size is an important parameter when designing oral drug delivery systems and, by extension, nutraceutical delivery systems.

Lipid delivery systems can be designed to produce μEs at the stomach/intestine, in which case they are referred to as self-microemulsifying delivery systems (SMEDS).

The drop sizes presented in Fig. 6.6 are relevant to SMEDS formulations. There is yet another approach where the final state is not a μE but an emulsion that, during the process of emulsification, produces a short-lived μE that produces ultra-low interfacial tensions, as predicted by equation 8. The ultra-low interfacial tension (~10^{-3} mJ/m^2) produced with these intermediate μEs reduces the energy required to produce the emulsion. Because of the minimal energy input required, these systems are known as self-emulsifying delivery systems (SEDS). The main features of the process of self-emulsification have been described in numerous articles (Gursoy & Benita 2004; Izquierdo et al. 2002; Miñana-Perez et al. 1986; Pérez et al. 2002; Salager et al. 2004; Solans et al., 2005). Fig. 6.7.A presents a simplified emulsification "map" that explains the formation of nano-emulsions (SEDS included) using self-emulsification. The y-axis of Fig. 6.7.A presents the HLD of the formulation and the x-axis the volume fraction of water (considering only oil and water with low surfactant content). Systems with positive HLD and/or high oil content are continuous in the oil phase. The transition from oil-continuous to water-continuous systems (negative HLD and/or high water content) can be accomplished by numerous paths. Fig. 6.7.A only presents two of the most common paths. In path A the oil and water content is fixed and the formulation conditions are changed (mainly temperature) to reach a transition point where HLD = 0, the interfacial tension, reaches a minimum, as shown in Fig. 6.7.B. The use of temperature (mainly for nonionic surfactants) as the method to change HLD, is known as the phase inversion temperature (PIT) emulsification method (Izquierdo et al., 2002). In path B, the formulation conditions (negative HLD) favor the formation of oil-in-water systems, but the high initial oil content forces the system to be water-in-oil. Upon dilution with water the system inverts from water-in-oil to oil-in-water.

The emulsification through path A can be understood by considering the equilibrium interfacial tension predicted by Equation 8, which is presented in Fig. 6.7.B for a typical μE system. The second consideration is the balance of inertial forces and surface tension forces. A review by Walstra (1993) summarizes the force balance equations that apply for laminar and turbulent flow conditions. This balance of forces is represented by the Weber number (shear/surface tension forces):

$$We = \mu Gr/\gamma \qquad [9]$$

where μ is the viscosity of the continuous phase, G is the shear rate, r is the drop size (radius), and γ is the oil-water interfacial tension. For laminar flow and when the viscosity of internal phase is similar to the viscosity of the external phase, the critical Weber number (beyond which the drops are not mechanically stable) is close to 1. Using the interfacial tensions predicted from Equation 8, the viscosity of water, and shear rate of 100 s^{-1} (Hz), Fig. 6.7.B presents emulsion drop sizes as a function of HLD. As shown in Fig. 6.7.B, as the system undergoes phase inversion the drop size tend to reach values of 1 micron or less. Unfortunately, we were not able to find measured values of shear rate in the stomach and the intestines. In fact, the peristaltic action in

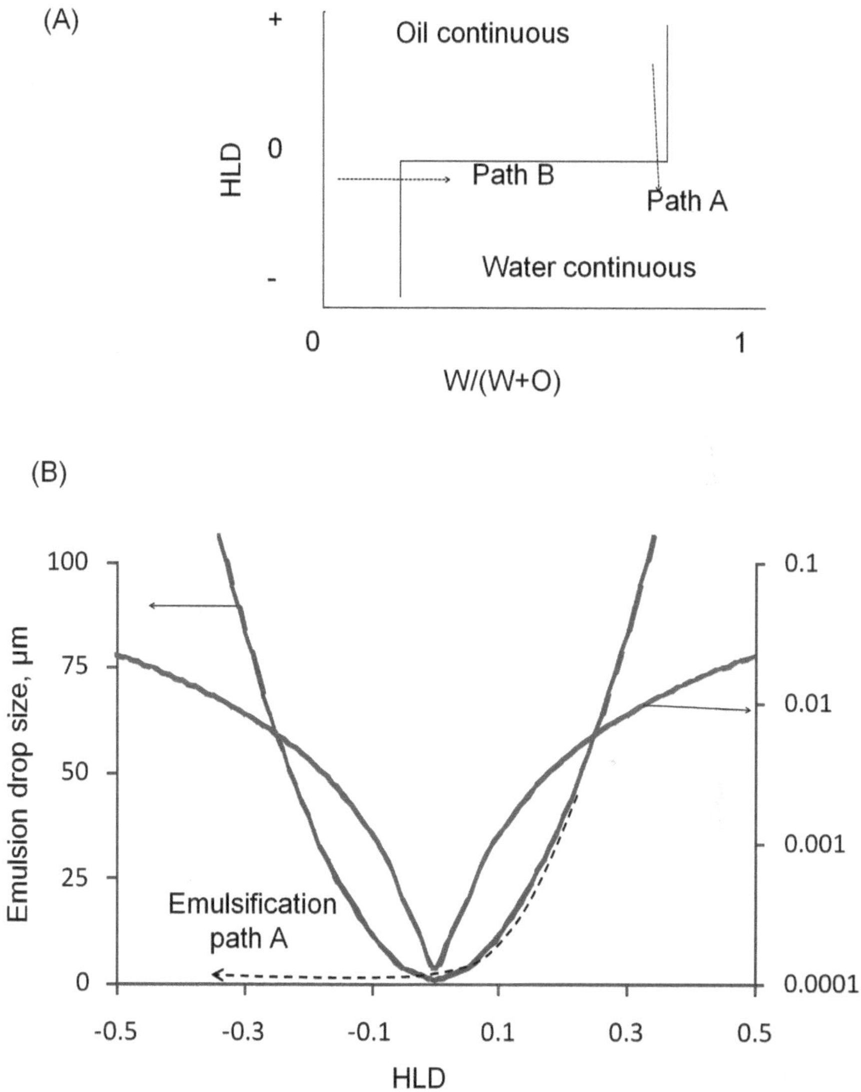

Fig. 6.7. Part A—Emulsification map showing the areas of HLD and oil/water ratios that produce inversion on the continuous phase of the emulsion. Adapted from Salager et al. (2004). Part B—Changes in interfacial tension predicted by the HLD-NAC model (equation 8) and drop size predicted by HLD-NAC + Walstra (Equation 9) for a shear rate of 100 Hz. The dashed line represents the expected change in drop size following the dilution path "A" indicated in Fig. 6.7.A.

the digestive tract is complex to measure experimentally. It has been estimated (not considering food particles) that the shear rate in the stomach, before reaching the pylorus, is close to 20 Hz (Kozu et al., 2010). This shear rate is expected to be higher in the pylorus and in the upper intestine as the peristaltic action creates flow restrictions and localized high shear.

While, according to Fig.6.7.B, path A could produce sub-micron particles, smaller drop sizes have been obtained using path B when with near zero HLDs (Pérez et al., 2002). The problem with both paths is that the minimum drop size needs to be retained after completing the phase inversion. The dilution method (path B) facilitates the stabilization because diluting the emulsion reduces the volume fraction of oil drops, thus decreasing the probability of collision and coalescence. For path A the stabilization can be accomplished if the change in HLD also coincides with the formation of liquid crystal phases that protect the emulsion. Liquid crystal phases also promote emulsion stability in systems obtained via path B.

To illustrate the fundamental difference between the SMEDS and SEDS approaches, Fig. 6.8 presents a hypothetical ternary phase diagram of a formulation with negative HLD (Type I μE). In the case of SMEDS (high surfactant content), the dilution with gastric/intestinal fluid (mainly water) of the preconcentrate (surfactant+oil) produces phase inversion that does not result in excess phases (i.e.,

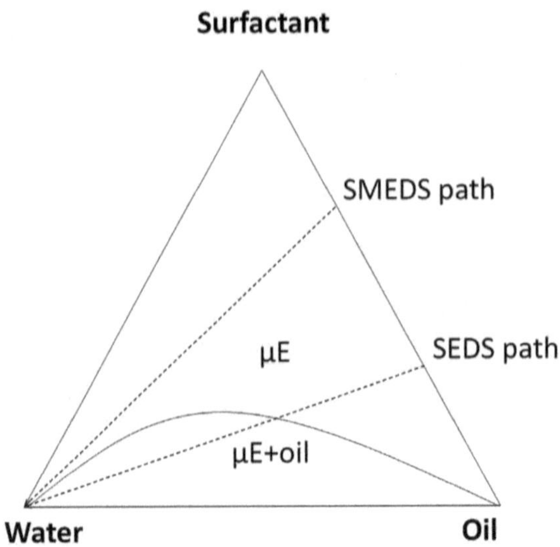

Fig. 6.8. Schematic of a ternary phase diagram for a Type I μE showing the preconcentrate dilution lines that lead to SMEDS (through single phase systems) and SEDS (crossing the 2-phase region) formulations.

no emulsion). Such formulations that can be fully diluted are also known as U-Type μEs (Garti & Yuli-Amar, 2008; Garti, 2003; Garti et al., 2001; 2004). The SEDS preconcentrate (surfactant+oil only) requires less surfactant than SMEDS, thus higher potential loading, but produces larger drop size, particularly at low surfactant concentration (Pérez et al., 2002). As the SEDS preconcentrate is diluted it undergoes a phase inversion, and excess oil is released at high dilution ratios, leading to the formation of the emulsion.

SMEDS AND SEDS IN PHARMACEUTICAL APPLICATIONS

The early μE formulations for oral drug delivery were based on lecithin and short and medium chain alcohol cosolvents (Malmsten, 1999). The idea, at that time, was that lecithin (mixture of phospholipids) is naturally occurring in plants and animals, thus posing a minimal risk. Furthermore, lecithin is an excellent emulsifier. However, lecithin on its own tends to form lamellar liquid crystals that prevent the formation of μEs (Bergenståhl & Fontell, 1983). Short and medium chain alcohols have been used to break the liquid crystal structure and yield μE systems (Tenjarla 1999). Unfortunately, medium (C4-C8) chain alcohols can produce undesirable side effects in oral formulations and are known to promote skin irritation in transdermal applications (McKarns et al., 1997; Yuan et al., 2008).

There are three well known examples SEDS/SMEDS-based delivery products in the market: Neoral®, Norvir®, and Fortovase®, all of which are used for the oral delivery of hydrophobic peptides. Most of these formulations were developed in the late 1990s using polymeric nonionic surfactants instead of the early lecithin-medium chain alcohol approach. Abbott's Laboratories ritonavir formulation (Norvir®), for example, was formulated using a polyethylene glycol (PEG-35) ester of castor oil (Gursoy & Benita, 2004). Delgado-Charro et al. (1997) showed that a mixture of polymeric surfactants Labrasol (PEG-8 C8/C10 fatty acid ester) and Plurol (polyglycerol ester of C16+ fatty acids) produced formulations that did not produce the adverse reactions observed with medium chain alcohols. These polymeric surfactant formulations are the current industry standard and have been reported in various articles (Djordjevic et al., 2004; Gursoy & Benita 2004; Kreilgaard, 2002; Spernath & Aserin, 2006; Tenjarla 1999).

Despite the early adoption of μE-based oral delivery, the terms self-microemulsifying drug delivery system (SMEDDS) and self-emulsifying drug delivery system (SEDDS) were introduced later by different groups (Khoo et al., 1998; Porter & Charman, 2001; Pouton, 2000). These groups warned that, at the time, various μE oral delivery systems were formulated to produce microemulsions at the bench, without simulating the conditions of the intestinal fluid where the μE is supposed to deliver the drug. They recommended designing lipid formulations to produce μEs (SMEDS) or emulsions (SEDS) under conditions that simulate digestion and the environment of the intestines.

It is important to clarify that the term SMEDS is used more broadly than the definition illustrated in Fig. 6.8. In many situations, if a formulation reaches a stable translucent emulsion with droplet size below 100 nm, it is still considered a SMEDS system (Narang et al., 2007). A recent review by He et al. (2010) describe oral SMEDS and SEDS formulations for various drugs and extracts, including itraconazole, paclitaxel, acycloguanosine (Acyclovir), ethyl apovincaminate (Vinpocetine), and silymarin. The oils used in these formulations included ethyl oleate, glyceryl monooleate, propylene glycol monocaprylate (Capryol 90™), and vegetable oils (e.g., soybean oil) containing a cosolvent like ethanol. Most of the formulations quoted by He et al. (2010) use PEG-castor oil surfactants (Cremophor® series) either on their own or in combination with other polymeric PEG-based surfactants such as PEG-sorbitan monooleate (Tween 80) and PEG-caproylocapryl (Labrasol™). Despite the popularity of PEG esters of castor oil (Cremophor) surfactants, these surfactants have been found to produce, when used in high concentrations, undesirable reactions including triggering allergic response, hyperlipidemia (contributing to transformation of HDL into LDL), neurotoxicity, and potentially undesired drug interactions (Gelderblom et al., 2001). These undesirable side effects have been observed in parenteral anticancer Paclitaxel formulations that, due to the low solubility of the drug, require 50 wt% or more of Cremophor. Gelderblom et al. (2001) proposed that the double bonds in the castor oil are responsible for some of these undesired effects. The fact that these surfactants are resistant to metabolic reactions explains why they tend to distribute, almost exclusively, in the plasma and interact with lipoproteins. More recent research has concentrated on lowering Cremophor concentration or finding alternative surfactants and delivery systems (Singh & Dash, 2009).

SMEDS AND SEDS IN FOOD APPLICATIONS

Cremophor formulations have also been used to deliver nutraceuticals. Fig. 6.9 presents the ternary phase diagram and accompanying plasma concentration profile for a curcumin SMEDDS formulation (Setthacheewakul et al., 2010). The surfactant mixture in Fig. 6.9 consists of PEG-35 castor oil (Cremophor EL) and PEG-8 C8/C10 fatty acids (Labrasol™). The oil phase is a mixture of propylene glycol monocaprylate (Capryol 90™) and di-ester C8/C10 propylene glycol (Labrafac PG). The authors selected these surfactants and oil because they offered the highest solubility of curcumin and, at the same time, the lowest possible surfactant content while producing a single phase region where the μE could be completely diluted with water while avoiding phase separation. In other words, the authors were able to produce a U-type formulation. The plasma concentration curves in Fig. 6.9 illustrate that SMEDS formulations, when compared to the suspension dose, produce a significant increase in plasma concentration and, therefore, produce a substantial increase in the bioavailability of curcumin. The plasma concentration data (Fig. 6.9) also illustrates

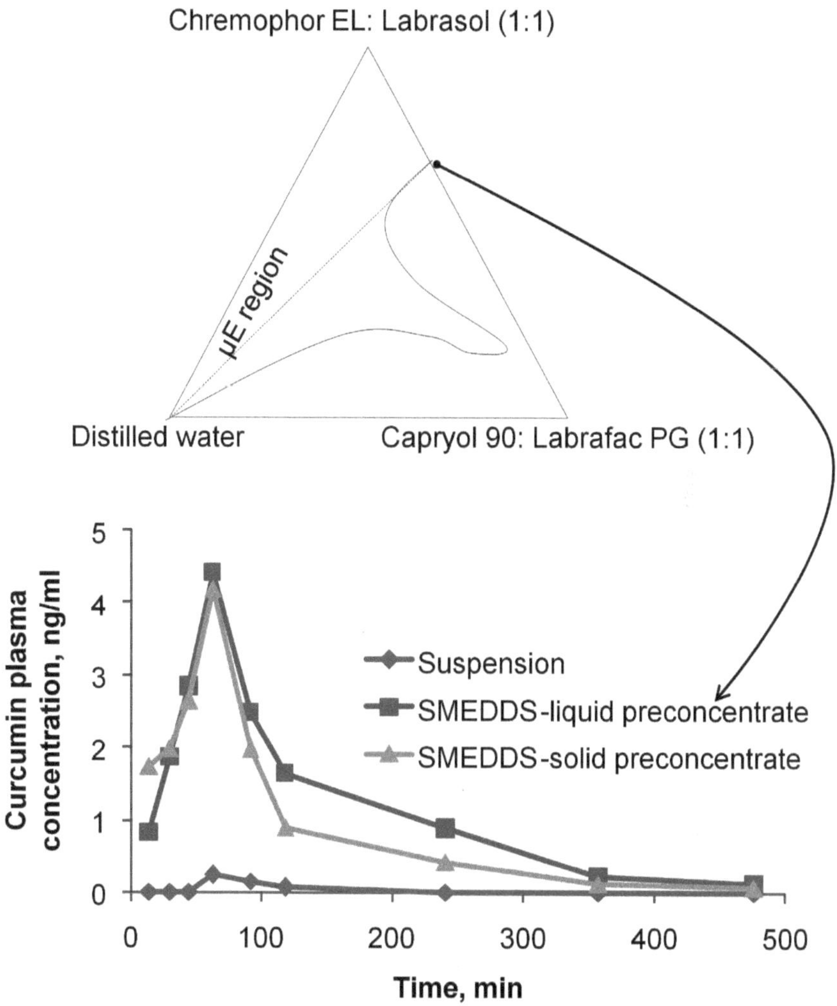

Fig. 6.9. Top—ternary phase diagram for a SMEDS formulation of curcumin, showing the expected dilution line of the SMEDS preconcentrate. Bottom—plasma concentration of curcumin fed to male Wistar rats in suspension, liquid preconcentrate and in solid preconcentrate forms. Adapted from Setthacheewakul et al. (2010).

that a liquid form of the preconcentrate (containing all the surfactant, solvent, oil, and curcumin but no water) produced the same bioavailability as a solid form of this preconcentrate. This solid form was obtained by mixing the liquid form with a premixed paste containing microcrystalline cellulose, internally cross-linked carboxymethyl cellulose (croscarmellose sodium), pregelatinized starch, glyceryl behenate, and silica. The paste impregnated with the liquid preconcentrate was then extruded and spheroidized into 2.5 mm pellets. Both liquid and solid forms were found to be stable for at least 6 months when stored at 45°C and 75% R.H. (relative humidity).

There are various processing techniques to transform liquid preconcentrates for SMEDS and SEDS into solid forms. A recent review by Kumar et al. (2010) summarizes the processes used to produce solid SEDS dose forms. These processing technologies include spray drying (only possible for gel or solid preconcentrates), adsorption on solid carriers, melt extrusion, and melt granulation/spheroidization. Kumar et al. (2010) emphasized that all of these processing technologies should ensure that the formation of the emulsion is facilitated, not hindered, by the solid matrix. The best substrates used thus far involve hydrophilic polymers, such as polysaccharides (e.g., starch and carboxymethyl cellulose used in the solid preconcentrate) and polyethylene glycol (Li et al., 2009). One interesting method of producing solid preconcentrates involves calcium pectinate shells produced by a vibrating nozzle (Zvonar et al., 2010). According to this method, a SMEDS preconcentrate containing PEG-8 caprylocaproyl glycerides (Labrasol®) and polyglyceryl-3 dioleate (Plurol® oleique) as surfactants and C8-C10 triglycerides (Miglyol® 812) as the oil phase were co-injected through a concentric syringe, with a shell-forming solution containing pectin. This 2-phase stream was atomized via a high frequency (3000 Hz) vibrating nozzle and the mist was further stabilized by an electric field. The mist was collected in a $CaCl_2$ solution where the pectin complexes with calcium ions harden the shell around the μE-preconcentrate core. The suspension can be washed and dried using various techniques. The loading efficiency for the μE-preconcentrate can be as high as 50%, which is slightly higher than the 36% loading obtained in the system of Setthacheewakul et al. (2010) and illustrated in Fig. 6.9. However, the method of Setthacheewakul et al. (2010) is simpler and as effective as the system of Zvonar et al. (2010).

The two formulations described above, while effective in producing SMEDS and compatible with encapsulated solid forms, still require the use of PEG-containing surfactants, which are either restricted or not approved for food applications. Various efforts have been made to move away from PEG-based surfactants. One of the early food grade μEs was reported by Patel et al. (2006). These authors formulated systems using lecithin as surfactant; propylene glycol, instead of water, as the hydrophilic phase; and various oils, including long-chain triglycerides such as soybean oil, medium chain triglycerides, a short chain triglyceride (tributyrin), and fatty acid esters (isopropyl myristate and ethyl oleate). The largest single phase regions were obtained with tributyrin. Unfortunately, SMEDS spontaneously emulsify in the aqueous environment in the intestine, making the use of propylene glycol as hydrophilic phase irrelevant. Even introducing a mixture of 9 parts propylene glycol and 1 part water was enough to significantly reduce the single phase region, mainly due to the increase of the liquid crystal region of the ternary phase diagram. Another alternative formulation introduced by Zhang et al. (2008) consists of a mixture of glycerol monolaurate and ethanol as a polar oil phase and Tween 80 (PEG-20 sorbitan monooleate, approved for food applications with restrictions) as a surfactant. Not surprisingly, the authors found that formulations enriched with ethanol (i.e., less glycerol monolaurate) produce large single phase regions. However, the use of the term μE to describe such systems is questionable. Later, Amsalem et al. (2009) sought to eliminate the need for Tween surfactants and ethanol and produce U-type μEs. These authors recognized that it is very difficult to solubilize medium and long chain triglycerides and decided to use a mixture of triacetin and decaglycerol monocaprylate/caprate as a polar oil phase and a mixture of lecithin and sucrose ester laurate as surfactants combined with propylene glycol as solvent. This basic system generated a wide range of compositions that produced U-Type μEs, especially for systems formulated with high C8/10-decaglycerol to triacetin ratios and high propylene glycol to surfactant ratios.

The previous examples illustrate the difficulty in formulating food grade μEs containing triglycerides as the oil phase and the efficacy of propylene glycol in incorporating triglycerides into single phase formulations. Surfactants containing alkyl—polypropylene glycol (PPG) and a hydrophilic group such as sulfate, carboxylate, polyglucosides, or PEGs have been introduced to formulate μEs with triglycerides (Miñana-Perez et al., 1995; Salager et al., 2005). These "extended surfactants" do not have a food grade or pharmaceutical grade status yet, but they have been found to produce large single phase regions using relatively low surfactant concentrations. It has been proposed that the intermediate propylene glycol group introduces a region of intermediate polarity at the interface that disrupts the local organization of triglyceride molecules, thus weakening the triglyceride-triglyceride interaction that hinders μE formulation (Salager et al., 2005).

Another alternative to producing food-grade μEs is the use of lipophilic and hydrophilic linkers (Acosta et al., 2005; Sabatini et al., 2003; Salager et al., 2005; Yuan et al., 2008). Lipophilic linkers are highly lipophilic surfactant-like additives with weak polar groups that segregate near the oil-water interface, between the surfactant alkyl groups, "linking" (only a physicochemical interaction) the surfactant to oil molecules in the core of the oil-swollen micelle (Graciaa et al., 1993). Hydrophilic linkers, on the other hand, are also surfactant-like molecules that, while adsorbing at the oil-water interface, are not capable of increasing oil solubilization as their interaction with the oil phase is weak (Acosta et al., 2002). One of the central differences between a linker and a surfactant is that linkers alone cannot form μEs as they are missing the interactions with one of the phases (oil or water) (Acosta et al., 2004). Because the hydrophilic-lipophilic interactions can be tuned by changing the ratio of linkers to surfactant, a wide range of μEs can be obtained using food-grade oils, lecithin, and linkers (Acosta et al., 2005). The resulting μEs have low toxicity and have been used for transdermal delivery (Yuan et al., 2008). Recently, lecithin-linker μEs have been used to produce SEDS formulations (Chu & Acosta, 2009). Fig. 6.10 presents a simplified ternary phase diagram using simulated intestinal fluid as the aqueous phase. The linkers used in Fig. 6.10 include PEG-caprylate (hydrophilic linker) and sorbitan monooleate as lipophilic linker. The region outlined as

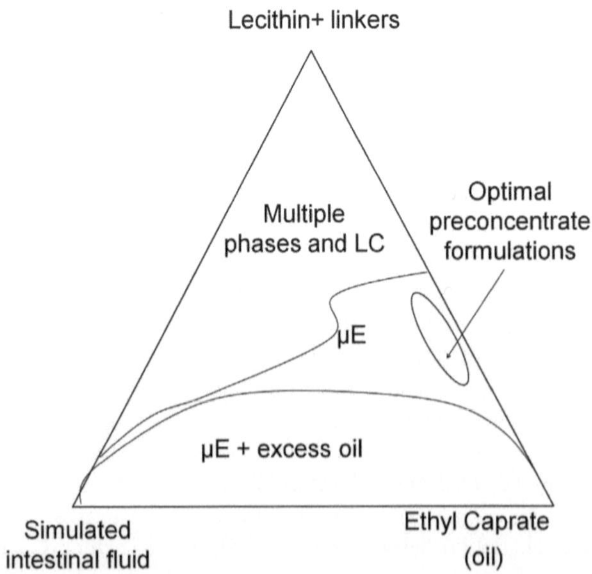

Fig. 6.10. Pseudoternary phase diagram for a system of lecthin (surfactant) + hydrophilic linker PEG-caprylate + lipophilic linker sorbitan monooleate. Dilution of single phase preconcentrate µEs results in stable SEDS systems with particle sizes of 300 nm or less, which further reduce towards 100 nm as the volume dilution ratio with fed-state intestinal fluid approaches 500. Adapted from Chu and Acosta (2009).

optimal preconcentrate formulations represent a series of compositions where, upon dilution, stable SEDS emulsions (less than 300 nm diameter) are produced. It is important to note that preconcentrate systems formulated in the liquid crystal region produced emulsions of large drop size, likely because the presence of liquid crystal prevented the formation of the intermediate µE systems. It is also important to note that, unlike PEG-based surfactant formulations, linker preconcentrates do not require total surfactant + linker content over 50%, and the main surfactant is food grade lecithin. The formulation of Fig. 6.10 is strictly a pharmaceutical grade system because of the use of PEG-based hydrophilic linkers. Work on food grade hydrophilic linkers is currently ongoing.

One advantage of the linker formulation over other µE formulations is that it is capable of making use of the phospholipids and bile salts of the intestinal fluid. In fact, at dilution ratios of 100 or more (aqueous phase to preconcentrate), the SEDS drop sizes are smaller when diluting with fed-state simulated intestinal fluid than the drop sizes obtained when diluting with a saline solution (Chu & Acosta, 2009). This interaction between SMEDS and dietary lipids to form mixed micelles systems has

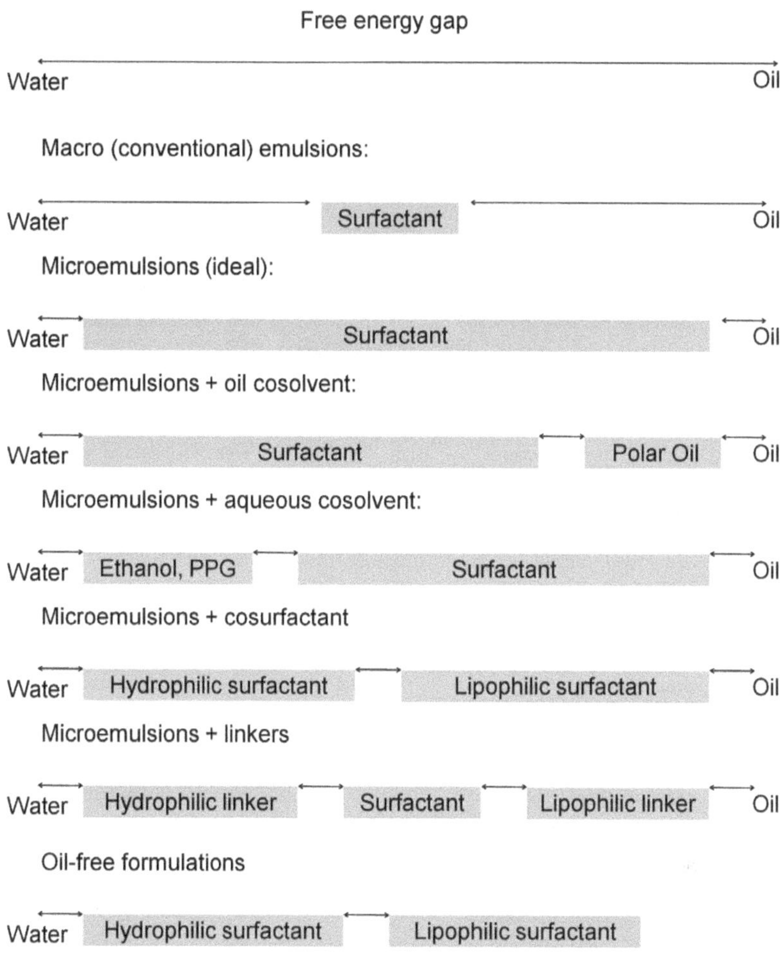

Fig. 6.11. Summary of formulation strategies used to produce SEDS and SMEDS.

been confirmed using various spectroscopic techniques (Rozner et al., 2010). However, there is little information available on the interaction of SMEDS with lipases other than the earlier comments about Cremophor and Tween systems.

Table 6-A presents a summary of nutraceuticals incorporated in microemulsion formulations, along with the loading and formulation strategy employed in each case. The formulation information in Table 6-A reveals that most of the existing μEs containing nutraceuticals are still formulated with PEG-based surfactants. Numerous formulations contain polar oils (e.g., fatty acid esters) and cosolvents such as ethanol and propylene glycol. To rationalize the formulation approaches described

Table 6-A. Selected Examples of μE Formulations Containing Nutraceuticals.

Reference	Formulation	Loading	Formulation approach
		Vitamin E	
Feng et al., 2009	Cremophor EL (PEG-35 castor oil), ethanol, ethyl butyrate	~20% in a preconcentrate form	Aqueous cosolvent and use of short polar oil (ethyl butirate), polymeric surfactant
Garti et al., 2004	Tween 60 (PEG-20 sorbitan monoesterate), ethanol, limonene	~17% in a preconcentrate form	Aqueous cosolvent, polymeric surfactant + "short" oil
Rozman et al., 2009	Tween 40 (PEG-20 sorbitan monopalmitate), Imwitor 308 (glyceryl monocaprylate), isopropyl myristate	1% (transdermal)	Polymeric surfactant + hydrophilic surfactant + polar oil
Chanamai, 2007	Tween 20 (PEG-20 sorbitan monolaurate), triglycerol monooleate, decaglycerol decaoleate	7%	Polymeric surfactants, free of solvent oil
Yuan et al., 2008	Lecithin, sorbitan monooleate, capric acid, sodium caprate, isopropylmyristate	5% (transdermal)	Lecithin + hydrophilic and lipophilic linkers
		Carotenoids	
Lopes et al., 2010	PEG-10 oleyl ester, propylene glycol mono- and tri-glycerides of medium chain fatty acids	500 ppm lycopene (transdermal)	Polymeric surfactant, aqueous cosolvent and polar oils
Garti et al. 2004	Tween 60 (PEG-20 sorbitan monoesterate), ethanol, limonene	~500 ppm lycopene in the preconcentrate form	Aqueous cosolvent, polymeric surfactant + "short" oil
		Steroids	
Garti et al., 2005	Tween 60 (PEG-20 sorbitan monoesterate), ethanol, limonene	~6%	Aqueous cosolvent, polymeric surfactant + "short" oil
Chu & Acosta, 2009	Lecithin, sorbitan monooleate, PEG-6 caprylate, ethyl caprate	~3% (pharmaceutical)	Lecithin + hydrophilic and lipophilic linkers
		Ascorbic acid	
Yi et al., 1991	Lecithin, sardine oil, vitamin E	0.04%	Used as antioxidant, W/O μE
Rozman et al., 2009	Tween 40 (PEG-20 sorbitan monopalmitate), Imwitor 308 (glyceryl monocaprylate), isopropyl myristate	0.4%	Polymeric surfactant + hydrophilic surfactant + polar oil

Table 6-A., cont., Selected Examples of μE Formulations Containing Nutraceuticals.

Reference	Formulation	Loading	Formulation approach
		Curcumin	
Setthacheewakul et al., 2010	Cremophor EL (PEG-35 castor oil), Labrasol (Caproyl-polyglyceride), capryol 90 (propylene glycol monocaprylate), labrafact PG (propylene glycol dicaprate)	4%	Polymeric surfactant + hydrophilic surfactant + polar oil; Similar to oil-free systems
Cui et al., 2009	Cremophor EL (PEG-35 castor oil), emulsifier OP (octylphenol ethoxylate) , ethyl oleate	5%	Polymeric surfactant+ polar oil
Wu et al., 2011	Cremophor RH (PEG-40 castor oil), ethanol , isopropyl myristate	5%	Polymeric surfactant+ solvent + polar oil

in Table 6-A as well as formulation approaches described earlier, Fig. 6.11 presents a schematic summary of SEDS/SMEDS formulation strategies. The primary issue is to overcome the hydrophilic-lipophilic gap between the oil and aqueous phase. The simplest method to close this gap is the use of a surfactant that adsorbs at the interface and interacts with the aqueous phase via hydrogen bonding (in the case of nonionic surfactants) or via ionic/polar interactions. The alkyl group of the surfactant then interacts with the oil phase primarily via Van der Waals interactions. If these interactions are weak, then the surfactant cannot solubilize oil and water and/or lower the interfacial tension enough in order to produce SEDS or SMEDS. Surfactants with greater surfactant-oil and surfactant-water interactions (e.g., extended surfactants) are capable of forming the desired SEDS and SMEDS as long as the surfactant-surfactant interaction is not strong enough to produce liquid crystal systems instead of μEs. As most surfactants do tend to produce liquid crystals (e.g., lecithin), one alternative is to bridge the gap by introducing polar co-solvent in the oil (e.g., mono- and di-glycerides) although some of these co-solvents can also behave as lipophilic linkers. Another alternative is to bridge the gap starting from the aqueous phase by introducing hydrophilic co-solvents such as ethanol and propylene glycol. A popular alternative is the use of cosurfactants, one slightly more hydrophilic (e.g., labrasol in Fig. 6.9) and one more lipophilic (e.g., cremophor in Fig. 6.9). An extreme version of the cosurfactant approach is the use of highly hydrophilic and highly lipophilic surfactant-like additives (linkers) that require the use of an intermediate surfactant to produce the μE system. Finally, one more option is to question the need to produce a conventional μE and instead produce surfactant-polar lipid mixtures without a dispersed oil phase, similar to the work of Zhang et al. (2008) described earlier. These oil-free formulations are related to Pouton's Type IV lipid formulations (note that this is not the same as Winsor's Type IV systems) (Pouton & Porter, 2008). The difference between Pouton's Type IV and the oil-free formulations in Fig. 6.11 is that Pouton's Type IV typically contains a hydrophilic cosolvent. Both oil-free systems and Pouton's Type IV tend to produce single phase micelle systems upon dilution in the intestine. Furthermore, surfactants are better solvents for drugs and nutraceuticals than most oils. The problem with oil-free systems containing hydrophilic cosolvents is that these solvents tend to dissolve in the aqueous phase upon dilution, changing the micellar environment and leading to drug/nutraceutical precipitation or the formation of liquid crystal phases (Pouton & Porter 2008).

CHALLENGES, OPPORTUNITIES, AND OUTSTANDING ISSUES FOR SMEDS AND SEDS

SMEDS and SEDS have been proven to improve the bioavailability of poorly-soluble drugs. Substantial advances have been made towards finding safer and more effective food-grade formulations. Also, there have been advances made towards the understanding of the thermodynamics of microemulsions and the mechanisms of self-emulsification. However, the challenge is to determine what is the best (optimal) formulation. The formulation of SEDS and SMEDS is in need of a more engineering-based approach to decide the best formulation conditions for a given drug or nutraceutical. The schematic of Fig. 6.12 illustrates such an engineering approach. It is important to start with understanding the drug or nutraceutical itself and then determining what molecular environment would be most beneficial to improve its dissolution. There are several possible approaches to this. Rane and Anderson (2008) reviewed several alternatives, including the use of predictive activity models such as UNIFAC (UNIversal Functional Activity Coefficient), regular solution theory and solubility parameters, Flory-Huggins theory, molecular dynamics, and Monte-Carlo models. One of the insights highlighted in that review is that triglycerides tend to form layered (organized) structures, as previously discussed when describing the role of propylene glycol groups in extended surfactants. The same structures were observed in earlier MD simulations by Sum et al. (2003). This local orientation in triglycerides has been confirmed using Raman spectroscopy (Da Silva & Rousseau, 2008). Pouton's Type I lipid formulations of mono-, di- and triglycerides in the presence of propylene glycol and water have also been evaluated using molecular dynamics (Warren et al., 2009). The authors found that at low water content the system tends to form reverse micelles and that propylene glycol tends to disrupt the structure of the lipids and water in favor of a more disorganized molecular structure containing water and lipids. Simplified methods that make use of simplified molecular dynamic simulations have also been introduced as an intermediate between UNIFAC and full molecular dynamics (Buggert et al., 2009). Another intermediate alternative to molecular dynamics is dissipative particle dynamic simulations that use a coarse segmentation of molecules into groups that allow for faster solutions of complex fluids, including lipid delivery systems (Shillcock & Lipowsky, 2006).

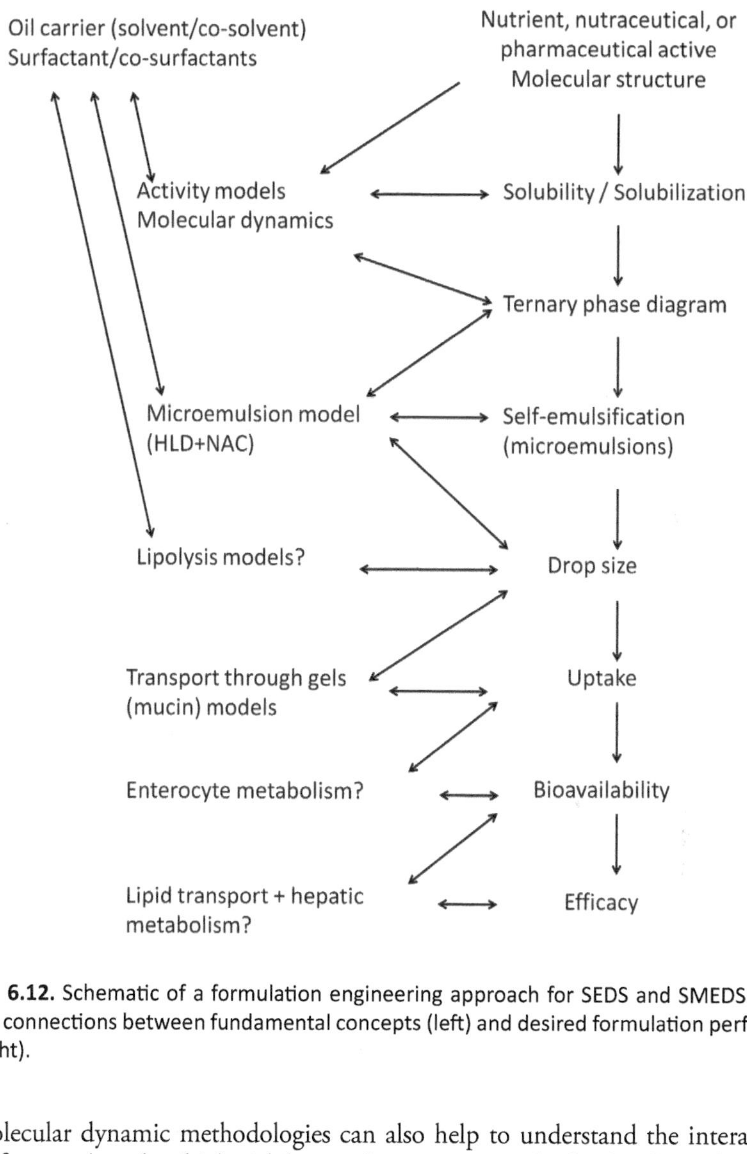

Fig. 6.12. Schematic of a formulation engineering approach for SEDS and SMEDS showing the connections between fundamental concepts (left) and desired formulation performance (right).

Molecular dynamic methodologies can also help to understand the interaction of surfactants (e.g., lectihin) with lipases, lipoproteins, and other lipid membranes that may affect the behavior of these proteins (Hsu & Yip, 2007).

The molecular and activity models mentioned above represent the first step in selecting the appropriate surfactants and solvents to maximize the solubility of the drug or nutraceutical of interest. The second step is the modeling of the microemulsion itself to determine the solubilization of oil, solvents, and drug or nutraceutical, and preparation of their ternary phase diagram. Currently the HLD-NAC model is perhaps the closest approach to an equation of state for surfactant-oil-water systems. Certainly, there are numerous constants in the model that have not been developed for surfactants used in drug delivery systems, but some of these constants will soon appear in the literature and are now available in a free version of the HLD-NAC software (Abbott, 2011). The HLD-NAC model can predict certain conditions of the ternary phase diagram but requires further refinements to predict the entire ternary phase diagram and the formation of liquid crystals. After predicting the properties of the microemulsion as well as the best combination of surfactant/solvents to maximize solubilization, the next step will involve predicting the drop size to be obtained during the self-emulsification process (also introduced earlier through the Weber number) and the uptake (partition) into the mucin layer and the permeation through the intestine layer. Thus far little is known about this process of permeation in SEDS and SMEDS other than that it is mediated by the mucin layer that lines the small intestine. At this point we enter the more complex prediction of the interaction of SEDS and SMEDS surfactants and solvents with lipoproteins and the lipid transport and metabolic pathways. While it is unlikely that all of the elements of Fig. 6.12 can be accomplished, we hope that these ideas are taken as a roadmap to guide efforts in closing the current gaps in SEDS and SMEDS technology.

CONCLUSION

Self-emulsifying delivery systems hold great potential as delivery systems in food applications. Food grade formulations are now possible thanks to polymeric-type surfactants, clever use of food grade cosolvents, and the introduction of linker formulations. Several formulations have been reported in research articles and patents, signaling that the technology has started to be transferred into commercial applications. Despite these successes, there is much that needs to be learned about how self-emulsifying systems interact with the body's lipid transport mechanisms. Integrating this knowledge with our current understanding of the thermodynamics and dynamic aspects of self-emulsification could produce safer and more efficient formulations.

REFERENCES

Abbott, S. HLD-NAC surfactant software. Accessed April 9, 2011. <http://www.stevenabbott.co.uk/HLD-NAC.html>

Acosta, E. Testing the effectiveness of nutrient delivery systems. In Delivery and controlled release of bioactives in foods and nutraceuticals. Garti, N. Ed.; CRC Press: Boca Raton, FL., 2008; pp. 53–106.

Acosta, E. Bioavailability of nanoparticles in nutrient and nutraceutical delivery. Curr. Opin. Colloid Interface Sci. 2009, 14, 3–15.

Acosta, E.; Uchiyama, H.; Sabatini, D.A.; Harwell, J.H. The role of hydrophilic linkers. J. Surfactants Deterg. 2002, 5, 151–157.

Acosta, E.; Szekeres, E.; Sabatini, D.A.; Harwell, J.H. Net-average curvature model for solubilization and supersolubilization in surfactant μEs. Langmuir. 2003, 19, 186–195.

Acosta, E.J.; Harwell, J.H.; Sabatini, D.A. Self-assembly in linker-modified microemulsions. J. Colloid Interface Sci. 2004, 274, 652–664.

Acosta, E.J.; Nguyen, T.; Witthayapanyanon, A.; Harwell, J.H.; Sabatini, D.A. Linker-based biocompatible μEs. Environ. Sci. Technol. 2005, 39, 1275–1282.

Acosta, E.J.; Yuan, J.S.; Bhakta, A.S. The characteristic curvature of ionic surfactants. J. Surfactants Deterg. 2008, 11, 145–158.

Amidon, G.L.; Lennernas, H.; Shah, V.P.; Crison, J.R. A theoretical basis for a biopharmaceutic drug classification: the correlation of in vitro drug product dissolution and in vivo bioavailability. Pharm. Res. 1995, 12, 413–420.

Amsalem, O.; Yuli-Amar, I.; Aserin, A.; Garti, N. Phospholipids embedded fully dilutable liquid nanostructures. Part 1: Compositions and solubilization capacity. Colloids Surf. B 2009, 73, 15–22.

Araya, H.; Tomita, M.; Hayashi, M. The novel formulation design of self-emulsifying drug delivery systems (SEDDS) type O/W microemulsion III: the permeation mechanism of a poorly water soluble drug entrapped O/W microemulsion in rat isolated intestinal membrane by the Ussing chamber method. Drug metab. Pharmacokinetics. 2006, 21, 45–53.

Bergenståhl, B.; Fontell, K. Phase equilibria in the system soybean lecithin/water. Prog. Colloid Polym. Sci. 1983, 68, 48–52.

Buggert, M.; Cadena, C.; Mokrushina, L.; Smirnova, I.; Maginn, E.J.; Arlt, W. COSMO-RS calculations of partition coefficients: different tools for conformational search. Chem. Eng. Technol. 2009, 32, 977–986.

Chanamai, R. μEs for use in food and beverage products. US patent application No. US 2007/0087104 A1. Asignee: Wild flavours, Inc. 2007.

Chau, C.-F.; Wu, S.-H.; Yen, G.-C. The development of regulations for food nanotechnology. Trends Food Sci. Technol. 2007, 18, 269–280.

Christiansen, A.; Backensfeld, T.; Weitschies, W. Effects of non-ionic surfactants on in vitro triglyceride digestion and their susceptibility to digestion by pancreatic enzymes. Eur. J. Pharm. Sci. 2010, 41, 376–382.

Christiansen, A.; Backensfeld, T.; Denner, K.; Weitschies, W. Effects of non-ionic surfactants on cytochrome P450-mediated metabolism in vitro Eur. J. Pharm. Biopharm. 2011, 78, 166–172.

Chu, J., Acosta, E. Enhancing oral delivery of lipophilic actives using linker-based microemulsions (2009) 8th World Congress of Chemical Engineering. Article available online http://archivos.labcontrol.cl/wcce8/offline/techsched/manuscripts/afeqqm.pdf Accessed April 9, 2011.

Cu, Y.; Saltzman, W.M. Mathematical modeling of molecular diffusion through mucus. Adv. Drug Delivery Rev. 2009, 61, 101–114.

Cui, J.; Yu, B.; Zhao, Y.; Zhu, W.; Li, H.; Lou, H.; Zhai, G. Enhancement of oral absorption of curcumin by self-microemulsifying drug delivery systems. Int. J. Pharm. 2009, 371, 148–155.

Da Silva, E.; Rousseau, D. Molecular order and thermodynamics of the solid-liquid transition in triglycerides via Raman spectroscopy. Phys. Chem. Chem. Phys. 2008, 10, 4606–4613.

De Jong, A.; Plat, J.; Mensink, R.P. Metabolic effects of plant sterols and stanols (Review). J. Nutr. Biochem. 2003, 14, 362–369.

Delgado-Charro, M.B.; Iglesias-Vilas, G.; Blanco-Méndez, J.; López-Quintela M.A.; Marty J.-P.; Guy, R.H. Delivery of a hydrophilic solute through the skin from novel microemulsion systems. Eur. J. Pharm. Biopharm. 1997, 43, 37–42.

Djordjevic, L.; Primorac, M.; Stupar, M., Krajisnik, D. Characterization of caprylocaproyl macrogolglycerides based microemulsion drug delivery vehicles for an amphiphilic drug. Int. J. Pharm. 2004, 271, 11–19.

Engström, S.; Larsson, K. μEs in foods. In Handbook of μE science and technology. Kumar, P. and Mittal, K. Eds; Marcel Dekker: New York, NY, 1999; pp. 789–796.

Feng, J.-L.; Wang, Z.-W.; Zhang, J.; Wang, Z.-N.; Liu, F. Study on food-grade vitamin E μEs based on nonionic emulsifiers. Colloids Surf. A 2009, 339, 1–6.

Garti, N.; Yaghmur, A.; Leser, M.E.; Clement, V.; Watzke, H.J. Improved oil solubilization in oil/water food grade μEs in the presence of polyols and ethanol. J. Agric. Food. Chem. 2001, 49, 2552–2562.

Garti, N. μEs as microreactors for food applications. Curr. Opin. Colloid Inter. Sci. 2003, 8, 197–211.

Garti, N.; Zakharia, I.; Spernath, A.; Yaghmur, A.; Aserin, A.; Hoffman, R.E.; Jacobs, L. Solubilization of water-insoluble nutraceuticals in nonionic μEs for water-based use. Prog. Colloid Polym. Sci. 2004, 126, 184–189.

Garti, N.; Spernath, A.; Aserin, A.; Lutz, R. Nano-sized self-assemblies of nonionic surfactants as solubilization reservoirs and microreactors for food systems. Soft Matter 2005, 1, 206–218.

Garti, N.; Yuli-Amar, I. Micro- and nano-emulsions for delivery of functional food ingredients. In Delivery and controlled release of bioactives in foods and nutraceuticals. Garti, N. Ed.; CRC Press: Boca Raton, FL., 2008; pp. 149–183.

Gelderblom, H.; Verweij, J.; Nooter, K.; Sparreboom, A. Cremophor EL: The drawbacks and advantages of vehicle selection for drug formulation. Eur. J. Cancer. 2001, 37, 1590–1598.

Giddings, J.C.; Kucera, E.; Russel, C.P.; Myers, M.N. Statistical theory for the equilibrium distribution of rigid molecules in inert porous network. Exclusion chromatography. J. Phys. Chem. 1968, 72, 4397–4408.

Graciaa, A.; Lachaise, J.; Cucuphat, C.; Bourrel, M.; Salager, J.L. Improving solubilization in μEs with additives. 2. Long chain alcohols as lipophilic linkers. Langmuir. 1993, 9, 3371–3374.

Gursoy, R.N.; Benita, S. Self-emulsifying drug delivery systems (SEDDS) for improved oral delivery of lipophilic drugs. Biomed. Pharmacother. 2004, 58, 173–182.

He, C.-X.; He, Z.-G.; Gao, J.-Q. Microemulsions as drug delivery systems to improve the solubility and the bioavailability of poorly water-soluble drugs. Expert Opin. Drug Delivery 2010, 7, 445–460.

Hernell, O.; Staggers, J.E.; Carey, M.C. Physical-chemical behavior of dietary and biliary lipids during intestinal digestion and absorption. 2. Phase analysis and aggregation states of luminal lipids during duodenal fat digestion in healthy adult human beings. Biochemistry. 1990, 29, 2041–2056.

Hoar, T.P.; Schulman, J.H. Transparent water-in-oil dispersions: The oleopathic hydro-micelle [1] Nature. 1943, 152, 102–103.

Hsu, J.C.Y.; Yip, C.M. Molecular dynamics simulations of indolicidin association with model lipid bilayers. Biophys. J. 2007, 92, L100–L102.

Hussain, N.; Jaitley, V.; Florence, A.T. Recent advances in the understanding of uptake of microparticulates across the gastrointestinal lymphatics. Adv. Drug Delivery Rev. 2001, 50, 107–142.

Izquierdo, P.; Esquena, J.; Tadros, Th.F.; Dederen, C.; Garcia, M.J.; Azemar, N.; Solans, C. Formation and stability of nano-emulsions prepared using the phase inversion temperature method. Langmuir. 2002, 18, 26–30.

Jantratid, E.; Janssen,N.; Reppas, Ch.; Dressman, J. Dissolution media simulating conditions in the proximal human gastrointestinal tract: an update. Pharm. Res. 2008, 25, 1663–1676.

Jinno, J.I.; Kamada, N.; Miyake, M.; Yamada, K.; Mukai, T.; Odomi, M. Toguchi, H.; Liversidge, G.G.; Higaki, K.; Kimura, T. Effect of particle size reduction on dissolution and oral absorption of a poorly water-soluble drug, cilostazol, in beagle dogs. J. Control. Release 2006, 111, 56–64.

Khanvilkar, K.; Donovan, M.; Flanagan, D. Drug transfer through mucus. Adv. Drug Delivery Rev. 2001, 48, 173–193.

Khoo, S.-M.; Humberstone, A.J.; Porter, C.J.H.; Edwards, G.A.; Charman, W.N. Formulation design and bioavailability assessment of lipidic self-emulsifying formulations of halofantrine. Int. J. Pharm. 1998, 167, 155–164.

Kipp, J.E. The role of solid nanoparticle technology in the parenteral delivery of poorly water-soluble drugs. Int. J. Pharm. 2004, 284,109–22.

Kiran, S.K.; Acosta, E.J. Predicting the morphology and viscosity of microemulsions using the HLD-NAC model. Ind. Eng. Chem. Res. 2010, 49, 3424–3432.

Kogan, A.; Aserin, A.; Garti, N. Improved solubilization of carbamazepine and structural transitions in nonionic microemulsions upon aqueous phase dilution. J. Colloid Inter. Sci. 2007, 315, 637–647.

Koynova, R.; Caffrey, M. An index of lipid phase diagrams. Chem. Phys. Lipids. 2002, 115, 107–219.

Kozu, H.; Kobayashi, I.; Nakajima, M.; Uemura, K.; Sato, S.; Ichikawa, S. Analysis of Flow Phenomena in Gastric Contents Induced by Human Gastric Peristalsis Using CFD. Food Biophys. 2010, 5, 330–336.

Kreilgaard, M. Influence of microemulsions on cutaneous drug delivery. Adv. Drug Delivery Rev. 2002, 54, S77–S98.

Kumar, A.; Sharma, S.; Kamble, R. Self emulsifying drug delivery system (SEDDS): future aspects. Int. J. Pharm. Pharm. Sci. 2010, 2, 7–13.

Lai, S.K.; O'Hanlon, D.E.; Harrold, S.; Man, S.T.; Wang, Y.-Y.; Cone, R.; Hanes, J. Rapid transport of large polymeric nanoparticles in fresh undiluted human mucus. Proc. Natl. Acad. Sci. U.S.A. 2007, 104, 1482–1487.

Lawrence, M.J.; Rees, G.D. µE-based media as novel drug delivery systems. Adv. Drug Delivery Rev. 2000, 45, 89–121.

Li, P.; Hynes, S.R.; Haefele, T.F.; Pudipeddi, M.; Royce, A.E.; Serajuddin, A.T.M. Development of clinical dosage forms for a poorly water-soluble drug II: Formulation and characterization of a novel solid microemulsion preconcentrate system for oral delivery of a poorly water-soluble drug. J. Pharm. Sci. 2009, 98,1750–1764.

Lopes, L.B.; Vandewall, H.; Li, H.T.; Venugopal, V.; Li, H.K.; Naydin, S.; Hosmer, J.; Levendusky, M.; Zheng, H.; Bentley, M.V.L.B.; et al. Topical delivery of lycopene using µEs: Enhanced skin penetration and tissue antioxidant activity. J. Pharm. Sci. 2010, 99, 1346–1357.

MacGregor, K.J.; Embleton, J.K.; Lacy, J.E.; Perry, E.A.; Solomon, L.J.; Seager, H.; Pouton, C.W. Influence of lipolysis on drug absorption from the gastro-intestinal tract. Adv. Drug Delivery Rev. 1997, 25, 33–46.

Malmsten, M. µEs in paharmaceuticals. In Handbook of µE science and technology. Kumar, P. and Mittal, K., Eds.; Marcel Dekker: New York, NY, 1999, pp. 755–772.

McKarns, S.C.; Hansch, C.; Caldwell, W.S.; Morgan, W.T.; Moore, S.K.; Doolittle, D.J. Correlation between hydrophobicity of short-chain aliphatic alcohols and their ability to alter plasma membrane integrity. Fund. App. Toxicol. 1997, 36, 62–70.

Miller, C.A. Behavior of emulsions and microemulsions: implications for nanosystems. In Nanoscience—colloidal and interfacial aspects. Starov, V.M., Ed.; CRC Press: Boca Raton, FL, 2010; pp. 493–514.

Miñana-Perez, M.; Jarry, P.; Perez-Sanchez, M.; Ramirez-Gouveia, M.; Salager, J.L. Surfactant-oil-water systems near the affinity inversion. part v: properties of emulsions. J. Dispersion Sci. Technol. 1986, 7, 331–343.

Miñana-Perez, M.; Graciaa, A.; Lachaise, J.; Salager, J.-L. Solubilization of polar oils with extended surfactants. Colloids Surf., A. 1995, 100, 217–224.

Musthaba, S.M.; Baboota, S.; Ahmed, S.; Ahuja, A.; Ali, J. Status of novel drug delivery technology for phytotherapeutics. Expert Opin. Drug Delivery. 2009, 6, 625–637.

Narang, A.S.; Delmarre, D.; Gao, D. Stable drug encapsulation in micelles and microemulsions. Int. J. Pharm. 2007, 345, 9–25.

Neethirajan, S.; Jayas, D.S. Nanotechnology for the Food and Bioprocessing Industries. Food Bioprocess Technol. 2011, 4, 39–47.

Oh, D.-M.; Curl, R.L.; Amidon, G.L. Estimating the fraction dose absorbed from suspensions of poorly soluble compounds in humans: a mathematical model. Pharm. Res. 1993, 10, 264–270.

Ohno, Y.; Naganuma, T.; Ogaw, T.; Muramoto, K. Effect of lectins on the transport of food factors in Caco-2 cell monolayers. J. Agric. Food Chem. 2006, 54, 548–553.

Patel, J.P.; Brocks, D.R. The effect of oral lipids and circulating lipoproteins on the metabolism of drugs. Expert Opin. Drug Metab. Toxicol. 2009, 5, 1385–1398.

Patel, N.; Schmid, U.; Lawrence, M.J. Phospholipid-based microemulsions suitable for use in foods. J. Agric. Food Chem. 2006, 54, 7817–7824.

Pérez, M.; Zambrano, N.; Ramirez, M.; Tyrode, E.; Salager, J.L. Surfactant-oil-water systems near the affinity inversion. XII. Emulsion drop size versus formulation and composition. J. Dispersion Sci. Technol. 2002, 23, 55–63.

Porter, C.J.H.; Charman, W.N. In vitro assessment of oral lipid based formulations. Adv. Drug Delivery Rev. 2001, 50, S127–S147.

Pouton, C.W. Lipid formulations for oral administration of drugs: Non-emulsifying, self-emulsifying and 'self-microemulsifying' drug delivery systems. Eur. J. Pharm. Sci. 2000, 11, S93–S98.

Pouton, C.W.; Porter, C.J.H. Formulation of lipid-based delivery systems for oral administration: Materials, methods and strategies. Adv. Drug Delivery Rev. 2008, 60, 625–637.

Rane, S.S.; Anderson, B.D. What determines drug solubility in lipid vehicles: is it predictable? Adv. Drug Delivery Rev. 2008, 60, 638–656.

Rao, A.V.; Janezic, S.A. The role of dietary phytosterols in colon carcinogenesis. Nutr. Cancer, 1992, 18, 43–52.

Rehfeld, J.F. The new biology of gastrointestinal hormones. Physiol. Rev. 1998, 78, 1087–1108.

Rozman, B.; Gasperlin, M.; Tinois-Tessoneaud, E.; Pirot, F.; Falson, F. Simultaneous absorption of vitamins C and E from topical μEs using reconstructed human epidermis as a skin model. Eur. J. Pharm. Biopharm. 2009, 72, 69–75.

Rozner, S.; Shalev, D.E.; Shames, A.I.; Ottaviani, M.F.; Aserin, A.; Garti, N. Do food microemulsions and dietary mixed micelles interact? Colloids Surf. B 2010, 77, 22–30.

Sabatini, D.A.; Acosta, E.; Harwell, J.H. Linker molecules in surfactant mixtures. Curr. Opin. Colloid Inter. Sci. 2003, 8, 316–326.

Salager, J.L.; Antón, R.E. Ionic μEs. In Handbook of μE science and technology. Kumar, P. and Mittal, K., Eds.; Marcel Dekker: New York 1999, pp. 247–280.

Salager, J.L.; Marquez, N.; Graciaa, A.; Lachaise, J. Partitioning of ethoxylated octylphenol surfactants in μE-oil-water systems: influence of temperature and relation between partitioning coefficient and physicochemical formulation. Langmuir. 2000, 16, 5534–5539.

Salager, J.L.; Antón, R.E.; Andérez, J.M.; Aubry, J.M. Formulation des micro-émulsions par la méthode HLD. Techniques De L'Ingénieur, Traité Des Procédés, 2001, 1–20.
Salager, J.-L.; Forgiarini, A.; Márquez, L.; Peña, A.; Pizzino, A.; Rodriguez, M.P.; Rondón-González, M. Using emulsion inversion in industrial processes (2004) Adv. Colloid Interface Sci., 108–109, pp. 259–272.
Salager, J.-L.; Antón, R.E.; Sabatini, D.A.; Harwell, J.H.; Acosta, E.J.; Tolosa, L.I. Enhancing solubilization in µEs - State of the art and current trends. J. Surfactants Deterg. 2005, 8, 3–21.
Schulman, J.; Stoeckenius, W.; Prince, L. Mechanism of formation and structure of micro emulsions by electron microscopy. J. Phys. Chem. 1959, 63, 1677–1680.
Setthacheewakul, S.; Mahattanadul, S.; Phadoongsombut, N.; Pichayakorn, W.; Wiwattanapatapee, R. Development and evaluation of self-microemulsifying liquid and pellet formulations of curcumin, and absorption studies in rats. Eur. J. Pharm. Biopharm. 2010, 76, 475–485.
Shen, H.; Howles, P.; Tso, P. From interaction of lipidic vehicles with intestinal epithelial cell membranes to the formation and secretion of chylomicrons. Adv. Drug Delivery Rev. 2001, 50, S103–S125.
Shillcock, J.C.; Lipowsky, R. The computational route from bilayer membranes to vesicle fusion. J. Phys. Condens. Matter. 2006, 18, S1191–S1219.
Shinoda, K.; Kunieda, H. Conditions to produce so-called microemulsions: factors to increase the mutual solubility of oil and water by solubilizer. J. Colloid Inter. Sci. 1973, 42, 381– 387.
Singh, S.; Dash, A.K. Paclitaxel in cancer treatment: Perspectives and prospects of its delivery challenges. Crit. Rev. Ther. Drug Carr. Syst. 2009, 26, 333–372.
Solans, C.; Izquierdo, P.; Nolla, J.; Azemar, N.; Garcia-Celma, M.J. Nano-emulsions. Curr. Opin. Colloid Interface Sci. 2005, 10, 102–110.
Solomon, L.J.; Embleton, J.K.; Pouton, C.W. Inhibition of lipolysis of non-ionic surfactants, a structure/activity study.(1996) In Formulation of poorly available drugs for oral administration. Couvreur, P., Duchene, D., and Kalles, I., Eds.; Editions de Sante: Paris; pp. 437–438.
Spernath, A.; Aserin, A. µEs as carriers for drugs and nutraceuticals. Adv. Colloid Interface Sci. 2006, 128–130, 47–64.
Srinivas, P.R.; Philbert, M.; Vu, T.Q.; Huang, Q.; Kokini, J.L.; Saos, E.; Chen, H.; Peterson, C.M.; Friedl, K.E.; McDade-Ngutter, C.; et al. Nanotechnology research: applications in nutritional sciences. J. Nutr. 2010, 140, 119–124.
Staggers, J.E.; Hernell, O.; Stafford, R.J.; Carey, M.C. Physical-chemical behavior of dietary and biliary lipids during intestinal digestion and absorption. 1. Phase behavior and aggregation states of model lipid systems patterned after aqueous duodenal contents of healthy adult human beings. Biochemistry. 1990, 28, 2028–2040.
Szentkuti, L. Light microscopical observations on luminally administered dyes, dextrans, nanospheres and microspheres in the pre-epithelial mucus gel layer of the rat distal colon. J. Control. Rel. 1997, 46, 233–242.
Sum, A.K.; Biddy, M.J.; De Pablo, J.J.; Tupy, M.J. Predictive molecular model for the thermodynamic and transport properties of triacylglycerols. J. Phys. Chem. B. 2003, 107, 14443–14451.
Tenjarla, S. Microemulsions: an overview and pharmaceutical applications. Crit. Rev. Ther. Drug Carr. Syst. 1999, 16, 461–521.
Torcello-Gómez, A.; Maldonado-Valderrama, J.; de Vicente, J.; Cabrerizo-Vílchez, M.A.; Gálvez-Ruiz, M.J.; Martín-Rodríguez, A. Investigating the effect of surfactants on lipase interfacial behaviour in the presence of bile salts. Food Hydrocolloids. 2011, 25, 809–816.
Tso, P.; Balint, J.A.; Bishop, M.B.; Rodgers, J.B. Acute inhibition of intestinal lipid transport by Pluronic L-81 in the rat. Amer. J. Physiol. - Gastrointestinal Liver Phys. 1981, 4, 487–497.

Velikov, K.P.; Pelan, E. Colloidal delivery systems for micronutrients and nutraceuticals. Soft Matter 2008, 4, 1964–1980.

Walstra, P. Principles of emulsion formation. Chem. Eng. Sci. 1993, 48, 333–349.

Wang, D.Q. H.; Carey, M.C. Complete mapping of crystallization pathways during cholesterol precipitation from model bile: influence of physical-chemical variables of pathophysiologic relevance and identification of a stable liquid crystalline state in cold, dilute and hydrophilic bile salt-containing systems. J. Lipid Res. 1996, 37, 606–630.

Warren, D.B.; Chalmers, D.K.; Pouton, C.W. Structure and dynamics of glyceride lipid formulations, with propylene glycol and water. Mol. Pharmaceutics. 2009, 6, 604–614.

Watkins, C. Nanotechnology and the fats and oils industry. INFORM 2003, 14, 168–171.

Watt, S.M.; Simmonds, W.J. The specificity of bile salts in the intestinal absorption of micellar cholesterol in the rat. Clin. Exp. Pharm. Physiol. 1976, 3, 305–322.

Williams, K.J. Molecular processes that handle—and mishandle—dietary lipids. J.Clin. Investigation 2008, 118, 3247–3259.

Winsor, P.A. Solvent properties of amphiphilic compounds. Butterworth: London, 1954.

Wu, X.; Xu, J.; Huang, X.; Wen, C. Self-microemulsifying drug delivery system improves curcumin dissolution and bioavailability. Drug Dev. Ind. Pharm. 2011, 37, 15–23.

Xu, L.; Liu, Y.; Bai, R.; Chen, C. Applications and toxicological issues surrounding nanotechnology in the food industry. Pure Appl. Chem. 2010, 82, 349–372.

Yi, O.S.; Han, D.; Shin, H.K. Synergistic antioxidative effects of tocopherol and ascorbic acid in fish oil/lecithin/water system. J. Am. Oil Chem. Soc. 1991, 68, 881–883.

Yuan, J.S.; Ansari, M.; Samaan, M.; Acosta, E.J. Linker-based lecithin μEs for transdermal delivery of lidocaine. Int. J. Pharm. 2008, 349, 130–143.

Yuan, J.S.; Yip, A.; Nguyen, N.; Chu, J., Wen, X.-Y.; Acosta, E.J. Effect of surfactant concentration on transdermal lidocaine delivery with linker microemulsions. Int. J. Pharm. 2010, 392, 274–284.

Zhang, H.; Lu, Z.; Wang, S.; Shen, Y.; Feng, F.; Zheng, X. Development and antifungal evaluation of a food-grade U-type microemulsion. J. Appl. Microbiol. 2008, 105, 993–1001.

Zolnik, B.S.; Sadrieh, N. Regulatory perspective on the importance of ADME assessment of nanoscale material containing drugs. Adv. Drug Delivery Rev. 2009, 61, 422–427.

Zvonar, A.; Berginc, K.; Kristl, A.; Gašperlin, M. Microencapsulation of self-microemulsifying system: improving solubility and permeability of furosemide. Int. J. Pharm. 2010, 388, 151–158.

7

APPLICATIONS OF NANOTECHNOLOGY IN PHARMACEUTICAL DEVELOPMENT

Moghis U. Ahmad, Shoukath M. Ali, and Imran Ahmad
Jina Pharmaceuticals Inc., 28100 N. Ashley Circle, Suite 103, Libertyville, IL 60048, USA

INTRODUCTION

Nanotechnology has given a novel perspective to human understanding and brought revolutionary changes in the field of life sciences, including drug delivery, diagnostics, food, and the production of biomaterials suitable for applications in the field of medicine. This has resulted in new therapies, drug delivery systems, and biomaterials for implants or prosthesis. Nanoparticles have important potential applications in administration of therapeutic agents and in diagnostic procedures (Karanth & Murthy, 2008). The nanoparticulates drug delivery system offers plenty of advantages over conventional dosage forms, for example reduced toxicity, enhanced biodistribution, and improved patient compliance. Pharmaceutical nanoparticles are subnano-size structures, which contain drug or bioactive substances, and are constituted of several tens or hundreds of atoms or molecules in crystalline, amorphous, spherical, or needles morphologies.

In recent years, the pharmaceutical industry has been looking at ways to minimize the development costs of drugs. The increase in production costs, competition from generic manufacturers, and increased failure rates are putting pressure on the pharmaceutical business sector. The use of nanotechnology in formulation development and lifecycle management can make drug development significantly cost effective. In addition, nanotechnology can target specific drugs, which can reduce toxicity and improve efficacy.

Nanotechnology-based delivery systems can also protect drugs from degradation. These properties may help in sustained release of drugs leading to reduction in the number of doses required and resulting in lower treatment cost. Nano-based systems allow the delivery of insoluble drugs, thus allowing the use of previously rejected drugs that were difficult to administer. At present, researchers are also utilizing nanotechnology systems for developing off-patent drugs. The nanostructures employed as drug delivery systems have multiple advantages that make them superior to conventional delivery systems. In this regard, extensive research has been undertaken into the development of nanostructures such as liposome, nanoemulsion, nanocapsule, solid lipid nanoparticles, and others for drug delivery. The materials used in the fabrication of nanostructures determine the type of nanostructures obtained, and these nanostructures determine the release characteristics of the incorporated drugs.

The market for nanotechnology-based drug delivery is set to have a cumulative growth rate and expected to be US $20.1 billion by the year 2012, (11% of total drug delivery market by 2012; Joseph & Moore, 2008).

LIPIDS IN DRUG DELIVERY

Nanoparticle-based drug delivery has a significant advantage over the conventional drug delivery systems in terms of high stability, high specificity, high drug carrying capacity, controlled release, possibility of use with different routes of drug administration, and the capability to transport both hydrophilic and hydrophobic molecules. Drugs may be enclosed in the sphere of the nanoparticles or associated with the surface. Once the nanoparticles reach the target site, the drug payload may be released from the nanoparticle by diffusion, swelling, erosion, or degradation. Active systems are also possible, such as drug release by the input of external energy (e.g., targeted ultrasound, light, or magnetic field). Lipid nanoparticles are a promising alternative to existing carriers in chemical or other drug delivery systems. Lipid nanoparticles are capable of protecting their contents from the conditions within the body that cause degradation. They can be used to deliver insoluble drugs and can also be targeted to specific sites by ligands.

LIPOSOMES

Liposomes are vesicular structures with an aqueous core surrounded by a lipid bilayer. This was first reported by Bangham et al. (1965). The use of liposomes in drug delivery emerged in the 1970s (Kingsley et al., 2006). Liposomes are small vesicles prepared by the extrusion of phospholipids such as phosphatidylcholine (PC), phosphatidylglycerol (PG), phosphatidylethanolamine (PE), and phosphatidylserine (PS). These phospholipids have been used in food, cosmetics, and pharmaceutical industries (Alonso et al., 2000; Banerjee, 2001; Moussaoui et al., 2002; Nenadis et al., 2003). The size of liposomes can vary from 15 nm to several μm. Liposomes with a nanometer-sized cavity are called nanoliposomes. The properties of liposomes are determined by the choice of lipids, their composition, method of preparation, size, and surface charge (Yih & Al-Fandi, 2006). The solutes, such as a drug in the core, cannot pass through the hydrophobic bilayer, although hydrophobic molecules can be adsorbed into the bilayer, enabling the liposome to carry both the hydrophilic and hydrophobic molecules. Drug solutions with high or low pH can be carried by the liposomes, and the drugs get diffused out at the targeted sites when pH is neutralized in the body. Liposomes are used as drug carriers because of their ability to prevent degradation of drugs, reduce side effects, and deliver drugs to the site of action. However, there are certain limitations of liposomes that include low encapsulation efficiency, rapid leakage of water soluble drug in the presence of blood components, and poor storage ability (Lim et al., 2008; Soppimath et al., 2001). These limitations can be overcome by surface modification, which can be achieved by attaching polymers such as polyethylene glycol (PEG) units to improve the circulation time of liposomes in the blood, or by conjugation to antibodies or ligands such as lectins for target-specific drug delivery and stability (Bakowsky et al., 2008; Lim et al., 2008; Sihorkar & Vyas, 2001). PEGylated liposomes that avoid clearance by the reticuloendotheelial systems (RES) are known as "stealth liposomes" (Allen et al., 1991). Early research in this area has shown that stealth liposomes remain in the tumor interstitial fluid in close vicinity to tumor vessels (Yuan et al., 1994). Liposomes have also been surface modified with active targeting ligands to improve delivery of therapeutics to target cells (Dagar et al., 2003a; Gabizon et al., 2003; Pakunlu et al. 2004; Park et al., 2002).

Liposomes are used to deliver a variety of biological and drug compounds. The lists of compounds that can be delivered using liposomes include antibiotics, antioxidants (such as retinoid, carotenoids, tamoxifen, glutathione, urate, etc), vitamins (vitamins A, C, and E), ATP, NSAIDs (indomethacin and naproxen), hemoglobin, and genetic materials (such as plasmid DNA) (Joseph & Moore, 2008). Several lipid formulations such as liposomal amphotericin B (Davidson et al., 1991) and daunorubicin (Guaglianone et al., 1994) have been successfully commercialized. Doxorubicin, a highly effective cancer drug, has been encapsulated in liposomes to reduce toxic side effects and tested for treatment. Alternate method of encapsulating doxorubicin into liposomes has also been reported (Abraham et al., 2005). Liposomes are currently investigated for a variety of additional therapeutic agents. These include anticancer agents such as paclitaxel (Zhang et al., 2005), camptothecin and its analogs (Giles et al., 2004; Proulx et al., 2001; Verschragen et al., 2004), cisplatin (Zamboni et al., 2004), antibiotics such as amikacin (Donald et al., 2001), vanomycin, and ciprofloxacin (Kadry et al., 2004), and biologics such as antisense oligonucleotide (Rudin et al., 2004). Presently, three lipid based amphotericin B formulations are available on the market mainly Abelcet®, Amphotec®/Amphocil®, and Ambisome® (Gates & Pinney, 1993; Gregoriadis, 1987; Swenson et al., 1998). In all three lipid-based or liposomal preparations of amphotericin B, organic solvents were used to solubilize lipids and drugs. Recently, a new lipid-based formulation of amphotericin B (nanosomal amphotericin B) was developed without using any organic solvent or detergent at any step of the entire manufacturing process; it showed significantly reduced toxicity both in vitro and in vivo (Sheikh et al., 2010). Liposomes are also used to entrap gases and drugs for ultrasound-controlled drug delivery and release (Huang, 2008). This technique can also be used to entrap photosensitizers. Ultrasound can enhance the pharmacological activity of certain drugs, improve transportation of drugs through tissues and cell membranes, and can create a hyperthermic condition that causes destruction of cancer cells (Joseph & Moore, 2008).

CATIONIC LIPOSOMES

Cationic liposomes are prepared from at least one cationic phospholipid. The positively charged lipid bilayers can form complexes with negatively charged nucleic acid by simple mixing. The resulting cationic liposomes-nucleic acid complexes, called lipoplexes, are formed by a combination of electrostatic attraction and hydrophobic interaction and have been extensively used as nonviral vectors for the intracellular delivery of therapeutic genes (Duzgunes et al., 2003; Simoes et al., 2005; Yotsuyanagi & Hazemoto, 1998). The nonviral vectors based on nanoparticles (50–500 nm) were already tested to transport plasmid DNA. Nanotechnology in gene therapy has been applied to replace the viral vectors with potentially less immunogenic nanosize gene carriers. Therefore, delivery of repaired genes or the replacements of incorrect genes are fields where nanoscale objects could be introduced successfully (Davis, 1997). A cationic liposome was developed based on synthetic cationic cardiolipin analogue (CCLA) to test the DNA transfection efficiency. The CCLA-based cationic liposome was also used to determine the therapeutic efficacy of c-raf small interfering RNA (siRNA). The results suggested that CCLA-based liposome can be used as a nucleic acid delivery reagent for both in vitro and in vivo studies (Chien et al., 2005).

MICELLES

Micelles are spherical lipid nanostructures, but they do not have a bilayer or inner cavity. The hydrophobic ends of the phospholipids point inwards and the hydrophilic ends face outside, forming a spherical structure. Reverse micelles have this polarity the opposite way. The size of micelles for pharmaceutical applications ranges from 10 to 80 nm and have a short circulation time within the body compared to liposomes (Jain et al., 2010). However, this property gives micelles the advantage of being able to enter into tumor cells more easily when compared to liposomes because of enhanced permeability and retention (EPR) effect.

Lipid-polymer conjugate micelles can also be prepared. PEG-conjugated phospholipids such as disteroyl phosphatidylethanolamine (DSPE)-PEG are water soluble and self assemble as nanosized micelles instead of bilayers. PEGylated phospholipids micelles have the added advantage of simple and reproducible preparation compared to liposomes. In addition, PEGylated phospholipids micelles can also avoid mononuclear phagocytic systems (MPS) uptake and have demonstrated prolonged circulation times (Lukyanov et al., 2002). Such micelles are called sterically stabilized micelles (SSM). A number of therapeutic agents can be solubilized and stabilized in SSM and can be used as nanomedicines. These include vascoactive intestinal peptide (Onyuksel et al., 1999), camptothecin (Kao et al., 2005), paclitaxel (Krishnades et al., 2003), and diazepam (Ashok et al., 2004). When drug solubilization in SSM is exceeded, a second population of drug self-aggregated particles (100–300 nm) other than drug SSM was also observed. These drug self-aggregated particles get stabilized by PEGylated phospholipids on their surface against further precipitation and remain suspended in the clear aqueous dispersion in nanosize range. These PEGylated lipid-coated particles have a promising future to deliver high drug concentration in a small volume and can be further developed as nanomedicine. The solubilization potential of SSM can be improved by adding water-insoluble phospholipids, such as phosphatidylcholine (PC) to form sterically stabilized mixed micelles (SSMMs) (Ashok et al., 2004; Krishnadas et al., 2003). The SSMMs have similar nanosize as SSM but greater solubilization potential due to the incorporation of PC. However, the size and solubilization potential of SSMMs vary with PEG chain length and PC content. PEGylated phospholipids micelles can be freeze-dried and reconstituted without the need of additional lyo- and cryo-protectants (Kao et al., 2005). SSMs can be actively targeted by the attachment of ligands to the distal end of PEG (Krishnadas et al., 2004; Torchilin et al., 2003; Torchilin, 2004).

Conjugated polymer micelles with ligands have been used for efficient delivery, such as micelles conjugated with transferrin to target cancer cells and deliver DNA (Jule et al., 2003). Similarly, folate residues attached to micelles have been used to deliver Adriamycin to cancer cells (Lee et al., 2003). Positively charged nanoparticles are also used to enhance drug uptake by cancer cells; hence, positively charged lipids have been attached to these micelles to enhance uptake.

Micelles of thermal or pH-sensitive polymers have also been prepared. The pH-sensitive micelles are suitable for oral delivery applications. Temperature-dependent micelles are reported to increase the drug release capacity (Liu et al., 2005). Ultrasound has been suggested as a non-invasive stimulus to micelles for triggering drug release (Rapoport et al., 2003). Micelles have also been used as contrast agents in diagnostic applications. Paramagnetic metals, such as gadolinium (Gd) or manganese (Mn) can be easily incorporated into micelles for imaging applications. The advantage of such agents is enhanced target penetration due to their smaller size and easy movement to the target location.

NANOEMULSIONS

Nanoemulsions are dispersion of oil and water where the dispersed phase droplet sizes are below 1μ but usually 20–200 nm (Tadros et al., 2004). In earlier cases, nanoemulsions have also been called microemulsions. Nanoemulsion droplets are in the nanosize range and stabilize with a surface active film composed of surfactant and co-surfactant (Brime et al., 2003; He et al., 2003; Podlogar et al., 2004; Santos-Magalhaes et al., 2000; Seki et al., 2004). Unlike microemulsions, which are white in color due to their light scattering ability, the nanoemulsions are often smaller than visible wavelength and are transparent (Solans et al., 2005). Nanoemulsions are biodegradable, biocompatible, easy to prepare, and attractive as pharmaceutical formulations. Nanoemulsions provide much longer oil-water contact area due to the nanosize droplet compared to classical emulsions, this facilitates drug release from the dispersed phase droplets. The nanosize range of the droplets prevent creaming or sedimentation from occurring on storage or droplet coalescence. Structure of the nanoemulsion can affect the rate of drug release. Nanoemulsions are also employed as a sustained release delivery system for depot formation via subcutaneous injection (Santos-Magalhaes et al., 2000). They enhance gastrointestinal absorption and reduce inter- and intra-subject variability for various drugs. They exhibit excellent drug release profile because of their large interfacial area (Brusewitz et al., 2007). Nanoemulsions have been developed for parenteral, oral, ocular, pulmonary, and dermal deliveries (Solans et al., 2005).

Nanoemulsions have a number of advantages over larger emulsions. They can be stabilized to increase time before creaming occurs. Nanoemulsions are easy to produce, thermodynamically stable, and optically transparent or translucent. They are currently used in commercially available cosmetics and have been used for imaging and delivery of poorly soluble drugs. When used for drug delivery, nanoemulsions have significant advantages over traditional formulations for poorly soluble drugs. They eliminate the use of toxic co-solvents and can be administered at a much higher dose (approximately 10-fold higher), decreasing the administration volume and time (Tadros et al., 2004). The components of nanoemulsions can be GRAS (generally recognized as safe) compounds and therefore considered safe systems that can break down to their safe components. Nanoemulsions allow transport through membranes and have been suggested for use in patches for administering drugs topically.

Nanoemulsions of vincristine have been investigated in vivo (Junping et al., 2003). The formulation was composed of water, PEG-lipid with cholesterol as surfactant, and the oil phase was a vitamin E solution of oleic acid and vincristine. Vincristine was loaded into the oil phase of the nanoemulsion by increasing pH of the formulation to increase its lipid stability. The emulsion mean particle size observed was 138 nm, and the preparation was shown to be stable. Lower acute toxicity was also observed when animals were given vincristine nanoemulsion compared to free vincristine (Junping et al., 2003). This explains why nanoemulsions have the ability to decrease toxicity of anticancer drugs by increasing targeting to tumor sites. In another example, a low-dose Amphotericin B nanoemulsion was administered intravenously to mice, rats, dogs, and monkeys at a dose of 1.0 mg/kg and compared to Fungizone® (Fukui et al., 2003). In contrast to Fungizone, the Amphotericin B nanoemulsion showed a predictable pharmacokinetics of Amphotericin B.

LIPID NANOCAPSULES

Lipid nanocapsules are considered a cross between liposomes and nanoemulsion particles. They are composed of a liquid, oily core (medium chain triglycerides) surrounded by hydrophilic and lipophylic surfactants. The outer wall of lipid nanocapsules are thicker than nanoemulsion particles, allowing functionalisation and more controlled delivery, mostly used to deliver anticancer drugs (Allard et al., 2008; Khalid et al., 2006; Peltier et al., 2006). Lipid nanocapsules have been used to deliver therapeutic molecules and radionucleotides across the blood-brain barrier by conjugation of antibodies or antibody fragments (Beduneau et al., 2007a; Beduneau et al., 2007b; Beduneau et al., 2008). A novel, efficient, solvent-free phase-inversion method for their preparation has been reported (Heurtault et al.,

2002). Stealth lipid nanocapsules have also been synthesized using PEG to improve circulation time. Sizes between 50 and 300 nm are preferred for drug delivery. Because of their high stability and low permeability, drugs may not be loaded into the capsules after formulation; also, the release of the drug at target site may be difficult. Thus to improve the permeability, they are made responsive to physiological factors such as pH.

TYPES OF NANOPARTICLES

Nanoparticles are solid polymeric, submicron colloidal systems with a size range of 5 to 300nm. The drug of interest is dissolved, entrapped, absorbed, attached, or encapsulated into the nanoparticle matrix (Müller & Ghola, 2000). Depending on the method of preparation, a nanoparticle, nanosphere, or nanocapsule can be obtained with different properties and release characteristics for the encapsulated drug. Nanoparticles hold much interest because in this range materials can have different and enhanced properties compared with the same materials of a larger size due to the following two major principal factors: surface area and use of biodegradable materials for preparation of nanoparticles.

The increased surfaces are of quantum effect. These factors can enhance properties such as reactivity, strength, electric characteristics and in vivo behavior, and a much greater surface area per unit mass compared with the larger particles, leading to greater reactivity. The advantage of using nanoparticles loaded with drugs is in their small size that can penetrate through small capillaries and are taken up by cells. This enables the release of the drug at the right rate and dose at specific sites in the body for a certain time to release the accurate delivery. This enhances the therapeutic effect and reduces the toxicity and side effects. The use of biodegradable materials for nanoparticle preparation allows sustained release within the target site over a period of days and sometime weeks.

Some of the nanoparticles used as a carrier for drug and diagnostic agents are as follows: polymeric nanoparticles, nanosuspensions and nanocrystals, polymeric micelles, ceramic nanoparticles, magnetic nanoparticles, nanoshells coated with gold, nanomers and carbon nanotubes, and solid lipid nanoparticles.

SOLID LIPID NANOPARTICLES. Solid lipid nanoparticles (SLNs) are particles of nanometer dimensions with a solid-lipid matrix. They are oily droplets made from lipids which are solid at room temperature and stabilized by surfactant. The advantage of SLNs is that there is no need for organic solvents in the preparation. They provide protection from water and can be used for controlled drug release. Research interest in SLNs increased in the last decade due to their scalability potential. The lipids used are well tolerated by the body. The important features of SLNs include good tolerability, site-specific targeting, stability (stabilized by surfactant or polymers), controlled drug release, and protection of drugs from degradation (Wissing et al., 2004). SLNs have been studied and developed for parenteral, oral, pulmonary, dermal, ocular, and rectal routes of administration (Casadei et al., 2006; Cavalli et al., 2002; Liu et al., 2008; Puglia et al., 2008). Stealth and non-stealth SLNs have been used to deliver cancer drug paclitaxel and doxorubicin (Cavalli et al., 2000; Zara et al., 2002). However, there are some drawbacks with SLNs. Their loading capacity is low and there is a tendency to expel the contents during storage. SLNs are known for drug expulsion after polymorphic transition (the tendency for the particle matrix to form a perfect crystal lattice when solid lipids are used) on storage. The high water content is also problematic.

NANOSTRUCTURED LIPID CARRIERS. To overcome the limitations of SLNs, a second generation of lipid particles has been developed by mixing solid lipids with liquid lipids; these are called nanostructured lipid carriers (NLCs). The NLCs usually have a distorted structure, which makes the matrix structure imperfect, creating space to accommodate therapeutic agents and increase stability on storage, thus reducing drug expulsion. They have been investigated for the topical delivery of drugs, including anti-fungal and non-steroidal anti-inflammatory drugs. NLCs also have applications in cosmetics and dermatological preparations (Cavalli et al., 2002; Pardeike et al., 2009; Wissing et al., 2004).

LIPID-DRUG CONJUGATE NANOPARTICLES. Lipid drug conjugate (LDC) nanoparticles have been developed to overcome the limitations of types of drugs incorporated in SLNs. Lipophilic drugs are usually incorporated in SLNs, but due to partitioning effects during production, only highly potent hydrophilic drugs are effective in low concentrations and get incorporated in SLNs (Wissing et al., 2004). LDC enables the incorporation of both hydrophilic and lipophilic drugs. The loading capacity of LDC was observed up to 33% (Olbrich et al., 2002).

NANOTECHNOLOGY IN IMAGING

Imaging plays an increasingly important role in disease detection and planning of therapy and surgery. Clinical trials also depend more on imaging data to provide noninvasive, objective measures of therapy response. Nanotechnology has the potential to transform the field of medicine because it offers novel opportunities for sensing clinically relevant markets by using various types of imaging techniques, including single photon emission computed tomography (SPECT), positron emission tomography (PET), magnetic resonance imaging (MRI), fluorescence microscopy, computed tomography, and ultrasound. The current drive in developing techniques such as the MRI is to enhance spatial resolution and contrast agents. In this chapter, we will focus on selected imaging techniques.

LIPOSOMES FOR IMAGING

Liposomes can be modified to incorporate a magnetic element for use in monitoring their movement within the body using MRI. Such liposomes enhance the efficacy for their use by allowing more external control on their movement. Magnetoliposomes with nanometer-sized magnetite cores, entrapped in the bilayer of phospholipids molecules, can be used as biocompatible MRI agents with potential applications in combined imaging and drug delivery (Soenen et al., 2008).

Liposomes have been developed as carriers for a variety of contrast agents and radiopharmaceuticals. The sterically stabilized liposomes (SSLs) encapsulating ^{99m}Tc-Hexamethylpropyleneamine Oxime (^{99m}Tc-HMPAO) with and without covalently attached vasoactive intestinal peptide (VIP) ligand accumulate more in breast cancer than normal tissues. In particular, actively targeted VIP-SSL encapsulating ^{99m}Tc-HMPAO accumulates significantly more in breast cancer than SSL without VIP. The data support VIP as an ideal targeting agent to increase in vivo specificity and accumulation of liposomes in tumor sites more than normal tissues. This is because normal vasculature is not leaky and VIP receptors exist only in the extravascular space; the imaging agent could not be distributed to the normal tissues to give high background signal (Dagar et al., 2003a).

Liposomes composed of dipalmitoylphosphatidylcholine (DPPC), cholesterol, and palmityl-D-glucuronide (PGlcUA) prepared in the presence of 2-[18F] fluoro-2-deoxyglucose were used to observe real-time liposomal trafficking by positron emission tomography (PET). The result indicates that the liposomes (size 200 nm or less) accumulate in the liver right after the injection but they re-enter the bloodstream with time. The majority of the liposomes remained in the bloodstream until they finally accumulated in the tumor tissues. The PGlcUA was found effective in increasing the circulation half-life of the liposomes. Compared to control liposomes without PGlcUA, these liposomes had lower liver and higher tumor accumulation (Oku et al., 1995). Other uses of liposomes in ultrasound and scintigraphic imaging were discussed by Dagar et al. (2003b).

MAGNETIC NANOPARTICLES FOR IMAGING

Magnetic nanoparticles are one of the most widely studied uses of nanotechnology. Applications of magnetic nanoparticles include targeted drug delivery as contrast agents in MRI, gene delivery, cell separation, and cell labeling. Colloidal iron oxide formulated with dextran is clinically used as MRI contrast agents (Gandon et al., 1991). Iron oxide nanoparticles are used because of their biodegradable nature, biocompatibility, and superparamagnetic properties suited for MRI applications. The iron oxide nanoparticles were coated with oleic acid and loaded with anticancer agents, doxorubicin and paclitaxel, with loading efficiency up to 95% (Jain et al., 2008). The drug release can be enabled by inducing hypothermia through an applied magnetic field.

A study showed that covalent coupling of insulin on the iron oxide nanoparticles surface decreased internalization of the nanoparticles (Gupta et al., 2003). This was attributed to attachment of the insulin-coated nanoparticles to the cell surface receptors, thus preventing internalization and reduced cellular toxicity. Multiple crystals of iron oxide have been embedded in nanoparticle matrices, such as polyacrylamide with surface PEG and with solid lipid nanoparticles (Moffat et al., 2003; Peira et al., 2003).

CARBON NANOSTRUCTURES IN DRUG DELIVERY

CARBON NANOTUBES

Carbon nanotubes (CNTs) have a wide range of possible biomedical applications due to their unique physical and chemical properties (Yang et al., 2007). CNTs, because of their size and shape, can enter living cells without causing cell death or damage to living cells. The hollow structure of CNTs allows encapsulation of drug molecules, but at present its use is very limited (Liu et al., 2009). CNTs solubilization requires covalent functionalization to prevent aggregation and to increase their solubility (Chattopadhay et al., 2006; Geordakilas et al., 2002; Holzinger et al., 2001; Lee et al., 2005; Yokoi et al., 2005) or non-covalent (Bandyopadhaya et al., 2002; Chen et al., 2001; Star et al., 2001; Star et al., 2003; Zheng et al., 2003). Several drugs have been delivered using CNTs, which are normally insoluble and toxic due to their tendency to aggregate, such as amphotericin B (Wu et al., 2005). Use of CNTs increases the solubility and low aggregation, therefore low toxicity and increased anti-fungal action. A smart bio-nanotube was also reported which has ends that can open and close through altering its charge (Raviv et al., 2005). This opens a new channel for controlled encapsulation and release. A recent review details biological applications of CNTs (Liu et al., 2009).

A potential drawback for their usage in biomedical applications is hydrophobicity-induced aggregation and how CNTs interact with human cells. It is important to understand how CNTs interact with the cellular membrane (lipid molecules which make up cell membranes) and detergent molecules that may be used to solubilize both CNTs and membrane proteins. A coarse-grained molecular dynamics to study the self assembly of CNTs with various amphiphilies, namely a phospholipid, dipalmitoylphosphatidylcholine (DPPC), and two species of detergents, such as dihexanoylphosphatidylcholine (DHPC) and lysophosphatidylcholine (LPC) was used. It is reported that for low amphiphile/CNT ratio, the DPPC, DHPC, and LPC all wrap around the CNT (Wallace & Sansom, 2009). Upon increasing the number of amphiphiles, a transition in absorption was observed; DPPC encapsulates the CNT within a cylindrical micelle, while both DHPC and LPC absorb onto CNTs in hemimicelles. This study highlights differences in the adsorption mechanisms of bilayer-forming lipids (phospholipids) and detergents onto CNTs which may be exploited to enable enhancement of CNT solubilization by amphiphiles while minimizing perturbation of cell membrane integrity.

Although CNTs show promise for drug delivery, there are a number of toxicity issues that have not been fully investigated. Currently, research is ongoing in this area, and it will take time to come to a conclusion on the effects of CNTs (Bottini et al., 2006; Tian et al., 2006).

OTHER NANOPARTICLES IN DRUG DELIVERY

Delivery of water-insoluble drugs or drugs not soluble both in water and nonpolar solvents or those difficult to formulate can be made by the dispersion of drug particles in nanosize in an aqueous solution. Nanosuspensions of drug particles are produced by two different methods. The first method involves the breaking down of bigger particles to nanosize using high pressure homogenization of drug suspension in the presence of a suitable surfactant (as for example, Tween 80 or Pluronic F68) (Kayser et al., 2003). The second method involves crystallization, building the nanoparticles up from the supersaturated solution state (Rabinow, 2004). The solid state of nanosuspension is chemically more stable and gives high drug weight per volume loading, and it is useful for therapeutic compounds that need high dosing. Mean diameters of drug nanoparticles are between 200 and 400 nm (Müller & Keck, 2004). Recently, in our laboratory, amphotericin B was formulated in lipids (Nanosomal amphotericin B) without using any detergent or toxic organic solvents during the preparation. Electron microscopy and particle size determination of nanosomal amphotericin B showed a homogeneous population of nanosized particles below 100 nm (Sheikh et al., 2010). Similarly, nanosomal tacrolimus was prepared in an aqueous system, which showed a homogeneous population of nanosized particles less than 100 nm. It contained no alcohol or surfactant such as polyoxyl 60 hydrogenated castor oil (HCO-60). Nanosomal tacrolimus presents a new and improved therapeutic approach for organ transplant patients compared to the marketed HCO-60 based tacrolimus product (Ali et al., 2010).

Drug nanocrystals have been formed for amphotericin B (Kayser et al., 2003), etoposide, camptothecin, and paclitaxel (Merisko-Leversidge et al., 1996). Nanoparticles of hydrophilic drugs such as dexamethasone phosphate have been produced using supercritical carbon dioxide and encapsulated into poly (lactide-co-glycolide) macroparticles. These drug nanoparticles of 150–200 nm size when microencapsulated provide sustained drug release for 700 hours without the initial burst release observed when larger unprocessed drug particles (50–100 μm) were encapsulated. This was attributed to the uniform dispersion of dexamethasone phosphate nanocrystals in the polymer matrix that allows slow release of the drug with polymer degradation. The super critical carbon dioxide process can be applied to produce nanocrystals of other hydrophilic drugs to be microencapsulated in sustained release formulations (Thote et al., 2005).

NANOGELS

Nanogels are composed of flexible hydrophilic polymers in the nanosize scale. They are soluble in water and allow spontaneous loading of drugs in aqueous media. Nanogels possess large surface area and a network to allow incorporation of drug molecules. The nanogel particles (>100 nm) made of polyethylene glycol (PEG) and polyethylenimine (PEI) have been used to cross the blood-brain barrier and deliver oligonucleotides to the brain. Cationic nanogels have been applied to encapsulate the negatively charged oligonucleotides spontaneously by ionic interaction (Vinogradov et al., 2004). It is observed that at least two-thirds of the oligonucleotides remained associated with the nanogels after being transported across bovine brain micro vessel endothelial cells. Permeability of the oligonucleotides was enhanced up to 6-fold when compared to free oligonucleotides at fixed oligonucleotide concentration of 5 μM. Further modification of nanogels with transferrin or insulin vector molecules resulted in 11- to 12-fold increase in permeability. Biodistribution in vivo indicated 2.7 to 5.3% of the intravenously injected oligonucleotide-loaded nanogels distributed to the intact brain compared to 0.2% with the free oligonucleotides. Nanogels have also been used for the release of pH-dependent drugs such as doxorubicin (Oishi et al., 2007) and incorporation of insoluble small molecule AQ10, an anticancer drug (Ganta et al., 2008).

ALBUMIN NANOPARTICLES

Albumin is a major protein component in serum. It has been modified to create novel nanostructures for applications in drug delivery. The surface of albumin has several amino and carboxylic groups which are available for covalent conjugation of biomolecules and drugs. Albumin-DNA-polyethylenimine (PEI) conjugates have been used for gene delivery with reduced toxicity, and irritation (Rhaese et al., 2003). Albumin nanoparticles can be prepared by a desolation/cross-linking technique, where dissolved albumin in water is desolated by drop-wise addition of ethanol and glutaraldehyde to induce albumin nanoparticle cross-linking (Zhang et al., 2004). Albumin nanoparticles can be modified to alter size, polydispersity, surface charge, drug loading, and drug release. Functionalized albumin nanoparticles have been shown to cross the blood-brain barrier (Mishra et al., 2006). Similarly, the bovine serum albumin nanoparticles loaded with sodium ferulate have been targeted towards the liver (Li et al., 2008).

The breakthrough in the clinical application of albumin nanoparticles has been a solvent-free formulation of paclitaxel (Abraxane®), which has been approved by the U.S. FDA for the treatment of metastatic breast cancer (Gradishar et al., 2005). The paclitaxel albumin nanoparticles (130 nm) was administered as a saline suspension at up to 10-fold higher concentration than previous formulations. It eliminated the solvent-derived side effects, need for pre-medication, and showed reduced toxicity. The overall response rate was 33% for Abraxane®, compared with 19% for paclitaxel (Taxol®) (Garber, 2004). The greater efficacy and lower toxicity of Abraxane® is due to passive targeting and greater retention of nanoparticles in cancer cells compared to free paclitaxel.

CHITOSAN-LIPID NANOPARTICLES

Chitosan is a natural polysaccharide derived from the shells of crustaceans. Chitosan has the ability to clot blood and is useful in bandages and other haemostatic agents. Nanoparticles of chitosan and egg phosphatidylcholine (egg PC) are reported for the delivery of anticancer drug paclitaxel (Grant et al., 2005). The chitosan-phospholipid nanoparticle was found to be very stable and biocompatible, depending on the ratio of the two materials. The activities of this system in mice were confirmed, but accumulation of paclitaxel was seen in the heart, which may cause complication (Dhanikula et al., 2005; Lim Soo et al., 2008) and need further research.

Chitosan is normally coated on the surface of lipid-based nanostructures to improve the adhesive properties and increase stability. Self-assembled spherical nanoparticles were formed by injection of an alcoholic lecithin solution into an aqueous chitosan solution (Sonvico et al., 2006). There is a possibility that these nanoparticles could encapsulate progesterone, however they had poor loading capabilities with hydrophilic drugs.

NANOTECHNOLOGY AND SAFETY ISSUES

Nanotechnology is considered to be an emerging, promising technology that will have a significant impact in all industrial sectors, particularly in the area of cancer research. Nanotechnology-based therapeutics is gaining significant interest and will have greater market share in the pharmaceutical industry. However, the novel properties and characteristics that bring new benefits in drug delivery also bring new challenges in risk management and toxicity. Some of these novel characteristics are either poorly understood or need further studies. Both pioneers of nanotechnology and its opponents are finding it difficult to argue their case because of the limited information available to support one side or the other. At present, nanoparticles prepared from natural or GRAS (generally recognized as safe) components, such as lipid-based or albumin nanoparticles appear to have the most potential due to their minimal toxicity. Similarly, nanoemulsions are usually GRAS components and therefore considered as safe systems which are easily biodegradable. Other systems such as CNTs are promising but more research is required to address the toxicity issues. Until then, CNT-based delivery systems are not likely to be accepted for therapies.

CONCLUSION

Nanoparticle-based drug delivery systems such as liposomes, micelles, nanoemulsions, nanocapsules, solid lipid nanoparticles, magnetic nanoparticles, nanogels, albumin nanoparticles, and so on have great potential to be therapeutically effective systems for human and animal diseases. Nanoparticle-based drug delivery systems hold great potential to overcome some of the barriers to efficient targeting of cells and molecules in inflammation and cancer. There is a possibility to overcome problems of drug resistance in target cells and to facilitate movements of drugs across the blood-brain barrier. The challenge, however, remains the precise characterization of molecular targets and ensuring that these nanoparticles are directed towards the targeted organs to prevent adverse effects on healthy tissues. Furthermore, because nanosystems increase the efficiency of drug delivery, recalibrations of doses are necessary. Finally, in order to take more nanotechnology-based products into the clinic, it is necessary to address the risk issues simultaneously, including any risks resulting from the nanoscale properties of the materials used.

REFERENCES

Abraham, S.A.; Waterhouse, D.N.; Mayer, L.D.; Cullis, P.R.; Madden, T.D.; Bally, M.B. The liposomal formulation of doxorubicin. Methods in Enzymology, 2005, 391, 71–97.

Ali, S.M.; Ahmad, A.; Sheikh, S.; Ahmad, M.U.; Rane, R.C.; Kale, P.; Paithankar, M.; Saptarishi, D.; Seghal, A.; Maheshwari, K.; et al. Polyoxyl 60 hydrogenated castor oil free nanosomal formulation of immunosuppressant tacrolimus: pharmacokinetics, safety, and tolerability in rodents and humans. Int. Immun. 2010, 10, 325–330.

Allard, E.; Passirani, C.; Garcion, E.; Pigeon, P.; Vessieres, A.; Jaouen, G.; Benoit, J.P. Lipid nanocapsules loaded with an organometallic tamoxifen derivative as a novel drug-carrier system for experimental malignant gliomas. J. Control. Rel. 2008, 130, 146–153.

Allen, T.M.; Hansen, C.; Martin, F.; Redemann, C.; Yau-Young, A. Liposomes containing synthetic lipid derivatives of poly (ethylene glycol) show prolonged circulation half-lives in vivo. Biochim. Biophys. Acta 1991, 1066, 29–36.

Alonso, A.; Goni, F.M.; Buckley, J.T. Lipids favoring inverted phase enhance the ability of aerolysm to permeabilize liposome bilayers. Biochemistry, 2000, 39, 14019– 14024.

Ashok, B.; Arleth, L.; Hjelm, R.P.; Rubinstein, I.; Onyuksel, H. In vitro characterization of PEGylated phospholipid micelles for improved drug solubilization: effects of PEG chain length and PC incorporation. J. Pharm. Sci. 2004, 93, 2476–2487.

Bakowsky, H.; Richter, T.; Kneuer, C.; Hoekstra, D.; Rothe, U.; Bendas, G.; Ehrhardt, C.; Bakowsky, U. Adhesion characteristics and stability assessment of lectin-modified liposomes for site-specific drug delivery. Biochim. Biophys. Acta 2008, 1778, 242–249.

Bandyopadhaya, R.; Nativ-Roth, E.; Regev, O.; Yerushalmi-Rozen, R. Stabilization of individual carbon nanotubes in aqueous solutions. Nano Letters 2002, 2, 25–28.

Banerjee, R. Liposomes: applications in medicine. J. Biomaterials Applications 2001, 16, 3–21.

Bangham, A.D.; Standish, M.M.; Watkins, J.C. Diffusion of univalent ions across the lamellae of swollen phospholipids. J. Mol. Biol. 1965, 13, 238–252.

Beduneau, A.; Salunier, P.; Hindre, F.; Clavreul, A.; Leroux, J.C.; Benoit, J.P. Design of targeted lipid nanocapsules by conjugation of whole antibodies and antibody Fab' fragments. Biomaterials 2007a, 28, 4978–4990.

Beduneau, A.; Saulnier, P.; Benoit, J.P. Active targeting of brain tumors using nanocarriers. Biomaterials 2007b, 28, 4947–4967.

Beduneau, A.; Hindre, F.; Clavreul, A.; Leroux, J.C.; Saulnier, P.; Benoit, J.P. Brain targeting using novel lipid nanovectors. J. Control. Rel. 2008, 126, 44–49.

Bottini, M.; Bruckner, S.; Nika, K.; Bottini, N.; Magrini, A.; Bergamaschi, A.; Mustelin, T. Multi-walled carbon nanotubes induce T lymphocyte apoptosis. Toxicol. Let. 2006, 160,121–126.

Brime, B.; Frutos, P.; Bringas, P.; Nieto,A.; Ballesteros, M.P.; Frutos, G. Comparative pharmacokinetics and safety of a novel lyophilized amphotericin B lecithin-based oil-water microemulsion and amphotericin B deoxycholate in animal models. J. Antimicrob. Chemother. 2003, 52, 103–109.

Brüsewitz, C.; Schendler, A.; Funke, A.; Wagner, T.; Lipp, R. Novel poloxamer-based nanoemulsions to enhance the intestinal absorption of active compounds. Int J. Pharm. 2007, 329, 1173–1181.

Casadei, M.A.; Cerreto, F.; Cesa, S.; Giannuzzo, M.; Feeney, M.; Marianecci, C.; Paolicelli, P. Solid lipid nanoparticles incorporated in dextran hydrogels: a new drug delivery system for oral formulations. Int. J. Pharm. 2006, 325,140–146.

Cavalli, R.; Caputo, O.; Gasco, M.R. Preparation and characterization of solid lipid nanospheres containing paclitaxel. Eur. J. Pharm. Sci. 2000, 10, 305–309.

Cavalli, R.; Gasco, M.R.; Chetoni, P.; Burgalassi, S.; Saettone, M.F. Solid lipid nanoparticles (SLN) as ocular delivery system for tobramycin. Int. J. Pharm. 2002, 238, 241–245.

Chattopadhyay, J.; Cortez, F.; Chakraborty, S.; Slater, N.; Billups, W. Synthesis of water soluble PEGylated single-walled carbon nanotubes. Chem. Mater. 2006, 18, 6864–6868.

Chen, J.; Rao, A.M.; Lyuksyutov, S.; Itkis, M.E.; Hamon, M.A.; Hu, H.; Cohn, R.W.; Eklund, P.C.; Colbert, D.T.; Smalley, R.E.; et al. Dissolution of full-length single-walled carbon nanotubes. J. Phys. Chem. 2001, b105, 2525–2528.

Chien, P-Y.; Wang, J.; Carbonaro, D.; Lei, S.; Miller, B.; Sheikh, S.; Ali, S.M.; Ahmad, M.U.; Ahmad, I. Novel cationic cardiolipin analogue-based liposome for efficient DNA and small interfering RNA delivery in vitro and in vivo. Cancer Gene Therapy 2005, 12, 321–328.

Dagar, S.; Krishandas, A.; Rubinstein, I.; Blend, M.J.; Onyuksel, H. VIP grafted sterically stabilized liposomes for targeted imaging of breast cancer: in vivo studies. J. Control. Rel. 2003a, 91,123–133.

Dagar, S.; Rubinstein, I.; Onyuksel, H. Liposomes in ultrasound and gamma scintigraphic imaging. Methods Enzymol. 2003b, 373, 198–214.

Davidson, R.N.; Croft, S.L.; Scott, A.; Maini, M.; Moody, A.H.; Bryceson, A.D. Liposomal amphotericin B in drug-resistant visceral leishmaniasis. Lancet, 1991, 337, 1061–1062.

Davis, S.S. Biomedical applications of nanotechnology-implications for drug targeting and gene therapy. Trends Biotechnol. 1997, 15, 217–224.

Dhanikula, A.B.; Singh, D.R.; Panchagnula, R. In vivo pharmacokinetics and tissue distribution studies in mice of alternative formulations for local and systemic delivery of paclitaxel: gel, film, prodrug, liposomes and micelles. Curr. Drug Del. 2005, 2, 35–44.

Donald, P.R.; Sirgel, F.A.; Venter, A.; Smit, E.; Parkin, D.P.; van deWal, B.W.; Mitchison, D.A. The early bactericidal activity of a low-clearance liposomal amikacin in pulmonary tuberculosis. J. Antimicrob Chemother. 2001, 48, 877–880.

Duzgunes, N.; De, C.; Harduya, T.; Simoes, S.; Zhdanov, R.I.; Konopka, K.; Pedroso de Lima, M.C. Cationic liposomes for gene delivery: novel cationic lipids and enhancement by proteins and peptides. Curr. Med. Chem. 2003, 10, 1213–1220.

Fukui, H.; Koike, T.; Saheki, A.; Sonoke, S.; Seki, J. A novel delivery system for amphotericin B with lipid nano-sphere (LNS). Int. J Pharm, 2003, 265, 37–45.

Gabizon, A.; Horowitz, A.T.; Goren, D.; Tzemach, D.; Shmeeda, H.; Zalipsky, S. In vivo fate of folate-targeted polyethylene-glycol liposomes in tumor-bearing mice. Clin. Cancer Res. 2003, 9, 6551–6559.

Gandon, Y.; Heautot, J.F.; Burnet, F.; Guyader, D.; Deugnier, Y.; Carsin, M. Superparamagnetic iron oxide: clinical time-response study. Eur. J Radiol. 1991, 12, 195–200.

Ganta, C.; Shi, A.; Battina, S.K.; Pyle, M.; Rana, S.; Hua, D.H.; Tamura, M.; Troyer, D. Combination of nanogel polyethylene glycol-polyethylenimine and 6-(hydroxymethyl)-1, 4-anthracenedione as an anticancer nanomedicine. J. Nanosci. and Nanotech. 2008, 8, 2334–2340.

Garber, K. Improved paclitaxel formulation hints at new chemotherapy approach. J. Natl. Cancer Inst. 2004, 96, 90–91.

Gates, C.; Pinney, R.J. Amphotericin B and its delivery by liposomal and lipid formulations. J. Clin. Pharm. Ther. 1993, 18, 27–38.

Geordakilas, V.; Kordatos, K.; Prato, M.; Guldi, D.M.; Holzinger, M.; Hirsch, A. Organic functionalization of carbon nanotubes. J. Am. Chem. Soc. 2002, 124, 760–761.

Giles, F.J.; Tallman, M.S.; Garcia-Manero, G.; Cortes, J.E.; Thomas, D.A.; Wierda, W.G.; Verstovsek, S.; Hamilton, M.; Barrett, E.; Albitar, M.; et al. Phase I and pharmacokinetics study of a low clearance, unilamellar liposomal formulation of lurtotean, a topoisomerase 1 inhibitor, in patients with advanced leukemia. Cancer 2004, 100, 1449–1458.

Gradishar, W.J.; Tjulandin, S.; Davidson, N.; Shaw, H.; Desai, N.; Bhar, P.; Hawkins, M.; O'Shaughnessy, J. Superior efficacy of albumin-bound paclitaxel, ABI-007, compared with polyethylated castor-oil-based paclitaxel in women with metastatic breast cancer: results of a phase III trial. J. Clin .Oncol. 2005, 23, 7794–7803.

Grant, J.; Blicker, M.; Piquette-Miller, M.; Allen, C. Hybrid films from blends of chitosan and egg phosphatidylcholine for localized delivery of paclitaxel. J. Pharm. Sci. 2005, 94, 1512–1527.

Gregoriadis, G. Overview of liposomes. J. Antimicrob. Chemother. 1987, 28, 39–48.

Guaglianone, P.; Chan, K.; DelaFlor-Weiss, E.; Hanisch, R.; Jeffers, D.; Sharma, D.; Muggia, F. Phase I and pharmacologic study of liposomal daunorubicin (DaunoXome). Invest. New Drugs 1994, 12, 103–110.

Gupta, A.K.; Berry, C.; Gupta, M.; Curtis, A. Receptor-mediated targeting of magnetic nanoparticles using insulin as a surface ligand to prevent endocytosis. IEEE Trans. Nanobioscience 2003, 2, 255–261.

He, L.; Wang, G. L.; Zhang, Q. An alternative paclitaxel microemulsion formulation: hypersensitivity evaluation and pharmacokinetic profile. Int J. Pharm. 2003, 250, 45–50.

Heurtault, B.; Saulnier, P.; Pech, B.; Proust, J.E.; Benoit, J.P. A novel phase inversion-based process for the preparation of lipid nanocarriers. Pharm. Res. 2002, 19, 6, 875–880.

Holzinger, M.; Vostrowsky, O.; Hirsch, A.; Hennrich, F.; Kappes, M.; Weiss, R.; Jellen, F. Sidewall functionalization of carbon nanotubes. Angew. Chem. Int. Ed. 2001, 40, 4002–4005.

Huang, S.L. Liposomes in ultrasonic drug and gene delivery, Adv. Drug Deliv. Rev. 2008, 60, 1167–1176.

Jain, N.; Jain, R.; Thakur, N.; Gupta, B.P.; Jain, D.K.; Banveer, J.; Jain, S. Nanotechnology: a safe and effective drug delivery system. Asian J. Pharm. Clin. Res. 2010, 3, 159–165.

Jain, T.K.; Richey, J.; Strand, M.; Leslie-Pelecky, D.L.; Flask, C.A.; Labhasetwar, V. Magnetic nanoparticles with dual functional properties: drug delivery and magnetic resonance imaging. Biomaterials 2008, 29, 4012–4021.

Joseph, T.; Moore, R. Drug delivery using nanotechnologies: report on technologies, markets and competitive environment, Inst. Nanotechnol.10 July 2008.

Jule, E.; Nagasaki, Y.; Kataoka, K. Lactose-installed poly (ethylene glycol)-poly (D, L- lactide) block copolymer micelles exhibit fast-rate binding and high affinity toward a protein bed simulating a cell surface. A surface plasmon resonance study. Bioconjugate Chem. 2003, 14, 177–186.

Junping, W.; Takayama, K.; Nagai, T.; Maitani, Y. Pharmacokinetics and antitumor effects of vincristine carried by microemulsions composed of PEG-lipid, oleic acid, vitamin E and cholesterol. Int J. Pharm. 2003, 251, 13–21.

Kadry, A.A.; Al-Suwayeh, S.A.; Abd-Allah, A.R.; Bayomi, M.A. Treatment of experimental osteomyelitis by liposomal antibiotics. J. Antimicrob Chemother. 2004, 54, 1103–1108.

Kao, O.; Rubinstein, I.; Onyuksel, H. Camptothecin in sterically stabilized phospholipid micelles: a novel nanomedicine. Nanomedicine 2005, 1, 77–84.

Karanth, H.; Murthy, R.S.R. Nanotechnology in brain targeting. Int. J. Pharm. Sci. Nanotech. 2008, 1, 10–21.

Kayser, O.; Olbrich, C.; Yardley, V.; Kiderlen, A.F.; Croft, S.L. Formulation of amphotericin B as nanosuspension for oral administration. Int. J. Pharm. 2003, 254, 73–75.

Khalid, M.N.; Simard, P.; Hoarau, D.; Dragomir, A.; Leroux, J.C. Long circulating poly (ethylene glycol)-decorated lipid nanocapsules deliver docetaxel to solid tumors. Pharm. Res. 2006, 23, 4, 752–758.

Kingsley, J.D.; Dou, H.; Destache, C.J. Nanotechnology: a focus on nanoparticles as a drug delivery system. J. Neuroimmune Pharmacol. 2006, 1, 340–345.

Krishnadas, A.; Rubinstein, I.; Onyuksel, H. Sterically stabilized phospholipid mixed micelles: in vitro evaluation as a novel carrier for water-insoluble drugs. Pharm. Res. 2003, 20, 297–302.

Krishnadas, A.; Rubinstein, I.; Sekosan, M.; Onyuksel, H. Targeted delivery of paclitaxel to breast cancer by vascoactive intestinal peptide conjugated sterically stabilized phospholipid mixed micelles. AAPS J. 2004, 6 (Suppl 1), M1191.

Lee, E.S.; Na, K.; Bae, Y.H. Polymeric micelle for tumor pH and folate-mediated targeting. J. Cont. Rel. 2003, 91, 103–113.

Lee, K.M.; Li, L.; Dai, L. Asymmetric end-functionalization of multiwalled carbon nanotubes. J. Am. Chem. Soc. 2005, 127, 4122–4123.

Li, F.Q.; Su,H.; Wang, J.; Liu, J.Y.; Zhu, Q.G.; Fei, Y.B.; Pan, Y.H.; Hu, J.H. Preparation and characterization of sodium ferulate entrapped bovine serum albumin nanoparticles for liver targeting. Int. J. Pharm. 2008, 349, 274–282.

Lim, H.J.; Cho, E.C.; Shim, J.; Kim, D-H.; An, E.J.; Kim, J. Polymer-associated liposomes as a novel delivery system for cyclodextrin-bound drugs. J. Colloid Inter. Sci. 2008, 320, 460–468.

Lim Soo, P.; Cho, J.; Grant, J.; Ho, E.; Piquette-Miller, M.; Allen, C. Drug release mechanism of paclitaxel from a chitosan-lipid implant system: effect of swelling, degradation and morphology. Eur. J. Pharm. Biopharm. 2008, 69, 149–157.

Liu, J.; Gong, T.; Fu, H.; Wang, C.; Wang, X.; Chien, Q.; Zhang, Q.; He, Q.; Zhang, Z. Solid lipid nanoparticles for pulmonary delivery of insulin. Int. J. Pharm. 2008, 356, 333–344.

Liu, S.Q.; Tong, Y.W.; Yang, Y.Y. Incorporation and in vivo release of doxorubicin in thermally sensitive micelles made from poly (n-isopropylacrylamide-co-n, n-dimethylacrylamide)-b-poly (D, L-lactide-co-glycolide) with varying compositions. Biomaterials 2005, 26, 5064–5074.

Liu, Z.; Tabakman, S.; Welsher, K.; Dai, H. Carbon nanotubes in biology and medicine: In vitro and in vivo detection, imaging and drug delivery. Nano Res. 2009, 2, 85–120.

Lukyanov, A.N.; Gao, Z.; Mazzola, L.; Torchilin, V.P. Polyethylene glycol-diacyllipid micelles demonstrate increased accumulation in subcutaneous tumors in mice. Pharm. Res. 2002, 19, 1424–1429.

Merisko-Liversidge, E.; Sarpotdar, P.; Bruno, J.; Hajj, S.; Wei, L.; Peltier, N.; Rake, J.; Shaw, J.M.; Pugh, S.; Polin, L.; et al. Formulation and antitumor activity evaluation of nanocrystalline suspension of poorly soluble anticancer drugs. Pharm. Res. 1996, 13, 272–278.

Mishra, V.; Mahor, S.; Rawat, A.; Gupta, P.N.; Dubey, P.; Khatri, K.; Vyas, S.P. Targeted brain delivery of AZT via transferrin anchored PEGylated albumin nanoparticles. J. Drug Target 2006, 14, 45–53.

Moffat, B.A.; Reddy, G.R.; McConville, P.; Hall, D.E.; Chenevert, T.L.; Kopelman, R.R.; Philbert, M.; Weissleder, R.; Rehemtulla, A.; Ross, B.D. A novel polyacrylamide magnetic nanoparticle contrast agent for molecular imaging using MRI. Mol. Imaging, 2003, 2, 324–332.

Moussaoui, N.; Cansell, M.; Denizot, A. Marinosomes, marine lipid-based liposomes: physical characterization and potential application in cosmetics. Int. J. Pharm. 2002, 242, 361–385.

Müller, R.H.; Gohla, S. Solid lipid nanoparticles (SLN) for controlled drug delivery–a review of the state of the art. Eur. J. Pharm. Biopharm. 2000, 50, 161–177.

Müller, R.H.; Keck, C.M. Challenges and solutions for the delivery of biotech drugs—a review of drug nanocrystal technology and lipid nanoparticles. J. Biotechnol. 2004, 113, 151–170.

Nenadis, N.; Zafiropoulou, I.; Tsimidou, M. Commonly used food antioxidants: a comparative study in dispersed systems. Food Chem. 2003, 82, 403–407.

Oishi, M.; Hayashi, H.; Michihiro, I.D.; Nagasaki, Y. Endosomal release and intracellular delivery of anticancer drugs using pH-sensitive PEGylated nanogels. J. Materials Chem. 2007, 17, 3720–3725.

Oku, N.; Tokudome, Y.; Tsukada, H.; Okada, S. Real-time analysis of liposomal trafficking in tumor-bearing mice by the use of positron emission tomography. Biochim. Biophys. Acta. 1995, 1238, 86–90.

Olbrich, C.; Geßner, A.; Kayser, O.; Müller, R.H. Lipid-drug conjugate (LDC) nanoparticles as novel carrier system for the hydrophilic antitrypanosomal drug diminazenediaceturate. J. Drug Target 2002, 10, 387–396.

Onyuksel, H.; Ikezaki, H.; Patel, M.; Gao, X.P.; Rubinstein, I. A novel formulation of VIP in sterically stabilized micelles amplifies vasodilation in vivo. Pharm. Res. 1999, 16, 155–160.

Pakunlu, R.I.; Wang, Y.; Tsao, W.; Pozharov, V.; Cook, T.J.; Minko, T. Enhancement of the efficacy of chemotherapy for lung cancer by simultaneous suppression of multidrug resistance and antiapoptotic cellular defense: novel multicompotent delivery system. Cancer Res. 2004, 64, 6214–6224.

Pardeike, J.; Hommoss, A.; Müller, R.H. Lipid nanoparticles (SLN, NLC) in cosmetic and pharmaceutical dermal products. Int J. Pharm. 2009, 366,170–184.

Park, J.W.; Hong, K.; Kirpotin, D.B.; Colbern, G.; Shalaby, R.; Baselga, J.; Shao, Y.; Nielsen, UB; Marks, JD; Moore, D.; et al. Anti-HER2 immiunoliposomes: enhanced efficacy attributable to targeted delivery. Clin. Cancer Res. 2002, 8, 1172–1181.

Peira, E.; Marzola, P.; Podio, V.; Aime, S.; Sbarbati, A.; Gasco, M.R. In vitro and in vivo study of solid lipid nanoparticles loaded with superparamagnetic iron oxide. J. Drug Target, 2003, 11, 19–24.

Peltier, S.; Oger, J.M.; Lagarce, F.; Couet, W.; Benoit, J.P. Enhanced oral paclitaxel bioavailability after administration of paclitaxel-loaded lipid nanocapsules. Pharm. Res., 2006, 23, 6, 1243–1250.

Podlogar, F.; Gasperlin, M.; Tomsic, M.; Jamnik, A.; Rogac, M.B. Structural characterization of water-Tween 40/Imwitor 308-isopropyl myristate microemulsions using different experimental methods. Int J. Pharm. 2004, 276, 115–128.

Proulx, M.E.; Desormeaux, A.; Marquis, J.F.; Olivier, M.; Bergeron, M.G. Treatment of visceral leishmaiasis with sterically stabilized liposomes containing camptothecin. Antimicorb. Agents Chemother. 2001, 45, 2623–2627.

Puglia, C.; Blasi, P.; Rizza, I.; Schoubben, A.; Bonina, F.; Rossi, C.; Ricc, M. Lipid nanoparticles for prolonged topical delivery: an in vitro and in vivo investigation. Int. J. Pharm. 2008, 357, 295–304.

Rabinow, B.E. Nanosuspensions in drug delivery. Nat. Rev. Drug Discov. 2004, 3, 785–796.

Rapoport, N.; Pitt, W.G.; Sun, H.; Nelson, J.L. Drug delivery in polymeric micelles: from in vitro to in vivo. J. Control. Rel. 2003, 91, 85–95.

Raviv, U.; Needleman, D.J.; Li, Y.; Miller, H.P.; Wilson, L.; Safnya, C.R. Cationic liposome-microtube complexes: pathways to the formation of two-state lipid-protein nanotubes with open or closed ends. Proc. Natl. Acad. Sci. 2005, 102, 111670–11172.

Rhaese, S.; von Briesen, H.; Rubsamen-Waigmann, H.; Kreuter, J.; Langer, K. Human serum albumin-polyethylenimine nanoparticles for gene delivery. J. Cont. Rel. 2003, 92, 199–208.

Rudin, C.M.; Marshall, J.L.; Haung, C.H.; Kindler, H.L.; Zhang, C.; Kumar, D.; Gokhale, P.C.; Steinberg, J.; Wanaski, S.; Kasid, U.N.; et al. Delivery of liposomal c-raf-1 antisense oligonucleotide by weekly bolous dosing in patients with advanced solid tumors: a phase I study. Clin. Cancer Res. 2004, 10, 7244–7251.

Santos-Magalhaes, N.S.; Pontes, A.; Pereira, V.M.W; Caetano, M.N.P. Colloidal carriers for benzathine penicillin G: nanoemulsions and nanocapsules. Int J. Pharm. 2000, 208, 71–80.

Seki, J.; Sonoke, S.; Saheki, A.; Fukui, H.; Mayumi, T. A nanometer lipid emulsion, lipid nano-sphere (LNS), as a parentral drug carrier for passive drug targeting. Int J. Pharm. 2004, 273, 75–83.

Sheikh, S.; Ali, S.M.; Ahmad, M.U.; Ahmad, A.; Mushtaq, M.; Paithankar, M; Mandal, J.; Saptarishi, D.; Sehgal, A; Maheshwari, K.; et al. Nanosomal amphotericin B is an efficacious alternative to Ambisome® for fungal therapy. Int. J. Pharm. 2010, 397, 103–108.

Sihorkar, V.; Vyas, S.P. Potential of polysaccharide anchored liposomes in drug delivery, targeting and immunization. J. Pharm. Sci. 2001, 4, 138–158.

Simoes, S.; Filipe, A.; Fanece, H.; Mano, M.; Penacho, N.; Duzgunes, N.; de Lima, M.P. Cationic liposomes for gene delivery. Expert Opin. Drug Deliv. 2005, 2, 237–254.

Soenen, S.J.H.; Cocquyt, J.; Defour, L.; Saveyn, P.; Van Der Meeren, P.; De Cuyper, M. Design and development of magnetoliposome-based theranostics. Mater. Manufac. Proc. 2008, 23, 611–614.

Solans, C.; Izquierdo, P.; Nolla, J.; Azemar, N.; Garcia-Celma, M.J. Nanoemulsion. Current opinion in Colloid and Interface Science. 2005, 10,102-110.

Sonvico, F.; Cagnani, A.; Rossi,A.; Motta, S.; Di Bari, M.T.; Cavatorta, F.; Alonso, M.J.; Deriu, A.; Colombo, P. Formation of self-organized nanoparticles by lecithin/chitosan ionic interaction. Int. J. Pharm. 2006, 324, 67–73.

Soppimath, K.S.; Aminabhavi, T.M.; Kulkarni, A.R.; Rudzinski, W.E. Biodegradable polymeric nanoparticles as drug delivery devices. J. Cont. Rel. 2001, 70, 1–20.

Star, A.; Stoddart, J.F.; Steuerman, D.; Diehl, M.; Boukai, A.; Wong, E.W.; Yang, X.; Chung, S.W.; Choi, H.; Heath, J.R. Preparation and properties of polymer-wrapped single-walled carbon nanotubes. Angew. Chem. Int. Ed. 2001, 40, 1721–1725.

Star, A.; Liu, Y.; Grant, K.; Ridvan, L.; Stoddart, J.F.; Steuerman, D.W.; Diehl, M.R.; Boukai, A.; Heath, J.R. Noncovalent side-wall functionalization of single-walled carbon nanotubes. Macromol. 2003, 36, 553–560.

Swenson, C.E.; Perkins, W.R.; Roberts, P.; Ahmad, I.; Stevens, R.; Steven, D.A. Janoff, A.S. In vitro and in vivo antifungal activity of amphotericin B lipid complex: are phopholipases important? J. Antimicrob. Chemother. 1998, 42, 767–771.

Tadros, T; Izquierdo, P.; Esquena, J.; Solans, C. Formation and stability of nanoemulsions. Adv. Colloid Interface Sci. 2004, 109, 303–318.

Thote, A.J.; Gupta, R.B. Formation of nanoparticles of a hydrophilic drug using supercritical carbon dioxide and microencapsulation for sustained release. Nanomed. 2005, 1, 85–90.

Tian, F.; Cui, D.; Schwarz, H.; Estrada, G.G.; Kobayashi, H. Cytotoxicity of single-wall carbon nanotubes on human fibroblasts. Toxicol. In Vitro. 2006, 20, 1202–1212.

Torchilin, V.P. Targeted polymeric micelles for delivery of poorly soluble drugs. Cell Mol. Life Sci. 2004, 61, 2549–2559.

Torchilin, V.P.; Lukyanov, A.N.; Gao, Z.; Papahadjopoulos-Sternberg, B. Immunomicelles: targeted pharmaceutical carriers for poorly soluble drugs. Proc. Natl. Acad. Sci. USA 2003, 100, 60039–6044.

Verschragen, C.F.; Gilbert, B.E.; Loyer, E.; Huaringa, A.; Walsh, G.; Newman, R.A.; Knight, V. Clinical evaluation of the delivery and safety of aerosolized liposomal 9-Nitro-20(s)-camptothecin in patients with advanced pulmonary malignancies. Clin. Cancer Res. 2004, 10, 2319–2326.

Vinogradov, S.V.; Batrakova, E.V.; Kabanov, A.V. Nanogels for oligonucleotide delivery to the brain. Bioconjug. Chem. 2004, 15, 50–60.

Wallace, E.J.; Sansom, M.S.P. Carbon nanotube self-assembly with lipids and detergent: a molecular dynamics study. Nanotechnol. 2009, 20, 045101.

Wissing, S.A.; Kayser, O.; Müller, R.H. Solid lipid nanoparticles for parenteral drug delivery. Adv. Drug Del. Rev. 2004, 56, 1257–1272.

Wu, W.; Wieckowski, S.; Pastorin, G.; Benincasa, M.; Klumpp, C.; Briand, J.P.; Gennaro, R.; Prato, M.; Bianco, A. Targeted delivery of amphotericin B to cells by using functionalized carbon nanotubes. Angew. Chem. Int. Ed. 2005, 44, 6358–6362.

Yang, W.R.; Thordarson, P.; Gooding, J.J.; Ringer, S.P.; Braet, F. Carbon nanotubes for biological and biomedical applications. Nanotechnology. 2007, 18, 412001.

Yih, T.C.; Al-Fandi, M. Engineered nanoparticles as precise drug delivery systems. J. Cell. Biochem. 2006, 97, 1184–1190.

Yokoi, T.; Iwamatsu, S.-I.; komai, S.-I.; Hattori, T.; Murata, S. Chemical modification of carbon nanotubes with organic hydrazines. Carbon. 2005, 43, 2869-2874.

Yotsuyanagi, T.; Hazemoto, N. Cationic liposomes in gene delivery. Nippon Rinsho. 1998, 56, 705–712.

Yuan, F.; Leunig, M.; Haung, S.K.; Berk, D.A.; Papahadjopoulous, D.; Jain, R.K. Microvascular permeability and interstitial penetration of sterically stabilized (stealth) liposomes in a human tumor xenograft. Cancer Res. 1994, 54, 3352–3356.

Zamboni, W.C.; Gervais, A.C.; Egorin, M.J.; Schllens, J.H.; Zuhowski, E.G.; Pluin, D.; Joseph, E.; Hamburger, D.R.; Working, P.K.; Colbern, G.; et al. Systemic and tumor disposition of platinum after administration of cisplatin or stealth liposomal-cisplatin formulations (SPI-077 and SPI-077 B103) in a preclinical tumor model of melanoma. Cancer Chemother. Pharmacol. 2004, 53, 329–336.

Zara, G.P.; Cavalli, R.; Bargoni, A.; Fundaro, A.; Vighetto, D.; Gasco, M.R. Intravenous administration to rabbits of non-stealth and stealth doxorubicin-loaded solid lipid nanoparticles at increasing concentrations of stealth agent: pharmacokinetics and distribution of doxorubicin in brain and other tissues. J. Drug Target, 2002, 10, 327–335.

Zhang, L.; Hou, S.; Mao, S.; Wei, D.; Song, X.; Lu, Y. Uptake of folate-conjugated albumin nanoparticles to the SKOV3 cells. Int. J. Pharm. 2004, 287,155–162.

Zhang, J.A.; Anyarambhatla, G.; Ma, L.; Ugwu, S.; Xuan, T.; Sardone, T.; Ahmad, I. Development and characterization of a novel Cremophor EL free liposome-based paclitaxel (LEP-ETU) formulation. Eur J. Pharm. Biopharm. 2005, 59, 177–187.

Zheng, M.; Jagota, A.; Semke, E.D.; Diner, B.A.; McLean, R.S.; Lusting, S.R.; Richardson, R.E.; Tassi, N.G. DNA-assisted dispersion and separation of carbon nanotubes. Nat. Mater. 2003, 2, 338–342.

8

LIPID NANO-VEHICLES BASED ON LYOTROPIC LIQUID CRYSTALS AS DRUG DELIVERY VEHICLES

Dima Libster, Abraham Aserin, and Nissim Garti
Casali Institute of Applied Chemistry, The Institute of Chemistry, The Hebrew University of Jerusalem, Edmond J. Safra Campus, Givat Ram, Jerusalem 91904, Israel

INTRODUCTION

Liquid crystals (LC) are self-assembled organized mesophases with properties intermediate to those of crystalline solids and isotropic liquids (Gin et al., 2008). In LC phases long-range periodicity exists, although the molecules exhibit a dynamical disorder at atomic distances, as is the case in liquids. Accordingly, these materials can also be considered ordered fluids (Larsson, 1989). Lyotropic LC (LLC) are materials that are composed from at least two molecules: an amphiphilic molecule and its solvent. A hydrophilic solvent, such as water, hydrates the polar moieties of the amphiphiles via hydrogen bonding, while the flexible aliphatic tails of the amphiphiles aggregate into fused hydrophobic regions based on van der Waals interactions. In addition to morphologic dependence on the chemical composition, LLC are also sensitive to external parameters, such as temperature and pressure (Gin et al., 2008; Larsson, 1989; Amar-Yuli, 2008). As a function of the molecular shape of the surfactants, packing parameters, and interfacial curvature energy considerations, LLC can be formed with aqueous domains ranging from planar bilayer lamellae to extended, cylindrical channels to three-dimensional interconnected channels and manifolds (Gruner, 1989). These mesophases are defined as lamellar (L_α), hexagonal (H), bicontinuous cubic [Q (or V)], and discontinuous cubic (I) phases, based on their symmetry (Seddon, 1990). In addition, most lyotropic mesophases exist as symmetric pairs; a "normal" (type I) oil-in-water system, consisting of lipid aggregates in a continuous water matrix; and a topologically "inverted" (type II) water-in-oil version. The headgroups hydrated by water are arranged within a continuous nonpolar matrix, which is composed of the fluid hydrocarbon chains (Seddon, 1990). In addition to their biological significance, inverse lipid phases could be useful as host systems for incorporation of food additives (Sagalowicz et al., 2006a; 2006b), the crystallization of membrane proteins for drug delivery (Sagalowicz et al., 2006b; Cherezov et al., 2006), and for inorganic synthesis (Boyd et al., 2007).

The lamellar structure does not possess any intrinsic curvature and is considered as the midpoint of an ideal, symmetrical LLC phase progression (Fig. 8.1) (Gruner, 1989; Seddon, 1990). The current review is mainly focused only on the inverted (W/O) mesophases (cubic and hexagonal), representing an important class of nanostructures for potential applications.

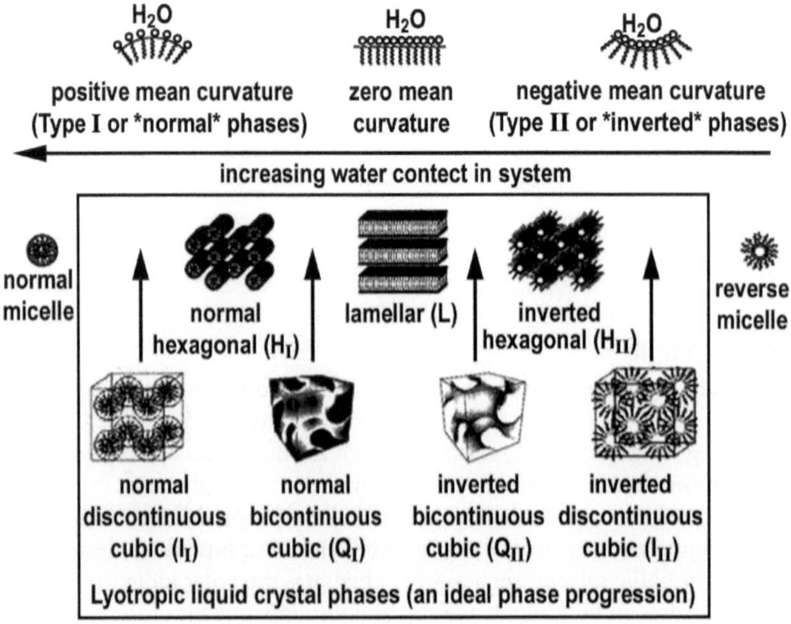

Fig. 8.1. Schematic representations of common LLC-phases formed by amphiphiles in water (Gin et al., 2008).

CUBIC PHASES

The bicontinuous cubic phase is the most complicated among all the LLC. Here the lipids are located in a complex, optically isotropic three-dimensional lattice. Cubic structures can be either normal (type I, O/W) or inverse (type II, W/O), and their topology is differentiated as bicontinuous or micellar, resulting in seven cubic space groups. Only three space groups are both inverse and bicontinuous structures (Fig. 8.2): the gyroid (G) type (*Ia3d*, denoted Q^{230}), the diamond (D) type (primitive lattice *Pn3m*, denoted Q^{229}), and the primitive (P) type (body-centered lattice *Im3m*, denoted Q^{224}) (Gruner et al., 1988). The aqueous channels of the G-surface consist of two separate, left-handed and right-handed helical channels. The aqueous channels can extend through the matrix, but the centers of the water channels never intersect. Hence they are connected and the structure adopts a helical arrangement. The structure D is characterized by a bilayer, which separates two interpenetrating aqueous channel systems forming a diamond lattice. In this configuration four aqueous channels of the D-surface meet at a tetrahedral angle of 109.5°. Structure P contains two aqueous channel systems that are separated by a bilayer. The unit cell possesses three mutually perpendicular aqueous channels, connected to contiguous unit cells, taking shape of a cubic array.

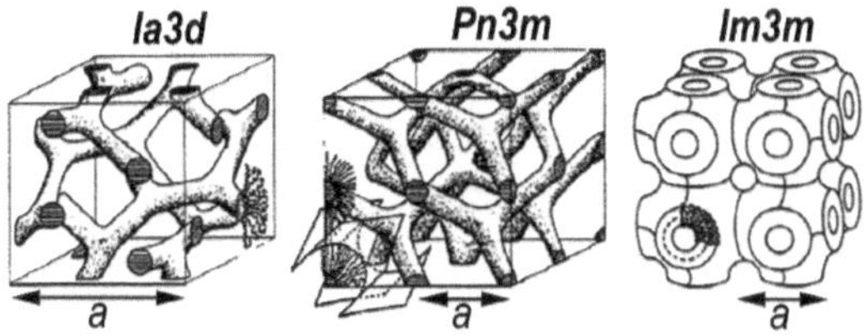

Fig. 8.2. Schematic illustrations of the inverse bicontinuous cubic phases *Ia3d* (gyroid type), *Pn3m* (diamond type), and *Im3m* (primitive type) (Seddon et al., 2000).

REVERSE HEXAGONAL MESOPHASE

The reverse hexagonal mesophase is of a primitive type (*P6mm*) and characterized by one cylinder per unit cell corner. These densely packed, straight, water-filled cylinders exhibit two-dimensional ordering. Each cylinder is surrounded by a layer of surfactant molecules that are perpendicular to the cylinder interface such that their hydrophobic moieties point outward from the water rods (Fig. 8.3). There is a growing indication that inverse hexagonal mesophases play structural and dynamic roles in biological systems (Gradzielski, 2004; Seddon, 1990; Yang et al., 2003). These systems are assumed to be active as transient intermediates in biological phenomena that require topological rearrangements of lipid bilayers, such as membrane fusion/fission and the trans-bilayer transport of lipids and polar solutes

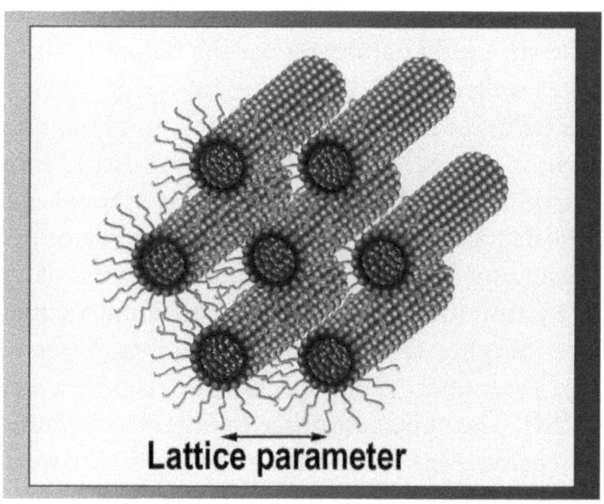

Fig. 8.3. Schematic presentation of H_{II} mesophase showing the packing of water-filled rods surrounded by lipid layers.

(Gradzielski, 2004; Seddon, 1990; Yang et al., 2003). H_{II} mesophases have recently been considered promising drug delivery vehicles, mainly owing to their unique structural features (Boyd et al., 2006; Boyd et al., 2007).

CUBIC MICELLAR MESOPHASES

Cubic micellar structures are less widespread than bicontinuous cubic. Discrete micellar mesophases are made of micellar amphiphilic nanostructures organized in 3-dimensional (3D) structures/lattices with cubic symmetry. The monolinolein/water system, upon tetradecane addition, was shown to exhibit an additional LC phase not present in the monolinolein/water binary mixture: an inverse discrete micellar cubic phase of the *Fd3m* type (Q_{227} group), as revealed by Yaghmur et al. (2006). The structure of the *Fd3m* micellar cubic phase was composed of two types of micelles with different sizes organized in a double diamond network. These micelles are not spherical but showed a dodecahedron and a hexakaidecahedron shape, respectively. In this way they minimized the high packing frustration constraints on the alkyl tails associated with the *Fd3m* arrangement (Pouzot et al., 2007). This mesophase was also observed in other systems upon oil addition, such as block copolymers (Alexandridis et al., 1996; Uddin et al., 2001), or phospholipid/fatty acid systems (Luzzatti et al., 1992). It is generally agreed that formation of an *Fd3m* micellar phase is promoted in systems where lipids have a negative preferred curvature when long alkyl hydrophobic tails are associated with weakly hydrated hydrophilic heads (Seddon et al., 1996).

Efrat and colleagues included ethanol (EtOH) in the binary GMO/water system and obtained a micellar mesophase of cubic symmetry with significantly lower viscosity (Efrat et al., 2007; 2008). The newly designed mesophase was termed Q_L—a fluid, liquid discontinuous cubic mesophase. It was found that EtOH influences the structural parameters of the bicontinuous cubic phases at a constant GMO/water weight ratio of 0.82 by increasing the area per lipid at the headgroup (a_0). The radius of the water channels (R_{wat}) also increased significantly as the ethanol content increased, but the mean curvature ($\langle H \rangle$) was reduced. These changes in the structural parameters indicated that ethanol mainly affected the polar region of the surfactant since it was accommodated around the headgroups. The rheological measurements showed that the new phase is of low viscosity (36.6 Pas in comparison to around 10^5 Pas of the bicontinuous cubic phase). The intermicellar structure complied with the so-called I_2 mesophase with primitive (P) symmetry. Its lattice parameter (the d-space) is *ca.* 105Å (based on synchrotron SAXS, and cryo-TEM). The new mesophase maintained its cubic order, although it exhibits a low viscosity, even lower than the values that were reported for I_2. Efrat and colleagues also studied the solubilization effects of a variety of bioactive guest molecules on the Q_L mesophase at various compositions and loads, including coenzyme (CoQ_{10}), cholesterol, phytosterols, and lycopene (Efrat et al., 2008; 2009).

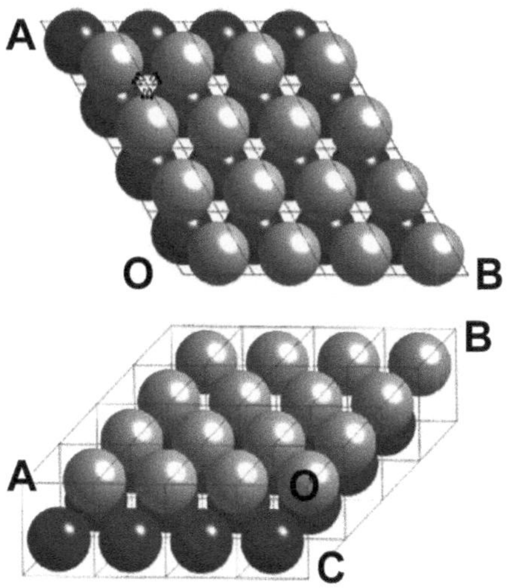

Fig. 8.4. Plan and perspective views of the schematic structure of the 3-D hexagonal inverse micellar phase. The spheres represent the polar regions (water cores plus lipid headgroups), and the remaining fluid volume is filled by the hydrophobic regions of the lipid molecules. The different shading of the two identical layers of spheres is purely for clarity (Shearman et al., 2006.).

NOVEL HEXAGONAL MESOPHASES

3D HEXAGONAL INVERSE MICELLAR PHASE

Another LC structure, based on a close packing of inverse micelles, is a 3-D hexagonal inverse micellar phase, of space group *P63/mmc*, recently discovered by Shearman et al. (2006). Ternary lipid mixtures comprising two lipids, dioleoylphosphatidylcholine (DOPC), dioleoylglycerol (DOG), and cholesterol, of molar ratios 1:2:1 and 1:2:2, in excess water induced formation of 3-D hexagonal mesophase over a temperature range of 16–52°C and a wide range of pressures (1–3000 bar). The new inverse micellar phase consists of an hexagonal closed-packed (hcp) packing of identical, spherical inverse micelles, with two lattice parameters, a and c, that were found to be a = 71.5 Å and c = 116.5 Å (Fig. 8.4). This is the first new inverse LLC phase, reported in two decades and it is the only known phase whose structure consists of a close packing of identical inverse micelles.

3-DOMAIN HEXAGONAL MESOPHASE

de Campo et al. (2011) reported on a novel lyotropic hexagonal mesophase formed by triphilic star-polyphiles, resulting in three distinct immiscible domains of hydrophilic/oleophilic/fluorophilic rods arranged in a 12.6.4. tiling. Triphilic star-

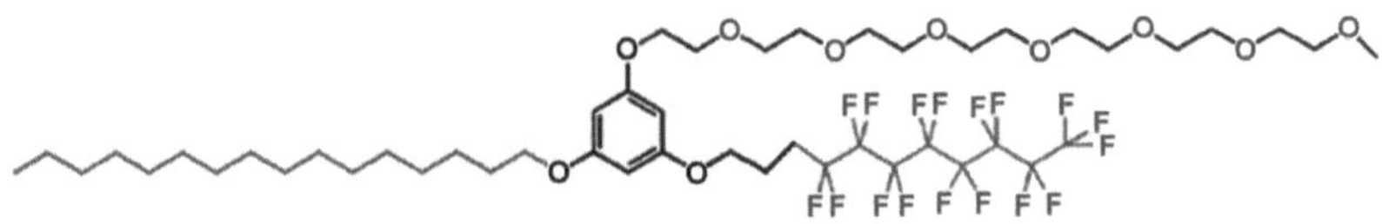

Fig. 8.5. Molecular structure of the star-polyphiles used in this study, abbreviated as B–O_a–H_b–F_c, where B denotes the aromatic molecular core, and O/H/F the methyl-capped oligoethyleneglycol, hydrocarbon and fluorocarbon chains of lengths *a*, *b*, and *c*, respectively (de Campo et al., 2011).

polyphiles are short-chain oligomeric molecules with a radial arrangement of hydrophilic, hydrocarbon, and fluorocarbon chains linked to a common centre (Fig. 8.5).

Fig. 8.6 shows the synchrotron SAXS patterns for the 25% water samples containing B–O_7–$(H_{16})_2$, B–O_7–H_{16}–F_{10} and B–O_7–H_{14}–F_8. From ≈25 to 35°C the surfactant, and from ≈30 to 42°C the star-polyphiles exhibit a lyotropic liquid crystalline phase that is consistent with 2D hexagonal symmetry (*p6mm*), evidenced by at least 3 peaks in the typical peak ratio of √1 √3 √4 √7 √9 √12 (de Campo et al., 2011). Microscopy images for all these hexagonal phases showed fan-like textures, typical of conventional hexagonal mesophases, when grown slowly in a flooding experiment. The underlying structure of B–O_7–$(H_{16})_2$ can be tentatively identified from the SAXS and optical texture data as a typical type-2 hexagonal phase (i.e., a 2D-columnar phase that is composed of hydrophilic cylinders hexagonally arranged in a hydrophobic matrix).

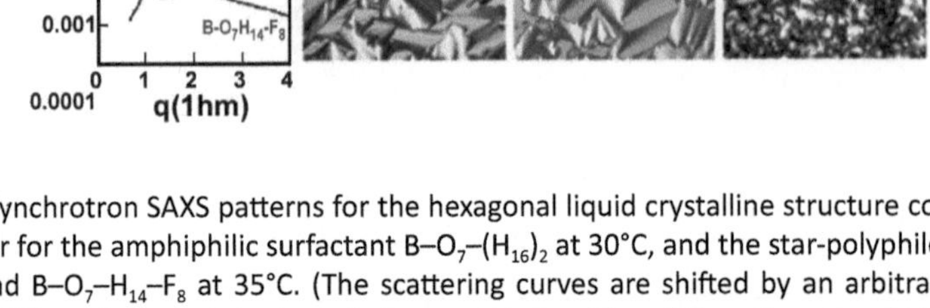

Fig. 8.6. Synchrotron SAXS patterns for the hexagonal liquid crystalline structure containing 25% water for the amphiphilic surfactant B–O_7–$(H_{16})_2$ at 30°C, and the star-polyphiles B–O_7–H_{16}–F_{10} and B–O_7–H_{14}–F_8 at 35°C. (The scattering curves are shifted by an arbitrary factor for clarity.) The microscopic textures between polarizers for all these hexagonal phases are typical fan-like patterns. B–O_7–$(H_{16})_2$ and B–O_7–H_{14}–F_8 liquid crystals were grown slowly in a water flooding experiment, while the B–O_7–H_{16}–F_{10} liquid crystal shown here resulted from re-crystallization from the melted state (de Campo et al., 2011).

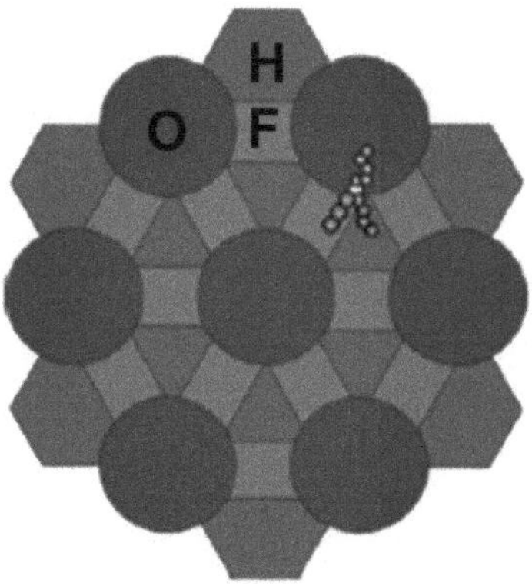

Fig. 8.7. Schematic picture of the star-polyphile molecular arrangement in the 12.6.4 tiling found for the star-polyphile hexagonal mesophases. The dimensions of the green fluorocarbon domains compared with the molecular dimensions requires some interdigitation of fluorocarbon chains (de Campo et al., 2011).

The dimensions of the hydrophilic channels estimated from the SAXS data suggested slight segregation within the hydrophilic domains. The $B–O_7–(H_{16})_2$ amphiphile hexagonal phase has a unit cell dimension of 7.7 nm, and with 46% v/v hydrophilic domain, the hydrophilic channel radius is estimated to be 2.7 nm. In contrast, SANS data for the star-polyphiles offered clear evidence that these systems are effectively segregated into three distinct domains at the mesoscale with spatially distinct hydrophilic, oleophilic, and fluorophilic domains (with minor additional SLD-inhomogeneities). The simplest hexagonal mesostructure consistent with the molecular star-architecture, and in very good agreement with the experimental data, is a hexagonal arrangement of infinite columns whose cross-section is the planar 12.6.4 tiling. It was inferred by the authors that the hydrophilic domain fills the large dodecagonal prismatic domains, the hydrocarbon chains fill the hexagonal prisms, and the fluorocarbon chains fill the quadrangular prisms (Fig. 8.7).

PHASE BEHAVIOR

Most of the common surfactants form direct (type I) phases, where the interface bends away from the polar solvent, toward the tail region. However, most biological amphiphiles (such as phospholipids) form "type II" LLC, where the interface curves toward the polar region. Only a few synthetic amphiphiles can mimic the behavior of biological lipids and form inverted mesophases. The unsaturated monoglycerides (monoolein and monolinolein) belong to this category (Drummond & Fong, 1999). Glycerol monooleate (GMO, monoolein) is a polar lipid that is commonly

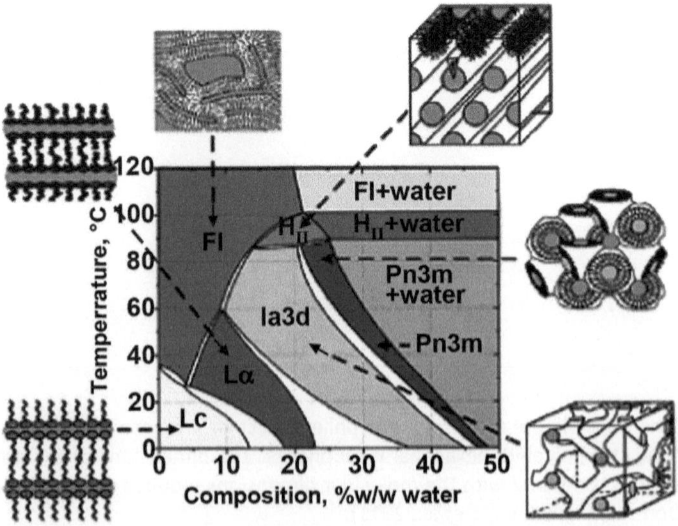

Fig. 8.8. Temperature-composition phase diagram of the monoolein/water system (up to 50 wt % water). A cartoon representation of the various phase states is included in which colored zones represent water. The mesophases are as follows: L_c-crystalline lamellar, L_α-lamellar, *Ia3d*-gyroid inverted bicontinuous cubic, *Pn3m*-primitive inverted bicontinuous cubic, H_{II}-inverted hexagonal, and F1-reverse micelles fluid phases (Cherezov et al., 2006).

used as a food emulsifier. Presently, GMO is the preferred amphiphile for formulating LC phases for scientific research and drug delivery. This is a non-toxic, biocompatible, and biodegradable lipid, which possesses low water solubility, but in excess water it swells and forms several LC phases.

Monoolein/water is one of the most extensively investigated types of lipid-based mesophases (Drummond & Fong, 1999). The phase diagram of this system (Fig. 8.8) revealed complex structural behavior. At room temperature the following phase sequence existed upon increasing hydration: lamellar crystalline phase (L_c) in coexistence with an L_2 phase, lamellar mesophase (L_α), and the inverted bicontinuous cubic mesophases-gyroid *Ia3d* and diamond *Pn3m*. Upon heating to about 85°C, the cubic phase is transformed into the H_{II} mesophase, followed by the micellar phase.

These concentration- and temperature-dependent structural transitions can be qualitatively explained in terms of effective critical packing parameter (CPP), as developed by Israelachvili et al. (1976). According to this theory, amphiphiles possess geometric parameters characterized by the critical packing parameter, CPP = V_s/a_0l, where V_s is the hydrophobic chain volume, a_0 is the polar headgroup area, and l is the length of the chain in its molten state. The packing parameter is useful in pre-

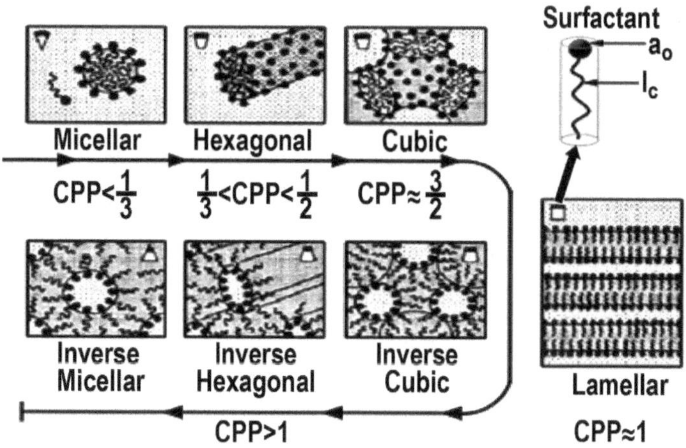

Fig. 8.9. Schematic presentation of classic mesophases, their corresponding CPP, and the "generic sequence" with increased water content (Efrat, 2009).

dicting which phases can be preferentially formed by a given lipid, since it connects the molecular shape and properties to the favored curvature of the polar–apolar interface and, therefore, the topology and shape of the aggregate. The main factors responsible for alterations of the mentioned parameters are the molecular shape of the surfactant, the chemical composition, and the temperature. According to the theory developed by Israelachvili and coworkers, inverse mesophases are formed by amphiphiles with CPP >1, which adopt inversed cone shaped geometry. Such lipids should possess a small headgroup area as compared to that of the tail region. In contrast, direct mesophases are preferred when the surfactant head cross-section is larger than that of the tail, resulting in CPP <1. In the case of lamellar mesophase these parameters are equal and, hence, result in CPP=1. The size of the polar headgroup area is dictated by both molecular shape and the degree of hydration.

Within the boundaries of a given mesophase the headgroup's area normally increases with increasing hydration and decreases upon temperature rise. The length and the volume of the lipophilic tail are also greatly affected by a temperature increase. Stronger thermal motion of the tail eventually decreases the length of the tails but increases their volume.

Usually the phase sequence follows a certain order; the "generic sequence" is reverse micellar → micellar cubic → reverse hexagonal → reverse bicontinuous cubic → lamellar → bicontinuous cubic → hexagonal → micellar cubic → micellar (Fig. 8.9) (Efrat, 2009).

It was shown for monoglyceride-based LLC that H_{II} mesophases are formed from amphiphiles with CPP~1.7, cubic phases with CPP~1.3, and lamellar structures with CPP=1 (Larsson, 1989). The lamellar phase is characterized by zero curvature since the cross sections of the polar heads and the lipophilic tails are similar. Upon increased hydration, cubic phases (*Ia3d* and *Pn3m*) with higher curvature and consequently higher CPP are preferred. The effective CPP theory can supply a reasonable explanation for the temperature-induced structural shifts from lamellar through cubic to reverse hexagonal phases, requiring higher curvature than in the lamellar phase. Increasing the thermal motion of both the hydrocarbon chains and the water molecules would increase the CPP values via expanding the volume of the lipophilic moiety but decreasing the chain length and the headgroup area. This leads to an increase in curvature and therefore induces the formation of cubic and hexagonal mesophases.

It should be noticed that the CPP concept may only be used as a rule of thumb to predict transitions between various forms of LLC. For example, it was shown that it fails to predict the appearance of the inverse bicontinuous phases and an intermediate phase (Shah & Paradkar, 2005; Shearman et al., 2006). This is the reason that a thermodynamic approach considering the total free energy of the LLC systems was applied by Seddon and coworkers (Seddon et al., 1996; 2000; Shearman et al., 2006), however, this is beyond the scope of this chapter.

H_{II} MESOPHASE COMPOSED OF GMO/TRIGLYCERIDE/WATER

It was shown that in the monoolein-based system the cubic phase is transformed into an H_{II} mesophase upon heating at ca. 85°C (Drummond & Fong, 1999). CPP theory can supply a reasonable explanation to the temperature-induced structural shifts from lamellar through cubic to reverse hexagonal phases, requiring greater curvature than in the lamellar phase. Increasing the thermal motion of both the hydrocarbon chains and the water molecules would increase the CPP values via expanding the volume of the lipophilic moiety but decrease the chain length and the headgroup area. This leads to an increase in curvature and therefore induces the formation of cubic and hexagonal mesophases. In addition, the hexagonal mesophase is characterized by greater packing cost than the cubic phase, but the opposite is true for curvature elastic energy. Therefore, elevated temperatures induced the tendency for interfacial curvature, which increased the curvature elastic costs of the bicontinuous cubic phase, stabilizing hexagonal symmetry (Gin et al., 2008; Israelachvili et al., 1976).

Bearing in mind the thermodynamic and structural considerations noted above, systematic research was conducted to decrease the cubic to hexagonal temperature transition and stabilize the GMO-based H_{II} mesophase at room temperature (Amar-Yuli & Garti, 2005). To achieve this goal, experiments relative to incorporation of triglycerides (TAGs) with various chain lengths to the binary GMO/water system were conducted. Amar-Yuli and Garti surmised that immobilization of a TAG between GMO tails would lead to a change in the geometry of monoolein molecules

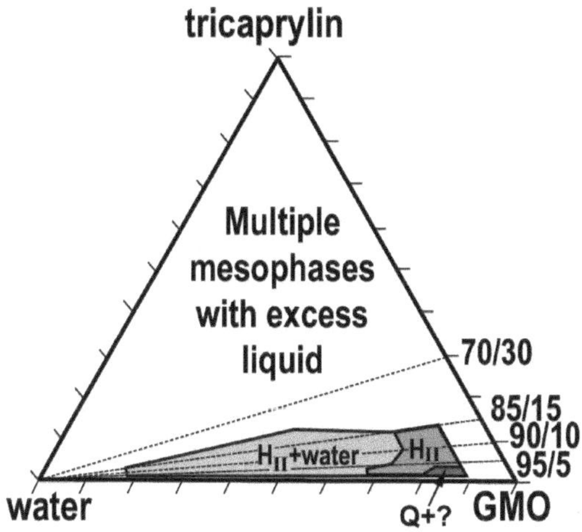

Fig. 8.10. The ternary phase diagram of GMO/tricaprylin/water at 25°C. The dilution lines represent the surfactant/oil weight ratio. (Amar-Yuli & Garti, 2005

from cylindrical to wedge-shaped and thereby to an increase in the CPP value of the system (Amar-Yuli & Garti, 2005). This should encourage transition from lamellar or cubic phases to hexagonal structures. In addition, the immobilization of TAG in the system was expected to reduce the packing frustration, stabilizing the hexagonal LLC at room temperature. These experimental results showed that a critical and optimal chain length of the triglyceride is required to induce the formation of H_{II} at room temperature. Among the examined TAGs, tricaprylin (C_8) attained the largest region of hexagonal structure in the phase diagram (Fig. 8.10), indicating its flexible accommodation between the tails of the GMO. It was suggested that the high mobility of tricaprylin in the surfactant aggregate provided direct transformation of the lamellar mesophase to the hexagonal phase, avoiding the formation of a cubic structure.

The structural properties of ternary hexagonal mesophases composed of GMO, tricaprylin, and water were extensively and systematically studied, as was shown in numerous publications (Amar-Yuli et al., 2007a; 2007b; 2008a; 2008b; 2009). Several additives, including dermal penetration enhancers, were solubilized to control the physical properties of these carriers (Amar-Yuli et al., 2007b; 2008a; 2008b), such as viscosity and thermal stability. For instance, the synergistic solubilization of two major hydrophilic (vitamin C, ascorbic acid, AA) and lipophilic (vitamin E, D-alpha-tocopherol, VE) antioxidants within H_{II} mesophases was reported by Bitan-Cherbakovsky et al. (Bitan-Cherbakovsky et al., 2009; 2010). This enabled expanding conditions to obtain stable H_{II} mesophase at room temperature. In addition, it

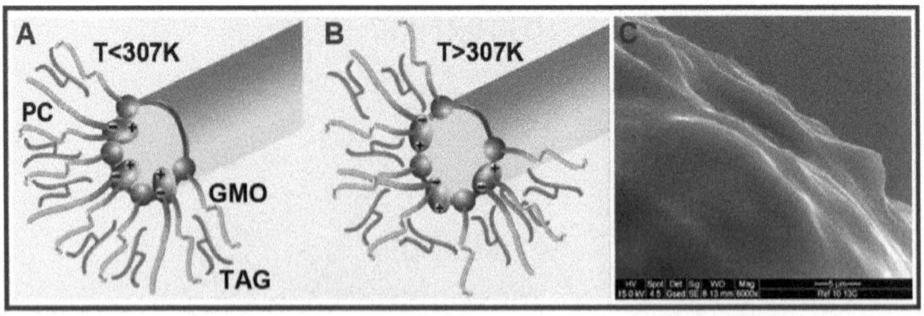

Fig. 8.11. (A) Schematic illustrations focused on one cylinder of the mixed surfactant (GMO and PC) hexagonal system organization. Note that GMO and PC polar moieties are hydrated but that TAG is located between the lipophilic chains of the surfactants. Below T = 307 K, interfacial water stabilizes the lipid layer via its interaction with the GMO head γ–OH group. Consequently, the TAG molecule remains intercalated between the tails. Dangling tails of the PC group give rise to counter ion movement as well as TAG dipolar relaxation. (B) Above T = 307 K, the interface is destabilized by the breakdown of the interfacial water layer. The perturbation of the PC heads leads to greater dangling, enabling diffusive movement of the TAG molecule through the tails (Ben Ishai et al., 2009). (C) ESEM micrographs of H_{II} mesophase of GMO/tricaprylin/PC (63, 7, and 10 wt %, respectively) containing 20 wt % water at 14°C and magnification ×8000 (Libster et al., 2008).

was shown that phosphatidylcholine (PC) can be embedded into the ternary GMO/TAG/water system (Ben Ishai et al., 2009; Libster et al., 2007; 2008; 2009a). PC was incorporated into the ternary mesophases, as it is known that phospholipid-based structures possess relatively high thermal stability and enhance transdermal drug permeation (Kurosaki et al., 1991; Spernath et al., 2006; Touitou et al., 1994) and transmembrane transport across the digestive tract (Brondsted et al., 1995; Liu et al., 1999).

Incorporation of PC into the ternary system (Fig. 8.11.A) caused competition for water binding between the hydroxyl groups of GMO and the phosphate groups of the phosphatidylcholine (PC), leading to dehydration of the GMO hydroxyls in favor of the phospholipid hydration (Libster et al., 2009a). On the macroscopic level, this was correlated with improvement of elastic properties and thermal stability of the H_{II} mesophase (Libster et al., 2007). Structural flexibility and stability are essential for a drug delivery system, especially for tuning its physical properties (such as viscosity) and composition. Dielectric spectroscopy analysis of this system showed a critical temperature, T_0 = 307 K, related to the dehydration of the GMO head groups (Ben Ishai et al., 2009). Effectively, this represented a breakdown of the interfacial layer of water. A physical picture emerged whereby at T_0 = 307 K, the “loosening ” of the GMO heads accentuates the dangling motion of the phosphatidylcholine (PC) tails, as evidenced by counterion motion along the PC head (Fig. 8.11.B). Furthermore, it precipitates the percolation of the large TAG molecules that are intercalated in the GMO and PC tails (Ben Ishai et al., 2009).

Libster et al. (2008) also explored and demonstrated correlations between the microstructural and mesostructural properties of the reverse hexagonal LLC using ESEM technique. It was shown that the mesoscopic organization of these systems is based on an alignment of polycrystalline domains. The topography of H_{II} mesophases, imaged directly in their hydrated state, as a function of aqueous phase concentration, was found to possess fractal characteristics, indicating a discontinuous and disordered alignment of the corresponding internal water rods on the mesoscale (Fig. 8.11.C). Fractal analysis indicated that the mesoscale topography of the H_{II} phase was likely to be influenced by the microstructural parameters and the water content of the samples (Libster et al., 2008). The fractal topology can potentially influence drug delivery through the large surface area of these carriers. Garti and coworkers also made considerable progress elucidating the solubilization of therapeutic peptides into the H_{II} mesophase (Cohen-Avrahami et al., 2010; Libster et al., 2007; 2009a; 2009b; Mishraki et al., 2005a; 2005b).

MAIN APPLICATIONS OF H_{II} MESOPHASES

SOLUBILIZATION AND DELIVERY OF BIOMACROMOLECULES

Solubilization of biomacromolecules into LLCs is of great practical interest. Mezzenga et al. (2005) investigated the effects of confining polymeric forms of glucose in the water domains of liquid crystals composed of monoglyceride and water solutions. Using rheological and SAXS methods the authors studied the effect of sugar concentration and molecular weight on cubic-to-H_{II} transitions by further solubilization of maltose and dextran I in the system. It was found that higher concentrations of sugars resulted in a decrease of the *Pn*3*m*–H_{II} transition temperature as a consequence of the Hofmeister effect. However, no dependence on molecular weight of the sugar was observed. In the case of a polymeric sugar, as soon as the end-to-end distance of the polymer approached the diameter of water channels its molecular weight made it more difficult for the guest molecule to fit within narrow water channels. Dextrans of larger molecular weights than dextran I did not fit into the water channels of the cubic phases, inducing a phase separation. Moreover, it was noticed that dextran I, solubilized into the *Ia*3*d* phase channels, induced the appearance of a *Pn*3*m* phase, which possesses larger water channels. Hence, when the chain end-to-end distance of the polysaccharide approached the diameter of the water channels, it induced phase transitions toward structures with different topologies, thus enhancing the solubilization capacity of large sugars.

Peptides and proteins are increasingly considered for the development of new therapeutic compounds (Daas et al., 2006; Kumar et al., 2006). Peptides of various sizes are currently tested for a broad spectrum of diseases, including skin cancer, acne, psoriasis, hypertension, hepatitis, and rheumatism. These biomacromolecules are usually found to be very effective, and in most cases low doses of them are needed for good medical treatment. Nevertheless, when delivered orally these peptides and proteins are susceptible to cleavage by various enzymes, mainly the human digestive proteases, leading to practically poor bioavailability and poor pharmaceutical efficacy. As a result of these critical limitations, excessively large doses of the peptide-based drugs are usually required to obtain therapeutic effects in vivo, which in turn often cause a wide range of hazardous side effects (Daas et al., 2006; Kumar et al., 2006). This is the main reason that the medical application of these otherwise high potential therapeutics is currently very limited and even those few that are used at the present time are far from being efficient. In this respect, LLCs seem to be promising candidates as alternative delivery means for various pharmaceuticals. Such vehicles can provide enhanced drug solubility, relative protection of the solubilized drugs, and controlled release of drugs while avoiding substantial side effects (Ericsson et al., 1991; Fong et al., 2009; Shah et al., 2001). The efficiency of the LLC delivery system can be further enhanced in transdermal delivery, where peptide degradation and low absorption rates are much less relevant.

Significant progress has been made during the last decade in the characterization of the interactions of peptides and proteins with LLCs, mainly for crystallographic and drug delivery purposes. LLCs have been shown to provide sustained release of drug molecules with a range of physicochemical properties (Drummond & Fong, 1999; Shah & Paradkar 2005; Shah et al., 2001). The cubic phase was shown to host a range of water-soluble biomacromolecules for use in controlled release applications. Clogston and Caffrey (2005) systematically examined biomacromolecules ranging from a single amino acid (tryptophan) to complex proteins and nuclear DNA. These investigators found that for a given cubic phase, the rate of diffusion depends on the molecular size of the specific diffusing molecule. Shah and Paradkar (2005) prepared *in situ* a cubic phase system of GMO. This system provided protection to the metalloenzyme seratiopeptidase (STP) in the gastric environment and gave delayed and controlled release with no initial burst after oral administration.

Although the cubic phase has been proposed and studied as a drug delivery vehicle, there is very little information about the interactions of peptides and proteins with the H_{II} mesophases and the therapeutic potential of these structures. Garti and coworkers explored and controlled the physical properties of H_{II} mesophases to use these systems as drug delivery vehicles for biologically active peptides and proteins. The results of this structural research enabled significant expansion of the application spectrum of hexagonal LLC, utilizing them for the solubilization of peptides and proteins (Libster et al., 2007; 2009a; 2009b; 2011) and low molecular weight drugs, such as gabapentin (Achrai et al., 2011), into this mesophase and its utilization as sustained drug delivery vehicle. Two model cyclic peptides, cyclosporin A (11 amino acids) and desmopressin (9 amino acids), of similar molecular weight but with very different hydrophilic and lipophilic properties were chosen to demonstrate the feasibility of using the H_{II} mesophase (Libster et al., 2011). In addition, a larger peptide, RALA (Cohen-Avrahami et al., 2010) (16 amino-acids), was solubilized into the H_{II} structures as a model skin penetration enhancer. Finally, a larger macromolecule,

LSZ protein (129 amino acids), was directly incorporated into a GMO-based H_{II} mesophase. (Mishraki et al., 2010a; 2010b)

DESMOPRESSIN SOLUBILIZATION AND DELIVERY

Desmopressin is a synthetic analogue of the antidiuretic hormone vasopressin, which is mainly used for treatment of enuresis in young children, central diabetes insipidus, hemophilia A, von Willebrand disease, and trauma-induced injuries (Pedersen et al., 2002). The location and mutual interactions of the desmopressin guest peptide with the host mesophase was elucidated by Libster et al. (2009b). The study also involved examination of the conformational and structural changes of the peptide caused by its immobilization within the carrier. In these experiments relatively high concentrations of the peptide (up to 10 wt%) were solubilized into the hexagonal mesophase system. The concentration-dependent SAXS measurements revealed a significant increase (up to 6 Å) in the lattice parameter of the mesophases as a result of the peptide incorporation. This chaotropic effect of the peptide was thought to be related to its interactions with hydroxyls of GMO in the outer interface region, as was inferred from the ATR-FTIR examination. It was found that neither the inner shell of the interface nor the acyl chains were affected by the peptide. Desmopressin molecules probably did not penetrate deeply enough into the interface region of the mesophase. As a result, they could not interact with the inner parts of the host GMO molecules, including the carbonyl groups, the ester moiety, and the lipophilic parts. Further, we tried to follow specific alterations in the peptide conformation, as induced by its interaction with GMO.

The desmopressin molecule consists of two parts: a cyclic part of the polypeptide chain involving a β-turn or a γ-turn structure and a highly flexible part consisting of the acyclic tail (Walse et al., 1998). This configuration gives this peptide a relatively large degree of flexibility. It was found that the dominant β-turn elements, unordered structures of the peptide and part of its "free" backbone carbonyls, remained almost unchanged as a result of desmopressin partitioning within the hexagonal mesophase. However, the interactions between desmopressin and the GMO were reflected in the conformation of the peptide when there was weakening or opening of the γ-turns and partial binding of the peptide free carbonyl groups.

It was indicated in the literature that the diffusion of a drug in the channels of the cubic mesophase is much faster than from the hexagonal phase (Boyd et al., 2006). This important issue was elaborated in the case of desmopressin. It was assumed that the diffusion of the peptide in the hexagonal phase, compared to water solution, was slowed by two factors. The first was the physical restriction of the peptide motion owing to the geometrical constraint of hexagonal architecture, more specifically to its diffusion within the water cylinders. The second factor is the chemical interactions of the peptide with the polar heads of monoolein. These two effects were separated and quantified by SD-NMR analysis (Fig. 8.12.A).

Renkin's model (Pluen et al., 1999) was used to calculate the theoretical diffusion coefficients of desmopressin within the channels of H_{II} mesophase, assuming

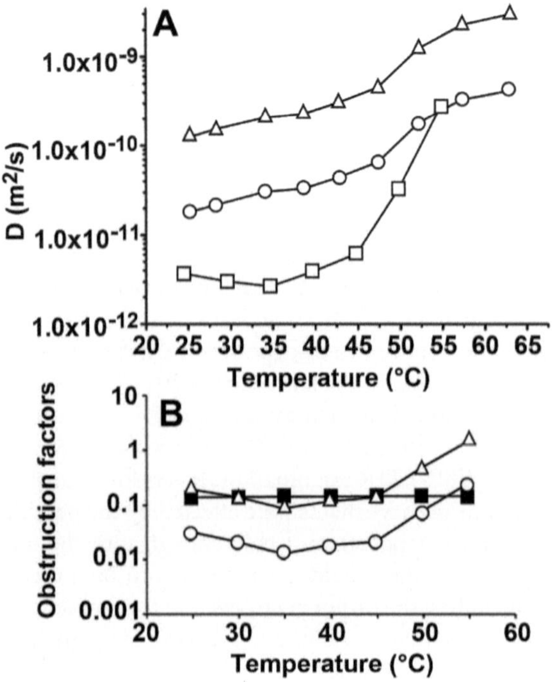

Fig. 8.12. (A) Desmopressin diffusion coefficients as a function of temperature in (Δ) 8 wt% in D_2O solution (O) according to the Rankin model, (□) 8 wt% in HII mesophase as a function of temperature. (B) Obstruction factors calculated from the SD-NMR analysis as a function of temperature. (O) The overall obstruction factor (λ); (■) β is the obstruction factor responsible for the physical restriction of the peptide in the hexagonal structure; (Δ) γ is the obstruction factor associated with the chemical interactions between the peptide and GMO (Libster et al., 2009b).

no interactions of the peptide with GMO and TAG. Using the theoretical diffusion coefficient of the drug and the measured diffusion values, the observed decrease in diffusion coefficients of the peptide in the H_{II} mesophase was clarified defining two obstruction factors. The obstruction factors (Fig. 8.12.B) enabled us to quantify the effects of both physical restriction (β) and chemical interactions (γ). With this approach, it turned out that between 25°C and 45°C both β and γ factors hold similar values, indicating that the physical restriction and the chemical interactions contributed almost equally to the decrease in the mobility of the peptide within the H_{II} mesophase compared to the water solution. However, above 45°C the increased values of γ-factor suggested considerable weakening of the GMO and desmopressin interactions and even repulsive interactions between the surfactant and the peptide at 55°C. This might trigger the observed decreased thermal stability of the loaded mesophase. Thus, it was concluded that the peptide dictated the phase behavior of the hexagonal structure, exhibiting the onset of the critical behavior at 45°C. These findings were compared with diffusion of this peptide in the aqueous channels of the cubic phase (Ericsson et al., 1991). Diffusion coefficients of desmopressin in the cubic phase at 40°C were smaller by about a factor of 9 in ^2H_2O solution, a difference that is larger than what is expected from pure obstruction effects; reduction by a factor of 3 was expected from inclusion of a solute in the water channel of a cubic phase. From the data of Libster et al. (2009b) it was found that at 40°C the desmopressin diffusion coefficient in the hexagonal mesophase was about a factor of 57 smaller than in ^2H_2O solution, while a reduction of 6.9 (Rankin model) was expected from the peptide physical restriction within the H_{II} mesophase. Comparing these data to those obtained for the cubic phase, it was clearly shown that desmopressin was retarded much more significantly within the hexagonal phase, compared to the cubic phase. As shown, one source of this diffusion impediment is that the peptide physical restriction within the H_{II} mesophase is more significant, compared to cubic phase (a reduction of 6.9 vs. 3). This is due to greater curvature and narrower aqueous channels of the hexagonal phase, containing less water. The second, more important and dominant reason is that the increased physical restriction of the peptide within the hexagonal structure imposed drastically stronger chemical interactions (via hydrogen bonding) of the peptide with monoolein. This eventually led to significant hindering of its diffusion within the H_{II} mesophase, compared to the cubic structure (a reduction of 57 vs. 9). Such difference may have remarkable implications for drug release.

The applicability and efficiency of the H_{II} mesophase as a carrier for transdermal delivery of desmopressin was also tested (Libster et al., 2009b). Such delivery is especially relevant since the great hydrophilicity of this peptide is the main reason for its uncontrolled fast release when it is used without a proper carrier. Therefore, it is highly desirable to achieve sustained release of this drug and to enable efficient treatment with relatively long duration. As shown in Fig. 8.13, reduced values of steady-state flux of the drug and its permeability coefficients through the skin were obtained compared to water solution. In other words, the peptide molecules penetrated more easily and freely through the skin from water solution than from hexagonal phase vehicles. Hence, a sustained release of desmopressin was obtained via the H_{II} mesophase. A different penetration pathway was suggested when using LLCs as drug vehicles when compared to direct delivery through water solution by Bender et al. (2008). The drug probably accumulated in micro-fissures and in a three-dimensional network of thin threads of the skin. Subsequently, the hydrophilic drug was supposed to diffuse into the surrounding intercellular lipid from the micro-fissure matrix, acting like a source for sustained release. Hence, it was proposed that the intercluster penetration pathway is preferable for delivery of hydrophilic compounds via LLC, in contrast to the intercellular pathway when using

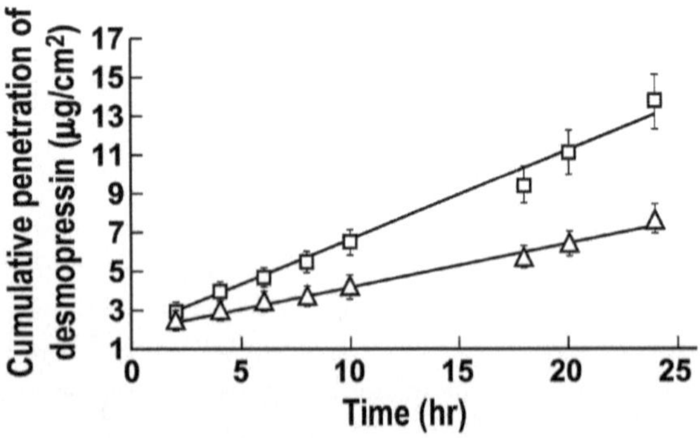

Fig. 8.13. Cumulative transdermal penetration of 1 wt% desmopressin (□) via water solution and (Δ) via H_{II} mesophase (Libster et al., 2009b).

the water solution. Several determining factors could explain the obtained sustained release of the peptide. The release is caused by diffusion of the peptide in LLC. According to the SD-NMR analysis, the release can be associated with the physical restriction of peptide diffusion within the tubes and the interactions of desmopressin with the monoglyceride hydroxyls, noting that the lowest self-diffusion coefficient of desmopressin was detected ca. 35°C. Another important factor is that the H_{II} mesophases possess dominating elastic properties, as well as proper microstructural (lattice parameter) and mesostructural parameters (alignment of discontinuous and anisotropic polycrystalline domains, possessing fractal characteristics), to allow the penetration of the desmopressin molecules into the microfissures.

"ON DEMAND" LLC-BASED DRUG DELIVERY SYSTEMS

Based on the principle that the H_{II} phases released more slowly model hydrophilic and hydrophobic drugs than the GMO cubic phase matrix did, Fong et al. (2009) designed phytantriol and GMO-based bicontinuous cubic (Q_2) and H_{II} nanostructures, designed to allow change to the nanostructure in response to an external change in temperature, with an intention of eventual control of drug release rates in vivo. Using glucose as a model hydrophilic drug, drug diffusion was shown to be reversible on switching between the H_{II} (very low release) and Q_2 nanostructures (high release) at temperatures above and below the physiological temperature, respectively (Fig. 8.14)

However, the matrix used by Fong et al. (2009) required the inclusion of a modifier (vitamin E acetate) to reduce the transition temperature to close to the physiological temperature for in vivo application, thereby enabling control over structure by application of, for example, a heat pack to the skin surface after subcutaneous administration. This is a major limitation, as there is no specificity in heat source; hence, exposure to extremes of temperature may unintentionally induce

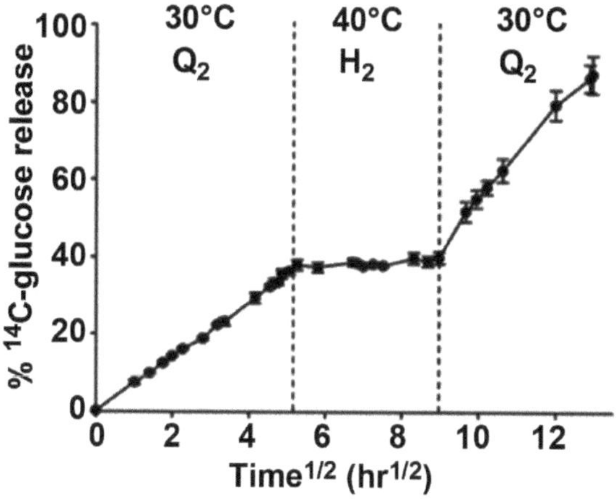

Fig. 8.14. Dynamic release profiles for glucose into phosphate buffered saline from phytantriol + 3% vitamin E acetate with changing temperature, plotted against square root of time. Temperature was switched from 30°C→40°C→30°C at the times indicated by the dashed lines in order to form the phase transition cubic→hexagonal→cubic (Fong et al., 2010).

drug release. Consequently, an alternative means to induce the phase transition was necessary that did not require direct heating, and did not require a reduction in the temperature at which the transition occurs, ideally occurring at the inherent transition temperature without additive (approximately 55°C for phytantriol), removing the potential for accidental activation.

In this context, toward the design of an advanced drug delivery system based on light-triggered phase transition of liquid crystalline phases, Fong et al. (2010) reported the design of novel liquid crystalline matrix—gold nanorod hybrid materials. Hydrophobized gold nanorods (GNRs) have been incorporated within the liquid crystalline matrix, composed of phytantriol and water, to provide remote heating and trigger the phase transitions on irradiation at close to their resonant wavelength. The surface of plasmonic metal nanoparticles delivers heat into surrounding material on exposure to an appropriate light source at the plasmon resonance. The application of near-infrared sensitive nanorods has significance in designing systems for ophthalmic, subcutaneous, or deeper tissue applications. The investigators used plasmonic nanoparticles to achieve reversible phase transitions, hence offering a novel practical solution. The presence of nanorods at concentrations up to 3 nM did not change the lattice parameter of cubic (v_2), hexagonal (H_{II}), and micellar (L_2) systems in the absence of laser irradiation.

Irradiation of the system (at 808 nm NIR diode laser) on inclusion of a low concentration of GNRs (0.3 nM) in the matrix did not induce a change in phase structure away from the *Pn*3*m* bicontinuous cubic phase. However, a small decrease

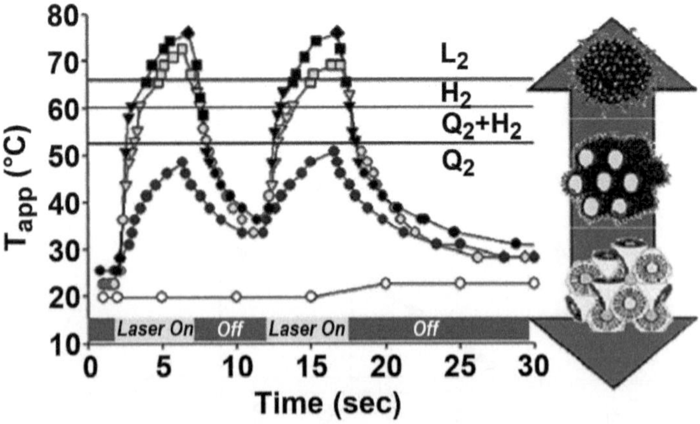

Fig. 8.15. Effect of laser irradiation on apparent temperature of the phytantriol + water matrix (T_{app}) with change in GNR concentration. [GNR] = 0 nM are empty circles, 0.3 nM filled diamonds, 1.5 nM filled triangles, and 3 nM filled squares. Circles indicate v_2 phase, triangles indicate $v_2 + H_2$, squares indicate $H_2 + L_2$, and diamonds indicate L_2. The cartoon on the right indicates the type of phase structure present with increasing temperature (Fong et al., 2010).

in lattice dimension occurred, indicating that heating of the matrix had occurred. The structure relaxed back to the original position when the laser was off. On repeated application of the 5 s laser pulse, the system again displayed the heating effect and relaxed back to the starting position when irradiation was complete (Fig. 8.15). This contraction and expansion of the lattice on heating and cooling (the "breathing mode") was accompanied by concurrent expulsion and uptake of water from the matrix to satisfy the changes in lattice dimension.

In the case of higher concentrations of GNR, the 5 s on pulse for both the 1.5 and 3 nM systems did induce a phase change to the H_2 and L_2 phase structures. At 3 nM, complete transformation to the L_2 phase occurred by the end of the 5 s irradiation, while the lower 1.5 nM concentration resulted in a mixed $H_2 + L_2$ phase, indicating a reduced heating effect (Fig. 8.15). Again, when the laser was switched off, the system ultimately returned to the initial v_2 (*Pn*3*m*) phase structure. Interestingly, on conversion from the L_2 or $L_2 + H_2$ state back to the v_2 structure, the "gyroid" bicontinuous cubic phase with *Ia*3*d* space group was encountered. The gyroid phase coexisted with the H_2 phase initially and then with the v_2 (*Pn*3*m*) phase for approximately 5–6 s after the laser was switched off.

In addition, Fig. 8.15 clearly demonstrates that in the case where no GNRs were added to the matrix there was no significant change in apparent sample temperature (T_{app}) upon laser irradiation. The reversibility of the heating effect in the presence of the GNR is evident from the T_{app} profiles for the three samples containing the nanorods. The sample containing 0.3 nM GNR heated to approximately 50°C, just below the transition above which coexisting H_2 phase occurs. Increasing the nanorod concentration to 1.5 nM induces heating to approx 70°C, while 3 nM GNR heated the matrix to an apparent temperature of 75°C. The repeat irradiation provided the same peak temperature within 1–2°C and reproducible kinetics of heating and cooling. The heating effect observed in these experiments is clearly a function of nanorod concentration, although the relationship between concentration and maximum temperature at differing irradiation conditions requires further investigation and is likely complicated by concurrent cooling. The "breaking wave" shape of the profiles indicated nonlinear heating/cooling gradients in the material.

The authors (Fong et al., 2010) concluded that GNRs embedded in liquid crystalline matrices produce localized plasmonic heating of the hybrid matrix, enabling fine control over nanostructure. The phase transitions resulting from photothermal heating were fully reversible and specific to the GNR/laser wavelength combination. Localized plasmonic heating of the liquid crystal did not compromise the integrity of the lipid molecules in any of the mesophases. Undoubtedly, this research represented a significant advance toward effective, light-activated drug delivery systems with potential to solve unmet medical needs.

MOLECULAR INTERACTIONS OF LLCS WITH PROTEINS AND NUCLEOTIDES

Amar-Yuli et al. (2011) investigated the relationship between the symmetry of lamellar, inverse bicontinuous cubic, and inverse hexagonal LLCs used as hosting for β-lactoglobulin (βlg) amyloid fibrils to exploit the LLC water reservoirs for encapsulation of large protein aggregates. The impact of the fibrils' confinement within the glycerol monolinoleate (GML) based LLCs on their secondary structure and spatial organization was studied by combining small angle X-ray scattering (SAXS), attenuated total reflectance Fourier transform infrared spectroscopy (ATR-FTIR), and atomic force microscopy (AFM) techniques.

Calculating the unit cell dimension of each mesophase revealed that encapsulation of 2 wt% fibrils only slightly modified the lattice parameter depending on the specific phase. Incorporation of fibrils within lamellar mesophase led to a decrease in the lattice parameter from 38.1 Å in the empty system to 35.5 Å. In contrast, the addition of the fibrils to the H_{II} and cubic mesophases increased the lattice parameters by more than 2 Å. Based on the hydrophilicity nature of the protein it was assumed that the βlg fibrils had been accommodated in the aqueous region, either in the LLC inner water domains or at the lipid–water interface. The moderate swelling phenomenon detected in the case of hexagonal and cubic structures suggests that the incorporation of the fibrils mostly results in their partitioning at the interface along the cylinders and not inside of the cylinders themselves, for which a kosmotropic effect and shrinking of the cylinders would then be expected.

FTIR measurements confirmed that the βlg fibrils were incorporated within the aqueous layers of the lamellar phase while in the cubic and hexagonal structures they

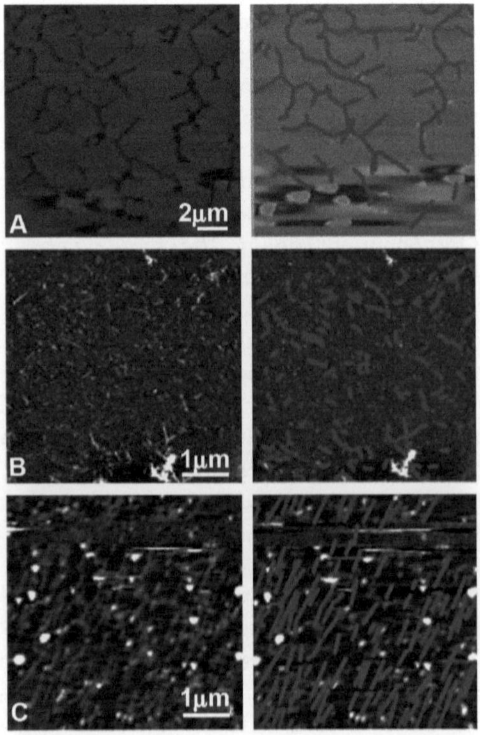

Fig. 8.16. AFM images of β-lactoglobulin fibrils assemblies, after 2 h sonication treatment, when confined in the lamellar (A), the cubic (B), and the hexagonal (C) phase. The right column depicts the fibrils tracking needed for the calculation of the 2D order parameter (Amar-Yuli et al., 2011).

were mostly located at the lipid–water interface, along the channels. Comparison of the secondary structure of fibrils between the three mesophases revealed that in the lamellar structure less unordered configuration of the fibrils (random coil+turns) was detected, compared to the hexagonal or cubic systems. The investigators assumed that upon fibrils embedment into hexagonal and cubic mesophases in the interface a larger unordered configuration of the protein was observed as a result of the simultaneous confinement and increased interaction with the GML interfacial region. Furthermore, to explore the relationship between the symmetry of the host mesophase and the fibrils' confinement the authors investigated the effects arising from encapsulating fibrils within these mesophases also by AFM (Fig. 8.16). The orientation distribution of the fibrils was presented via an average 2D order parameter (S), $S = \langle 2\cos^2\theta - 1\rangle$ where θ is the angle between an individual amyloid fibril and the common director vector existing in the case of a non-random distributed fibrils orientation. The order parameter described the orientational distribution or fibrils centered around the director ($\theta = 0$ condition). The values range between 0 and 1, where the former corresponds to an isotropic structure and the latter represents a perfect alignment along the director.

Fig. 8.16.A–C depict AFM images of βlg fibrils that were sonicated for 2 h embedded in the three mesophases considered and incubated for one week. In the lamellar phase, within the water-filled lamellae, the fibrils exhibited no specific orientation upon encapsulation, which is in agreement with the expected randomly oriented configurations for fibrils confined within planar topologies; this is clearly concluded by inspection of the original AFM image as shown in Fig. 8.16.A. The order parameter was calculated with respect to randomly selected directors ($\theta = 0$) and found to be as low as 0.05. This value clearly indicates randomly oriented fibrils confined within the planar lamellar topologies.

Similarly, when the fibrils were embedded within the isotropic cubic (*Pn3m*) (Fig. 8.16.B), lack of orientation was also demonstrated by the AFM image and corresponding fast Fourier transform (FFT) analysis (not shown). The corresponding order parameter is low (<0.2), which is expected based on the isotropic 3D structure of the bicontinuous cubic phase.

Incorporation of the fibrils in the anisotropic, unidimensional reverse hexagonal phase revealed a strong templating effect of the anisotropic mesophase on the fibrils' alignment (Fig. 8.16.C). This was witnessed by the nearly perfect orientation of the fibrils in correspondence to the hexagonally packed water channels (Fig. 8.4.C). The order parameter values reached a value as high as 0.97, which indicated a nearly perfect alignment. It was concluded that a random orientation of the fibrils was found when these were confined within the *Pn3m* mesophases, whereas encapsulation in the H_{II} structure caused nearly perfect orientation of the fibers.

These trends were further exploited by Amar-Yuli et al. (2011) to induce orientation of the LLCs by decorating the encapsulated amyloid fibrils with magnetic nanoparticles capable of responding to an external magnetic field stimulus. It was demonstrated by SANS measurements that this specific anisotropy coupling mechanism among the mesophases and the amyloid fibrils can also be used to indirectly affect the order of the two anisotropic LLC mesophases considered (i.e., lamellar and hexagonal) by modifying the amyloid fibrils with magnetic nanoparticles and tuning their orientation with an external magnetic field of moderate intensity.

Undoubtedly, the results of this study offered a new appealing route to control, store, and modify alignment in anisotropic mesophases or to use the same as templating media for orienting highly anisotropic particles such as protein fibrils. A new direction was suggested, proposing structural control of stimuli-responsive complex fluids.

Murgia et al. (2009) entrapped several nucleotides into cubic and lamellar monoolein-based mesophases in order to protect them and enable their release. Nucleoside and nucleotide analogues represent a class of innovative therapeutics for the

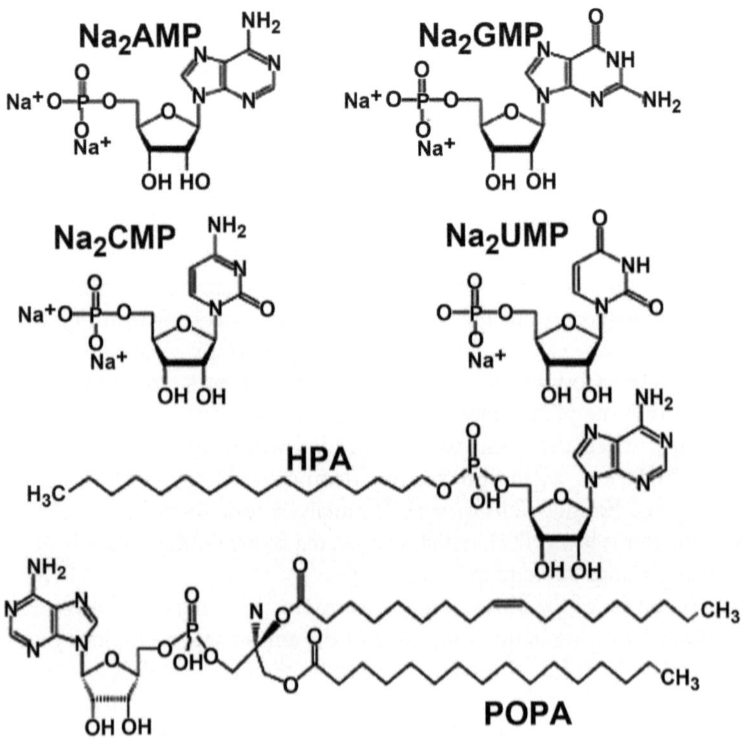

Fig. 8.17. Mononucleotides disodium salts and nucleolipids (Amar-Yuli et al., 2011).

treatment of viral diseases and cancer. Simultaneously, nucleotides can constitute a promising choice to promote molecular recognition. However, these substances need to be protected since they can be easily recognized and degraded by different extracellular nucleases, resulting in poor in vivo pharmacokinetic properties. The usually severe cytotoxic side effects of nucleoside and nucleotide analogues, particularly myelotoxicity, can be partially relieved by lipid derivatization of the phosphate group of the active nucleotide. The resulting drug is amphiphilic and can be delivered to the therapeutic target by incorporation in lipid-based nanovectors.

In this regard these investigators used simple models for both hydrophilic and hydrophobic modified/functionalized nucleotide-based drugs, AMP, GMP, UMP, and CMP (XMPs), along with two hydrophobically functionalized nucleotides (nucleolipids) [i.e., the 1-palmitoyl-2-oleoyl-*sn*-glycerol-3-phosphoadenosine (POPA) and the hexadecylphosphoadenosine (HPA)] (Fig. 8.17). In these nucleolipids a nucleic base is enzymatically exchanged with the choline headgroup of a lipid precursor. The resulting systems were investigated mainly through SAXS and ^{31}P-NMR techniques.

The GMO-based LC phases used here can be easily regarded as membrane models as a result of lipid type and bilayer occurrence. These lipid LC systems have been shown to constitute suitable matrices to entrap either hydrophilic nucleotides or amphiphilic nucleolipids and are, therefore, promising lipid-based nanovectors to protect and deliver nucleotide-analogue drugs. However a drastically different behavior regarding the long-term stability of the nanodevice is displayed. All these molecules contain a phosphate moiety. It has been ascertained that in the case of the amphiphilic additives (POPA and HPA), which locate at the GMO polar–apolar interface, the LC systems were found endowed of high stability—more than 2 years. Differently, the hydrophilic XMP molecules, which are located in the aqueous domain of the GMO LC phases, undergo a hydrolysis process at the phosphoester bond due to a preferential orientation with respect to the MO interface. The XMPs' degradation, within about 4 months, causes an impressive alteration of the interface arrangement, resulting in structural transition from cubic to hexagonal mesophase. It was demonstrated that the relatively small amount of HPO_4^{2-} anion resulting from the XMPs' hydrolysis was sufficient to induce a cubic-to-hexagonal phase transition as a result of different, and specific, interactions at the polar–apolar interface that favored a reverse curvature. The authors concluded that the presence of unesterified phosphate groups in fully hydrophilic drug molecules (for instance XMPs) may have dramatic effects on the stability of the lipid bilayer of a biological membrane.

CONCLUSION

This chapter demonstrated that LC mesophases with complex architectures were rationally designed and prepared for the solubilization and potential administration of bioactive molecules. It was shown that comprehensive understanding of the structural properties of the carriers is imperative for rational and successful tailoring of delivery vehicles. The principal strategy was to first characterize the different levels of organization of these special materials and then to explore the detailed relationships and possible correlations between them, their structures, and the macroscopic properties.

The results lately obtained, although very interesting and promising, definitely do not cover all the aspects and do not answer all the questions addressed by the fascinating field of lyotropic liquid crystals and their applications as drug delivery systems. Clearly, many additional experiments need to be carried out in order to clarify the detailed structure, the exact properties, and the specific potential of these systems. We hope that we at least opened a new window and provided new thoughts and interest into this rapidly growing field of research.

REFERENCES

Achrai, B.; Libster, D.; Aserin, A.; Garti, N. Solubilization of gabapentin into HII mesophases. J. Phys. Chem. B, 2011, 115, 825–835.

Alexandridis, P.; Olsson, U.; Lindman, B. A reverse micellar cubic phase. Langmuir 1996, 12, 1419–1422.

Amar-Yuli, I. Hexagonal liquid crystals and hexosomes structural modifications and solubilization. Ph.D. dissertation; The Hebrew University of Jerusalem, Israel, 2008.

Amar-Yuli, I.; Garti, N. Transitions induced by solubilized fat into reverse hexagonal mesophases. Colloids Surf. B 2005, 43, 72–82.

Amar-Yuli, I.; Wachtel, E.; Ben Shoshan, E.; Danino, D.; Aserin, A.; Garti, N. Hexosome and hexagonal phases mediated by hydration and polymeric stabilizer. Langmuir 2007a, 23, 3637–3645.

Amar-Yuli, I.; Wachtel, E.; Shalev, D.; Moshe, H.; Aserin, A.; Garti, N. Thermally induced fluid reversed hexagonal (H-II) mesophase. J. Phys. Chem. B 2007b, 111, 13544–13553.

Amar-Yuli, I.; Wachtel, E.; Shalev, D.; Aserin, A.; Garti, N. Low viscosity reversed hexagonal mesophases induced by hydrophilic additives. J. Phys. Chem. B 2008a, 112, 3971–3982.

Amar-Yuli, I.; Aserin, A.; Garti, N. Solubilization of nutraceuticals into reverse hexagonal mesophases. J. Phys. Chem. B 2008b, 112, 10171–10180.

Amar-Yuli, I.; Libster, D.; Aserin, A.; Garti, N. Solubilization of food bioactives within lyotropic liquid crystalline mesophases. Curr. Opin. Colloid Inter. Sci. 2009, 14, 21–32.

Amar-Yuli, I.; Adamcik, J.; Lara, C.; Bolisetty, S.; Vallooran, J.J; Mezzenga, R. Templating effects of lyotropic liquid crystals in the encapsulation of amyloid fibrils and their stimuli-responsive magnetic behavior. Soft Matter 2011, 7, 3348–3357.

Ben Ishai, P.; Libster, D.; Aserin, A.; Garti, N.; Feldman, Y. Molecular interactions in lyotropic reverse hexagonal liquid crystals: a dielectric spectroscopy study. J. Phys. Chem. B 2009, 113, 12639–12647.

Bender, J.; Simonsson, C.; Smedh, M.; Engström, S.; Ericson, M.B. Lipid cubic phases in topical drug delivery: visualization of skin distribution using two-photon microscopy. J. Control. Rel. 2008, 129, 163–1

Bitan-Cherbakovsky, L.; Amar-Yuli, I.; Aserin, A.; Garti, N. Structural rearrangements and interaction within HII mesophase induced by cosolubilization of vitamin E and ascorbic acid. Langmuir 2009, 25, 13106–13113.

Bitan-Cherbakovsky, L.; Amar-Yuli, I.; Aserin, A.; Garti, N. Solubilization of vitamin E into HII LLC mesophase in the presence and in the absence of vitamin C. Langmuir 2010, 26, 3648–3653.

Boyd, B.J.; Whittaker, D.V.; Khoo, S.-M.; Davey, G. Lyotropic liquid crystalline phases formed from glycerate surfactants as sustained release drug delivery systems. Int. J. Pharm. 2006, 309, 218–226.

Boyd, B.J.; Khoo, S.-M.; Whittaker, D.V.; Davey, G.; Porter, C.J.H. A lipid-based liquid crystalline matrix that provides sustained release and enhanced oral bioavailability for a model poorly water soluble drug in rats. Int. J. Pharm. 2007, 340, 52–60.

Brondsted, H.; Nielsen, H.M.; Hovgaard, L. Drug-delivery stufies in caco-2 monolayers. 3. Intestinal transport of various vasopressin analogs in the presence of lysophosphatidylcholine. Int. J. Pharm. 1995, 114, 151–157.

Cherezov, V.; Clogston, J.; Papiz, M.Z.; Caffrey, M. Room to move: crystallizing membrane proteins in swollen lipidic mesophases. J. Mol. Biol. 2006, 357, 1605–1618.

Clogston, J.; Caffrey, M. Controlling release from the lipidic cubic phase. Amino acids, peptides, proteins and nucleic acids. J. Control. Rel. 2005, 107, 97–111.

Cohen-Avrahami, M.; Aserin, A.; Garti, N. H-II mesophase and peptide cell-penetrating enhancers for improved transdermal delivery of sodium diclofenac. Colloids Surf. B 2010, 77, 131–138.

Daas, C.R.; Choong, P.F.M. Biophysical delivery of peptides: applicability for cancer therapy. Peptides 2006, 27, 3479-3488.

de Campo, L; Varslot, T; Moghaddam, M.J.; Kirkensgaard, J.J.K.; Mortensenand, K; Hyde, S.T.; A novel lyotropic liquid crystal formed by triphilic star-polyphiles: hydrophilic/oleophilic/fluorophilic rods arranged in a 12.6.4. tiling. Phys. Chem. Chem. Phys. 2011, 13, 3139-3152.

Drummond, C.J.; Fong, C. Surfactant self-assembly objects as novel drug delivery vehicles. Curr. Opin. Colloid Interface Sci. 1999, 4, 449–456.

Efrat, R. Stable cubic mesophases and nanoparticles (cubosomes and micellosomes) for solubilization of active molecules. Ph.D. dissertation; The Hebrew University of Jerusalem, Israel, 2009.

Efrat, R.; Aserin, A.; Kesselman, E.; Danino, D.; Wachtel, E.J.; Garti, N. Liquid micellar discontinuous cubic mesophase from ternary monoolein/ethanol/water mixtures. Colloids Surf. A 2007, 299, 133–145.

Efrat, R.; Aserin, A.; Garti, N. On structural transitions in a discontinuous micellar cubic phase loaded with sodium diclofenac. J. Colloid Interface Sci. 2008, 321, 166–176.

Efrat, R.; Kesselman, E.; Aserin, A.; Garti, N.; Danino, D. Solubilization of hydrophobic guest molecules in the monoolein discontinuous Q(L) cubic mesophase and its soft nanoparticles. Langmuir 2009, 25, 1316-1326.

Ericsson, B.; Eriksson, P.O.; Löfroth, J.E.; Engström, S. Cubic phases as delivery systems for peptide drugs. In Polymeric drugs and drug delivery systems, American Chemical Society, Washington, DC; 1991, 251-265.

Fong, W.K.; Hanley, T.; Boyd, B.J. Stimuli responsive liquid crystals provide 'on-demand' drug delivery in vitro and in vivo J. Controlled Rel. 2009, 135, 218–226.

Fong, W.K.; Hanley, T.; Thierry, B.; Kirby, N.; Boyd, B.J. Plasmonic nanorods provide reversible control over nanostructure of self-assembled drug delivery materials. Langmuir, 2010, 26, 6136–6139.

Gin, D.L.; Pecinovsky, C.S.; Bara, J.E.; Kerr, L. Functional lyotropic liquid crystal materials. Struct. Bond 2008, 128, 181-222.

Gradzielski, M. Investigations of the dynamics of morphological transitions in amphiphilic systems Curr. Opin. Colloid Interface Sci. 2004, 9, 256–263.

Gruner, S.M. Stability of lyotropic phases with curved interfaces. J. Phys. Chem. 1989, 93, 7562-7570.

Gruner, S.M.; Tate, M.W.; Kirk, G.L.; So, P.T.C.; Turner, D.C.; Keane, D.T.; Tilcock, C.P.S.; Cullis, P.R.; X-ray-diffraction study of the polymorphic behavior of N-methylated dioleoylphosphatidylethanolamine. Biochem. 1988, 27, 2853-2866.

Israelachvili, J.N.; Mitchell, D.J.; Ninhan, B.W. Theory of self-assembly of hydrocarbon amphiphiles into micelles and bilayers. J. Chem. Soc. 1976, 72, 1525-1568.

Kumar, T.R.S.; Soppimath, K.; Nachaegari, S.K. Novel delivery technologies for protein and peptide therapeutics. Curr. Pharm. Biotech. 2006, 7, 261-276.

Kurosaki, Y.; Nagahara, N.; Tanizawa, T.; Nishimura, H.; Nakayama, T.; Kimura, T. Use of lipid disperse systems in transdermal delivery: comparative study of flufenamic acid permeation among rat addominal skin, silicon rubber membrane and stratum corneum sheet isolated from hamster cheek pouch. Int. J. Pharm. 1991, 67, 1-9.

Larsson, K. Cubic lipid-water phases: structures and biomembrane aspects. J. Phys. Chem. 1989, 93, 7304-7314.

Libster, D.; Aserin, A.; Wachtel, E.; Shoham, G.; Garti, N. An HII liquid crystal-based delivery system for cyclosporin A: physical characterization. J. Colloid Interface Sci. 2007, 308, 514-524.

Libster, D.; Ben Ishai, P.; Aserin, A.; Shoham, G.; Garti, N. From the microscopic to the mesoscopic properties of lyotropic reverse hexagonal liquid crystals. Langmuir 2008, 24, 2118–2127.

Libster, D.; Ben Ishai, P.; Aserin, A.; Shoham, G.; Garti, N. Molecular interactions in reverse hexagonal mesophase in the presence of Cyclosporin A. Int. J. Pharm. 2009a, 367, 115–126.

Libster, D.; Aserin, A.; Yariv, D.; Shoham, G.; Garti, N. Concentration- and temperature-induced effects of incorporated desmopressin on the properties of reverse hexagonal mesophase. J. Phys. Chem. B. 2009b, 113, 6336–6346.

Libster D.; Aserin, A.; Garti, N. Interactions of biomacromolecules with reverse hexagonal liquid crystals: drug delivery and crystallization applications. J. Colloid. Interface Sci. 2011, 356, 375–386.

Liu, D.-Z.; LeCluyse, E.L.; Thakker, D.R. Dodecylphosphocholine-mediated enhancement of paracellular permeability and cytotoxicity in Caco-2 cell monolayers. J. Pharm. Sci. 1999, 88, 1161–1168.

Luzzatti, V.; Vargas, R.; Gulik, A.; Mariani, P.; Seddon, J.M.; Rivas, E. Lipid polymorphism: a correction. The structure of the cubic phase of extinction coefficient Fd–consists of two types of disjointed reverse micelles embedded in a three-dimensional hydrocarbon matrix. Biochemistry 1992, 31, 279–285.

Mezzenga, R.; Grigorov, M.; Zhang, Z.; Servais, C.; Sagalowicz, L.; Romoscanu, A.I. Polysaccharide-induced order-to-order transitions in lyotropic liquid crystals. Langmuir 2005, 21, 6165–6169.

Mishraki, T.; Libster, D.; Aserin, A.; Garti, N. Lysozyme entrapped within reverse hexagonal mesophases: physical properties and structural behavior. Colloids Surf. B 2010a, 75, 47–56.

Mishraki, T.; Libster, D.; Aserin, A.; Garti, N. Temperature-dependent behavior of lysozyme within the reverse hexagonal mesophases (HII). Colloids Surf. B 2010b, 75, 391–397.

Murgia, S.; Lampis, S.; Angius, R.; Berti, D.; Monduzzi, M. Orientation and specific interactions of nucleotides and nucleolipids inside monoolein-based liquid crystals. J. Phys. Chem. B, 2009, 113, 9205–9215.

Pedersen, T.B.; Frokjaer, S.; Mouritsen, O.G.; Jørgensen, K. A calorimetric study of phosphocholine membranes mixed with desmopressin and its diacylated prodrug derivative (DPP). Int. J. Pharm., 2002, 233, 199–206.

Pluen, A.; Netti, P.A.; Jain, R.K.; Berk, D.A. Diffusion of macromolecules in agarose gels: comparison of linear and globular configurations. Biophys. J. 1999, 77, 542–552.

Pouzot, M.; Mezzenga, R.; Leser, M.; Sagalowicz, L.; Guillot, S.; Glatter, O. Structural and rheological investigation of Fd3m inverse micellar cubic phases Langmuir 2007, 23, 9618–9628.

Sagalowicz, L.; Mezzenga, R.; Leser, M.E. Investigating reversed liquid crystalline mesophases. Curr. Opin. Colloid Interface Sci. 2006a, 11, 224–229.

Sagalowicz, L.; Leser, M.E.; Watzke, H.J.; Michel, M. Monoglyceride self-assembly structures as delivery vehicles. Trends Food Sci. Technol. 2006b, 17, 204–214.

Seddon, J.M. Structure of the inverted hexagonal (HII) phase, and non-lamellar phase-transitions of lipids. Biochim. Biophys. Acta 1990, 1031, 1–69.

Seddon, J.M.; Zeb, N.; Templer, R.H.; McElhaney, R.N.; Mannock, D.A. An Fd3m lyotropic cubic phase in a binary glycolipid/water system. Langmuir 1996, 12, 5250–5253.

Seddon, J.M.; Robins, J.; Gulik-Krzywicki, T.; Delacroix, H. Inverse micellar phases of phospholipids and glycolipids. Phys. Chem. Chem. Phys. 2000, 2, 4485–4493.

Shah, M.H.; Paradkar, A. Cubic liquid crystalline glyceryl monooleate matrices for oral delivery of enzyme Int. J. Pharm. 2005, 294, 161–171.

Shah, J.C.; Sadhale, Y.; Dakshina, M.C. Cubic phase gels as drug delivery systems. Adv. Drug Deliv. Rev. 2001, 47, 229–250.

Shearman, G.C.; Ces, O.; Templer, R.H.; Seddon, J.M. Inverse lyotropic phases of lipids and membrane curvature. J. Phys. Condens. Matter 2006, 18, S1105–S1124.

Spernath, A.; Aserin, A.; Garti, N. Phase transition induced by water dilution in phospholipid U-type food-grade microemulsions studied by DSC. J. Colloid Interface Sci., 2006, 299, 900–909.

Touitou, E.; Levi-Schaffer, F.; Dayan, N.; Alhaique, F.; Riccieri, F. Modulation of caffeine skin delivery by carrier design: liposomes versus permeation enhancers. Int. J. Pharm. 1994, 103, 131–136.

Uddin, M.H.; Rodriguez, C.; Watanabe, K.; Lopez-Quintela, A.; Kato, T.; Furukawa, H.; Harashima, A.; Kunieda, H. Phase behavior and formation of reverse cubic phase based emulsion in water/poly(oxyethylene) poly(dimethylsiloxane) surfactants/silicone oil systems. Langmuir 2001, 17, 5169–5175.

Walse, B.; Kihlberg, J.; Drakenberg, T. Conformation of the Vasopressin analogue DDAVP in aqueous solution. Eur. J. Biochem. 1998, 252, 428–440.

Yaghmur, A.; de Campo, L.; Salentinig, S.; Sagalowicz, L.; Leser, M.E.; Glatter, O. Oil-loaded monolinolein-based particles with confined inverse discontinuous cubic structure (Fd3m). Langmuir 2006, 22, 517–521.

Yang, L.; Ding, L.; Huang, H.W. New phases of phospholipids and implications to the membrane fusion problem. Biochem. 2003, 42, 6631-6635.

9

LIPID NANOPARTICULATE DRUG DELIVERY AND NANOMEDICINE

Heidi M. Mansour[1], Yun-Seok Rhee[2], Chun-Woong Park[1], and Patrick P. DeLuca[1]
[1]University of Kentucky College of Pharmacy, Department of Pharmaceutical Sciences-Drug Development Division, 789 S. Limestone St., Lexington, KY 40536-0596, USA and [2]Gyeongsang National University (GNU), College of Pharmacy, Jinju, Gyeongnam 660-751, Republic of Korea

INTRODUCTION

Nanomedicine, as a convergence of nanoscience with molecular pharmaceutics and medicinal science, is currently one of the fastest growing aspects of nanotechnology and will be for the foreseeable future. It is essential for physiological understanding of living targets based on biological science and for developing multifunctional nanomaterials as drug delivery carriers with novel nanotechnology. Nanomedicines can be designed for solubilization of lipophilic drugs that are poorly water-soluble, thereby improving drug absorption in the body. Nanomedicines can also be designed for drug-targeting at the cellular and molecular level, which offers enhanced therapeutic effects and decreased side effects.

Recent trends in nanomedicine include cellular and intracellular drug targeting related to the interactions of subcellular components, in addition to imaging and diagnostic applications. These nanomedicines have a submicron size (below 1 μm) and can be obtained either by particle size reduction of larger particles, thereby forming nano-sized particles (top-down approach), or by building up particles by precipitation of dissolved molecules (bottom up approach). The approaches that have progressed rapidly in nanomedicine and drug delivery are the lipid-based nanoparticulate drug delivery systems, such as solid lipid nanoparticles (SLNs), nanostructure lipid carriers (NLCs), and liposomal drug delivery carriers. This is due in large part to biocompatibility with relatively low human toxicity.

SLNs are colloidal carriers developed at the beginning of the 1990s as an alternative colloidal carrier system to emulsions, liposomes, and polymeric nanoparticles (Muller et al., 2000; Siekmann & Westesen, 1992). SLNs are colloidal particles of a lipid matrix that are solid at body temperature (Joshi & Muller, 2009; Muchow et al., 2008) and are made from solid lipids (i.e., solid lipids at room temperature), surfactant(s), and water (Mehnert & Mader, 2001). SLNs offer various advantages, including improved bioavailability of poorly water soluble drugs, high biocompatibility and biodegradability, low toxicity, protection of incorporated labile drugs, good storage stability, controlled release property, low cost of ingredients, sterility, ease of preparation and scale-up, high dispersibility in an aqueous medium, and wide potential application spectrum (e.g., parenteral, oral, ocular, dermal).

Similarly, nanostructure lipid carriers (NLCs) are regarded as the second-generation of lipid nanoparticles (LNs) and have been developed in recent years to overcome the limitations of SLNs. In contrast to SLNs being produced from solid lipids only, NLCs are made of a solid matrix that entraps liquid lipid nano-compartments. There are some potential limitations associated with SLNs including limited drug loading capacity, low drug entrapment efficiency, risk of gelation, and drug expulsion during storage (due to the crystallization of lipid matrix or lipid polymorphism). The advantage of NLCs over SLNs results from the liquid lipid, which is present in the solid matrix, avoiding the drug expulsion during storage caused by lipid polymorphic transformations from unstable to more stable configurations. Liquid lipids usually show a higher solubility for drugs than solid lipids. This means an improvement of drug loading capacity, reduced drug expulsion during storage, and a possibility of modulation of the drug release profile of NLCs compared to SLNs.

LIPID NANOPARTICLES WITH A SOLID MATRIX

PREPARATION TECHNIQUES

Several classical methods are discussed in this section at the basic principle level, although there are many reports on modified preparation methods to produce the lipid nanoparticle. Several production processes using cold, hot high pressure homogenization techniques, multiple emulsion technique, and hot homogenization method followed by ultrasonication are shown in Fig. 9.1.

HIGH PRESSURE HOMOGENIZATION. The high pressure homogenization technique has been widely used as a reliable and powerful technique for preparation of SLNs. In contrast to other techniques, scale-up can be possible without problems in most cases. Two general approaches to the homogenization step (i.e., the hot and cold homogenization techniques) can be used in the production of SLNs.

Hot Homogenization Technique. The lipid is melted at 5°C above its melting point and dispersed in a surfactant solution of identical temperature using a high speed stirrer to form a coarse pre-emulsion. The pre-emulsion thus obtained is then homogenized by varying the homogenization pressure, homogenization time and homogenization cycles. The obtained hot oil-in-water nanoemulsion is then cooled down to room temperature and the solid lipid nanoparticles (SLNs) are produced after crystallization.

Cold Homogenization Technique. The cold homogenization technique has been developed to overcome the following three problems of hot homogenization (Mehnert & Mader, 2001): (1) drug distribution into the aqueous phase during homogenization, (2) temperature-induced drug degradation, and (3) complexity of the

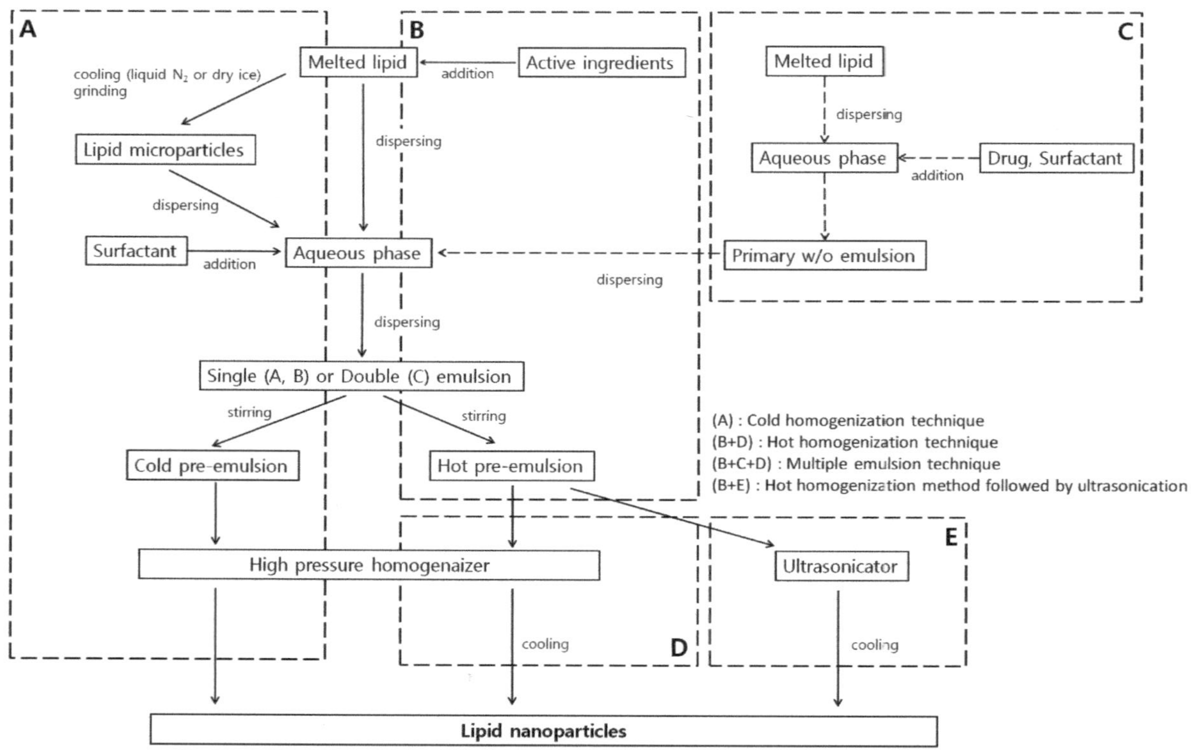

Fig. 9.1. Production processes of lipid nanoparticles using cold, hot high pressure homogenization techniques, multiple emulsion technique, and hot homogenization method followed by ultrasonication.

crystallization step of the nanoemulsion leading to several modifications and/or supercooled melts.

In cold homogenization, the drug containing the lipid melt is cooled by pouring this mixture into liquid nitrogen or dry ice to obtain drug-loaded lipid. The produced solid lipid is ground to lipid microparticles (approximately 50–100 μm). These lipid microparticles are suspended in a cold aqueous surfactant solution and homogenized at or below room temperature to obtain the required particle size.

HOT HOMOGENIZATION METHOD FOLLOWED BY ULTRASONICATION. Drug and lipids are melted by heating at 5°C above the melting point of the lipid. An aqueous phase is prepared by dissolving surfactants (poloxamer 188) in water and heated to the same temperature of the oil phase. The hot aqueous phase is added to the oil phase and homogenization is carried out using a homogenizer for a few minutes. To prevent recrystallization during homogenization and ultrasonication, the production temperature can kept at least 5°C above the lipid melting point. The coarse hot o/w emulsion thus obtained is sonicated using a probe sonicator for 20–30 min. The obtained nanoemulsion (o/w) is cooled down at room temperature or in an ice bath to form SLNs (Aji Alex et al., 2010; Kumar et al., 2007; Manjunath & Venkateswarlu, 2006; Siddiqui et al., 2010; Suresh et al., 2007; Venkateswarlu & Manjunath, 2004).

WARM O/W MICROEMULSION TECHNIQUE. The o/w microemulsion technique to obtain SLNs requires two steps: (1) preparation of warm o/w microemulsion and (2) formation of the lipid nanoparticles by dispersing the warm microemulsion in a cold aqueous medium.

A clear microemulsion consisting of drug, melted lipids (such as fatty acids/triglycerides), surfactants, and co-surfactants is obtained by gentle stirring at a temperature ranging from 60 to 80°C (slightly above the melting point of lipids). The prepared microemulsions are subsequently dispersed in cold water (2–3°C) under mechanical stirring. The obtained nanosuspensions are successively filtered and lyophilized with cryoprotectants producing a dry powder of SLNs (Aquilano et al., 1993; Bondi et al., 2003; Brioschi et al., 2008; Cavalli et al., 1995; 1997; 2001; Fontana et al., 2005; Jain et al., 2005; Kuo & Chen, 2009; Ma et al., 2009; Pedersen et al., 2006; Sanjula et al., 2009; Tiyaboonchai et al., 2007; Ugazio et al., 2002; Wong et al., 2004).

SOLVENT DIFFUSION METHOD. The drug and lipids are dissolved completely either in acetone or ethanol (or a mixture) using a water bath at a temperature ranging from 50 to 70°C. The resultant organic solution is poured into an acidic aqueous phase containing surfactant under mechanical agitation at room temperature. The pH value of the above obtained SLN or NLC dispersion is adjusted to about pH 1 by adding 0.1M hydrochloric acid to form aggregation of the nanoparticles. The entire dispersion is then centrifuged and re-suspended in distilled water (Hong et al., 2006;

Hu et al., 2002; 2004a; 2004b; 2005; Paliwal et al., 2009). Cold aqueous solution (0–8°C) can be used to allow for the hardening of the SLNs instead of the acidic aqueous phase (Hu et al., 2008; Pandey et al., 2005; Ye et al., 2008).

EMULSIFICATION-SOLVENT EVAPORATION TECHNIQUE. The drug and lipids are dissolved in a water-immiscible organic solvent (e.g., cyclohexane, dichloromethane, chloroform). This is then added into an aqueous solution containing surfactants with homogenization. Afterwards, it is stirred mechanically for several hours. Upon solvent evaporation, a nanoparticle dispersion is formed by precipitation of the lipid in the aqueous medium. The nanoparticles are retrieved by centrifugation and washed several times with deionized water to remove residual surfactant (del Pozo-Rodriguez et al., 2008; Mehnert & Mader, 2001; Rahman et al., 2010; Siekmann & Westesen, 1996; Sjöström et al., 1993a; 1993b; 1993c).

EMULSIFICATION-SOLVENT DIFFUSION TECHNIQUE. Partially water-miscible organic solvents (e.g., benzyl alcohol or butyl lactate) and water are mutually saturated before use in order to ensure initial thermodynamic equilibrium of both liquids. Lipids are dissolved in the water-saturated solvent and the organic solution is emulsified with the solvent-saturated aqueous solution containing the emulsifier mixture, using a homogenizer. The lipid nanoparticles are precipitated out by quickly adding a sufficient amount of water into the initial emulsion to extract the organic solvent into the continuous phase. Stirring is continued until the dispersion cools, and the suspensions are then purified (Shah et al., 2007; Trotta et al., 2001; 2003).

MULTIPLE EMULSION TECHNIQUE. A double emulsion w/o/w method is a useful technique for hydrophilic drugs or peptide/protein drugs due to its hydrophilic inner phase. The drug is dissolved in an aqueous solution (inner aqueous phase) with or without surfactants (e.g., Tween 80). The oil phase is prepared by dissolving lipids with or without surfactants (i.e., phospholipids). The inner aqueous phase is mixed with the oil phase and sonicated using an ultasonicator to obtain the primary w/o emulsion. The primary emulsion obtained is then poured into the surfactant (i.e., poloxamer, PVA) solution (outer aqueous phase) at room temperature under mechanical agitation to remove the organic solvent and form SLNs. The suspension is then filtered and used (Gallarate et al., 2009; Garcia-Fuentes et al., 2002; Liu et al., 2007a; 2008; 2009; Lv et al., 2009; Sarmento et al., 2007; Singh et al., 2010)

SOLVENT INJECTION METHOD. The solvent injection method, which is usually employed for the preparation of liposomes (Deamer & Bangham, 1976) was developed for the preparation of SLNs. The drug and lipids are dissolved in a specified volume of water-miscible organic solvent (e.g., ethanol, acetone, isopropanol) by heating to the melting temperature of the lipids. The resulting solution is rapidly injected into a stirred aqueous phase containing surfactant (e.g., poloxamer 407, Tween 80). The resulting dispersion is then filtered with a paper filter in order to remove any excess lipid (Gulbake et al., 2010; Gupta et al., 2007; Pandita et al., 2009; Schubert & Muller-Goymann, 2003; Shah & Pathak, 2010; Stevens et al., 2004; Wang et al., 2010).

OTHER TECHNIQUES.

Supercritical Fluid Extraction of Emulsions. Oil-in-water (o/w) emulsions are prepared by dissolving the drug and lipids in chloroform with a soy lecithin surfactant, dispersing this solution into an aqueous phase of sodium glycocholate and passing the mixture through a high pressure homogenizer in order to form a fine emulsion with a mean droplet size ranging between 30 and 100 nm. SLN suspensions are obtained using a continuous extraction method. The o/w emulsion is introduced into an extraction column from the top at a constant flow rate. Simultaneously, supercritical CO_2 is introduced counter-currently at a constant flow rate. The operating pressure and temperature conditions are selected to minimize product loss. The residence time required for producing pure aqueous suspensions of SLN is approximately 2 minutes with the product continuously removed from the bottom of the extraction column. The solvent-laden supercritical CO_2 is carried out using an extraction column for solvent recycling (Chattopadhyay et al., 2007).

Membrane Contactor. The melted lipidphase-containing drug is pressed, at a temperature above the melting point of the lipid, through the membrane pores allowing the formation of small droplets. The lipid droplets are detached from the membrane pores by the aqueous phase streaming tangentially to the membrane surface. SLNs are formed by cooling the preparation to a temperature below the lipid melting point (Charcosset et al., 2005; El-Harati et al., 2006).

Coacervation Method. Coacervation is a solvent-free technique to produce SLNs of fatty acids by acidification of a micellar solution of their alkaline salts. Fatty acid sodium salt is dispersed in a solution of polymeric stabilizer and the mixture is then heated under stirring just above the Krafft point of fatty acid sodium salt to obtain a clear solution. A selected acidifying solution (i.e., the coacervating solution) is then added drop-wise until pH 4.0 is reached. The obtained suspension is then cooled in a water bath with stirring until 15°C temperature is reached (Battaglia et al., 2010).

CHARACTERIZATION TECHNIQUES

The particle size and polydispersity index (PI) of lipid nanoparticles (LNs) are important factors (Mehnert & Mader, 2001; Muller & Heinemann, 1992) when considering parenteral administration of LNs and in manufacturing processes (e.g., sterilization, filtration). The average particle size and the PI of LNs is determined by photon correlation spectroscopy (PCS) and the laser diffraction (LD) method. The particle sizes obtained by static (LD) and dynamic light scattering (PCS) methods are usually in the same range, but particle sizes obtained by LD are always a bit smaller than those of PCS. In general, differences in the sizes by both methods result from the different measurement principles. PCS determines the hydrodynamic radii (i.e., an effective size), which are usually larger than the solid radii of spheres.

Zeta potential is the measurement of net charge acquired by particles in a particular medium and is considered to be one of the benchmarks in predicting the stability of colloidal systems. Particles will repel one another if the particles have high positive or negative values of zeta potential. A system having a value of ±30 mV is considered a stable formulation if dispersed in a liquid as colloidal dispersion. Both particle size and zeta potential measurements are critical in order to evaluate the physical stability of LNs over a period of time (Souto et al., 2005).

Differential scanning calorimetry (DSC) studies give insight into the melting behavior of crystalline material. The breakdown of the crystal lattice of the nanoparticles yields information about polymorphism and crystal ordering. During the cooling process of the LN preparation step, the lipid recrystallizes from the melting stage and may form polymorphic crystal. An important parameter affecting drug incorporation is the polymorphic modification of the lipid particle matrix. In general, the preparation process of the nanoparticles can change the type of modification of their respective fraction (Muller et al., 2008). The degree of crystallinity in LNs is analyzed by DSC (Freitas & Muller, 1999).

X-ray powder diffraction (XRPD) is employed to investigate the lamellae arrangement of the lipid molecules and the crystallinity of the fatty acid chains in the acylglycerols of lipids (palmitic and stearic acids) (Muller et al., 2008).

FTIR studies are conducted to find any interaction between the drug and the lipids (Rahman et al., 2010). Moreover, Raman spectroscopy is a useful tool for determining the chemical reaction between the drug and the lipids (Varia et al., 2008). Raman spectroscopy detects vibrations of molecules after excitation by an intensive laser beam, and it is used to obtain information about the structure and properties of molecules from their vibrational transitions (Mansour & Hickey, 2007). An investigation via Raman spectroscopy gives insight into the conformational order of hydrocarbon chains (i.e., if and how oil loading changes the lipid chain arrangement). Raman spectra of lipids are sensitive to conformational, packing, and dynamical changes involving hydrocarbon chains (Jores et al., 2005; Wartewig & Neubert, 2005). Raman spectroscopy is a particularly useful technique as it involves no sample preparation and most importantly allows measurements in the presence of water (Saupe et al., 2006).

Nuclear magnetic resonance (NMR) investigations are used to obtain information about the mobility and structure of both the nanoparticle system and the incorporated drug (Morel et al., 1998). The width of NMR lines is related to the phenomenon of relaxation, thus decreased mobility of the molecules leads to a decrease in the relaxation time which results in considerable line broadening. NMR is, therefore, an appropriate tool to distinguish between mobile liquid and immobilized solid states (Wissing et al., 2004b; Zimmermann et al., 2005).

Electron paramagnetic resonance (EPR) spectroscopy is a powerful tool for investigating dynamic phenomena and the characteristics of nanocompartments in colloidal lipid dispersions. The position and distribution of different spin probes within LN dispersions can be determined by EPR spectroscopy. EPR can also provide important information about the influence of structure and lipophilicity on the distribution and mobility of the compound. Moreover, the use of relevant theoretical models can provide a thorough understanding of the distribution processes involved in nanostructured systems (Pegi et al., 2003).

Electron microscopic technique is one of the most useful characterization methods used in colloidal drug delivery systems since it images the particles for batch imaging, particle size, particle shape, and surface roughness. The microstructure of the lipid nanoparticles can investigated by electron microscopy studies.

Several electron microscopic techniques such as scanning electron microscopy (SEM), transmission electron microscopy (TEM), cryo-field emission scanning electron microscopy (Cryo-FESEM), and cryo-transmission electron microscopy (Cryo-TEM) have been applied to imaging LNs. The advantage of SEM, compared to the other microscopy techniques, is its lower cost and the opportunity to examine large areas of dispersed particles in a short time (Hatziantoniou et al., 2007). TEM can reveal the shape, size, and uniformity of LNs; mechanical adhesion between LNs; and drug crystallization from LNs (Li et al., 2009; Zhao et al., 2010; Zhuang et al., 2010). Particle size information, which is measured by laser light scattering methods, can be confirmed by TEM and SEM studies (Liu et al., 2007a). One advantage of using Cryo-FESEM or Cryo-TEM is that the liquid dispersion can be frozen and viewed in the frozen state. It is not necessary to separate the nanoparticles from the water matrix, and thus the samples are investigated close to their natural state (Saupe et al., 2006). However, cryo-electron micrographs tend to be decisively biased towards small particles as a result of the sample preparation technique. Therefore, a comparison of particle sizes measured in cryo-TEM with the results of particle sizing techniques such as PCS or LD should be performed with great caution (Jores et al., 2004).

Atomic force microscopy (AFM) has been recently employed as a major microscopic tool to observe size and morphology of LNs (Aji Alex et al., 2010; Shahgaldian et al., 2003; zur Muhlen et al., 1996). This is due to AFM's many advantages for the detection and characterization of heterogeneous populations with regards to the amount of material as well as to the level of information.

PHARMACEUTICAL APPLICATIONS OF LIPID NANOPARTICLES

PARENTERAL DRUG DELIVERY

Lipid nanoparticles designed for the parenteral route of administration (intravenous) offer direct access to the systemic circulation and molecular targeting of specific organ and tissue sites such as the brain, lung and liver. Moreover, lipid nanoparticles have demonstrated great potential in improving parenteral delivery of hydrophobic agents (Joshi & Muller, 2009; Wissing et al., 2004a). Lipid nanoparticles have been regarded as an alternative to liposomes and nanoemulsions due to various advantages such as ease of manufacture, relatively high drug loading, and modulation of drug release.

Furthermore, the aqueous nature of lipid nanoparticles, their nanostructure, and biocompatibility of the excipients can provide parenteral delivery of the drugs without harsh surfactants (Joshi & Muller, 2009; Lu et al., 2008b; 2009; Martins et al., 2007; Muller et al., 2000; Zhuang et al., 2010).

To protect LNs against rapid elimination and degradation after parenteral administration, surface-modified strategies have been investigated for LNs by creating a protective barrier around the LNs themselves through steric hindrance, hydrodynamic volume effects, and charge repulsion. Surface modification of nanoparticles with polyethylene glycol (PEG), a polymer of hydrophilic nature, showed promising results. Stealth SLNs containing doxorubicin were prepared by adding stearic acid-PEG 2000 as a stealth agent to melted stearic acid (Fundaro et al., 2000; Zara et al., 2002). Prolonged circulation time was compared to the doxorubicin solution (Fundaro et al., 2000), and the increase in the stealth properties affected the doxorubicin transported into the brain (Zara et al., 2002). Stealth SLNs can evade phagocytosis of macrophages to some extent (Wang & Wu, 2006; Wu et al., 2010; Zhang et al., 2008).

Many LNs studies showed that SLNs have a good brain targeting efficiency in vivo (Blasi et al., 2007; Goppert & Muller, 2005; Hsu et al., 2010; Kaur et al., 2008; Wang et al., 2002; Yang et al., 1999b), and SLNs are taken up readily by the brain because of their lipidic nature (Kaur et al., 2008). Parenteral dexamethasone acetate-loaded SLNs were examined for lung-targeting delivery (Xiang et al., 2007), and enhanced liver targeting by SLNs was also investigated in animal models (mice, rabbits) (Lu et al., 2008a; Yu et al., 2003). Injectable actarit-loaded SLNs were investigated to improve therapeutic efficacy and reduce side effects such as nephrotoxicity and gastrointestinal disorders commonly associated with oral formulations of actarit. The results indicate that injectable actarit-loaded SLNs were promising in passive targeting therapeutic agents for rheumatoid arthritis (Ye et al., 2008).

ORAL DRUG DELIVERY

Incorporation of drugs into nanoparticles can enhance bioavailability and reduce variability in plasma levels (Kaur et al., 2008). Advantages of oral LNs include their ability to protect against chemical or enzymatic degradation in the gastrointestinal tract, possible large-scale production, controlled-release, and improved bioavailability. Furthermore, the carrier itself can be taken up to a certain extent by epithelial cells or lymphoid tissues in Peyer's patches (Garcia-Fuentes et al., 2002).

LN formulations loaded with poorly water-soluble drugs or insulin for oral administration have been investigated to improve oral bioavailability and demonstrated a promising strategy for oral delivery (Hu et al., 2010a; 2010c; Li et al., 2009; Muchow et al., 2008; Muller et al., 2006; Ugazio et al., 2002; Yang et al., 1999a; 2009). SLNs containing apomorphine showed 12- to 13-fold higher bioavailability than the reference solution (Tsai et al., 2010). The NLC formulation remarkably improved the oral bioavailability of vinpocetine (Luo et al., 2006; Zhuang et al., 2010).

Recently, LNs have been studied as potential carriers for oral intestinal lymphatic delivery (Aji Alex et al., 2010; Paliwal et al., 2009; Sanjula et al., 2009; Trevaskis et al., 2008). Several mechanisms of delivering drugs to the intestinal lymphatics following oral administration have been investigated and reported. These include paracellular mechanism (Hiroshi et al., 1981; Muranishi et al., 1997), transcellular mechanism (Caliph et al., 2000; Khoo et al., 2001), and transport through the M cells of Peyer patches (Clark et al., 2001; Hussain et al., 1997). Among these, the most promising mechanism is transcellular absorption, which operates for dietary lipids. LNs enhance the stimulation of chylomicron formation by enterocytes, which promote the absorption of lipid carriers into intestinal lymphatics and organs (Porter & Charman, 2001; Trevaskis et al., 2008).

Methotrexate-loaded SLNs improved the oral bioavailability of methotrexate, presumably following lipid digestion and co-absorption through lymphatic transport and route (Paliwal et al., 2009). SLN formulations effectively improved the lymphatic uptake of carvedilol, thus enhancing its oral bioavailability (Sanjula et al., 2009). From the intestinal lymphatic transport study it became evident that SLNs increased the cumulative percentage dose of lopinavir secreted into the lymph, which was 4.91-fold higher when compared with a conventional drug solution (Aji Alex et al., 2010).

PULMONARY DRUG DELIVERY

Pulmonary drug delivery, as inhaled aerosols, is effective in the targeted treatment of respiratory diseases including chronic obstructive pulmonary disease, asthma, cystic fibrosis, infectious diseases, tuberculosis, and lung cancer (Pison et al., 2006). Pulmonary administration via inhaled aerosols is also receiving increased attention as an effective noninvasive alternative for systemic drug delivery due to a thin epithelial barrier, high blood flow, avoidance of first-pass metabolism, more rapid onset of therapeutic action, and the availability of a huge surface area (Patton & Byron, 2007; Yang et al., 2008a). Nanocarrier systems in pulmonary drug delivery offer many important advantages (Mansour et al., 2009). Nanocarrier systems can provide sustained release of drugs in the lung tissue, resulting in a reduction in dosing frequency and improved patient compliance (Pandey & Khuller, 2005), while providing uniformity in drug dose distribution among the alveoli (Yang et al., 2008b). Moreover, by formation of drug nanoparticles, one can achieve a dose that is higher than that of a solution, which is thermodynamically limited by the aqueous solubility of the drug (Sung et al., 2007).

Recent investigations demonstrate that SLNs can be used in non-invasive pulmonary delivery of therapeutic drugs (Hu et al., 2010b; Li et al., 2008; 2010). SLNs have been used as a potential carrier for pulmonary delivery of insulin by improving both in vitro and in vivo stability (Liu et al., 2008). Pulmonary administration of microparticles containing thymopentin-loaded SLNs showed enhanced drug absorption and sustained drug release (Li et al., 2010). Epirubicin-loaded SLNs were prepared as an inhalable formulation for the treatment of lung cancer. The drug concentrations in the lungs and in plasma after inhalation of epirubicin-SLNs were much higher than those after administration of epirubicin solution (Hu et al., 2010b).

DERMAL AND TRANSDERMAL DRUG DELIVERY

LNs are attractive carriers for topical cosmetic and pharmaceutical products (Wissing & Muller, 2002; 2003a; 2003b). LNs have a high affinity for the stratum corneum, and therefore an enhanced penetration of the encapsulated active substances to the skin can be achieved. LNs enhance the penetration and transport active substances especially lipophilic compounds, and thus, increase the level of these compounds in the skin (Jenning et al., 2000a; Muller et al., 2002). The incorporation of chemically labile active substances into LNs has been shown to protect them against degradation giving new chances to the use of cosmetic ingredients that suffer chemical instability in conventional formulations (Dingler et al., 1999; Jenning & Gohla, 2001; Kaur et al., 2007). LNs are suitable for use on inflamed and damaged skin since LNs are composed of non-irritative and non-toxic lipids (Jenning et al., 2000a; Mei et al., 2003; Wissing & Muller, 2003b). Moreover, LNs have noticeable occlusive effects due to the formation of an intact nanofilm on the skin surface upon drying, which reduces trans-epidermal water loss and leads to wrinkle smoothing and enhancing the penetration of drugs through the stratum corneum (Uner, 2006). Besides the occlusion effect, the highly specific surface area of nanometer sized LNs enhances the contact of loaded-drug with stratum corneum (Jenning et al., 2000a; Schafer-Korting et al., 2007).

LNs have been intensively investigated to improve the skin and dermal uptake of several drugs such as vitamin E (Dingler et al., 1999), vitamin A (Castro et al., 2007; Jee et al., 2006; Jenning et al., 2000a; 2000b; Pople & Singh, 2006), vitamin K (Iscan et al., 2005), glucocorticoids (Jensen et al., 2010; Maia et al., 2000), triptolide (Mei et al., 2003), podophyllotoxin (Chen et al., 2006), isotretinoin (Liu et al., 2007b), econazole nitrate (Bhalekar et al., 2009; Sanna et al., 2007), psoralen (Fang et al., 2008), cyclosporin A (Kim et al., 2009), penciclovir (Lv et al., 2009), minoxidil (Silva et al., 2009), Siberian pine seed oil (Averina et al., 2010), oxybenzone(Gulbake et al., 2010), calcipotriol and methotrexate (Lin et al., 2010), buprenorphine (Wang et al., 2009), and non-steroidal anti-inflammatory drugs (NSAIDs) (Jain et al., 2005; Liu et al., 2010).

OCULAR DRUG DELIVERY

LN delivery is highly promising in ophthalmic drug delivery when compared with other conventional drug delivery systems. LNs can be formulated as nanosuspension in aqueous media due to the physical stability of LNs as liquid dosage forms (Mehnert & Mader, 2001). Nanoparticles can protect the drug against degradation, improve tolerance, and increase corneal uptake and intraocular half-lives (Kassem et al., 2007). Colloidal suspension of LNs can be administered as ophthalmic eye drop formulation with no blurred vision side effect (Mehnert & Mader, 2001). Many hydrophobic drugs have been successfully formulated as SLNs or NLCs containing tobramycin (Cavalli et al., 2002), pilocarpine (Cavalli et al., 1995), diclofenac sodium (Attama et al., 2008), flurbiprofen (Luo et al., 2010), timolol (Attama et al., 2009), cyclosporine A (Basaran et al., 2010; Gokce et al., 2009), or gatifloxacin (Kalam et al., 2010).

CYTOSOLIC DRUG DELIVERY

Cytosolic drug delivery refers to the internalization of nanocarriers to specific cell populations. The targets can be located in the endo-lysosome, cytoplasm, nucleus, mitochondria, or other organelles within the cell (Panyam & Labhasetwar, 2004). To exert its pharmacological effects, the drug should be delivered into the cytosol. However, cytosolic access of macromolecules (i.e., proteins, peptides, recombinant proteins, plasmid DNA) is challenging due to the low permeability of macromolecules through the cell membrane, high susceptibility to enzymatic degradation, undesirable toxicity to non-target cellular organelles, and extensive efflux from the cell by efflux transporters (i.e., multidrug resistance proteins and P-glycoproteins) (Chavanpatil et al., 2007; Panyam & Labhasetwar, 2004; Vasir & Labhasetwar, 2007). Most of the nanocarriers associated with the cell membrane are internalized into cells by means of endocytosis (Vasir & Labhasetwar, 2007). Various cationic nano-carrier systems including liposomes and poly-ethylene-imine (PEI) have been used for cytosolic drug delivery of genes. Among cationic nanocarrier systems, LNs have been recently reported to be useful for gene delivery due to their low toxicity, good storage stability, ease of scalability, and commercialization (del Pozo-Rodriguez et al., 2007; 2008; 2009). Surface charge modification of LNs with cationic moiety can assist in enhancing the cellular uptake of LNs. Cationic LNs are being extensively investigated for efficient intracellular delivery of macromolecules, such as genes, small interfering ribonucleic acids, and oligonucleotides (Basaran et al., 2010; Bondi & Craparo, 2010; Choi et al., 2008; Kim et al., 2008; Kuo & Chen, 2009; Kuo & Lin, 2008; Rudolph et al., 2004; Tabatt et al., 2004; Vighi et al., 2007; 2010a; 2010b).

PRODUCT DEVELOPMENT AND RELATED ISSUES IN LIPID NANOPARTICLES

Lipid-based nanoparticles were introduced in the 1990s as an alternative colloidal drug carrier (Joshi & Muller, 2009; Muchow et al., 2008; Muller et al., 2000). SLNs were used as colloidal carriers for intravenous/parenteral administration routes initially, but gradually interest shifted to oral administration and then dermal administration. One of the reasons is that the regulatory restrictions are relatively fewer for dermal products, allowing a faster launch to the market (Muller et al., 2007; Pardeike et al., 2009). Despite various advantages of LNs such as versatility, ease of scale-up, and low cost of solid lipids, there is no pharmaceutical product on the market yet (Joshi & Muller, 2009; Muchow et al., 2008; Pardeike et al., 2009). However, the time for product development and market introduction for cosmetic products is much shorter; consequently, several cosmetic LN products have been launched in many countries (Muller et al., 2007; Pardeike et al., 2009). LNs in cosmetic and pharmaceutical dermal products have been recently reviewed in depth by Muller et al. (2007) and Pardeike et al. (2009). Considering the relatively short history and potential advantages of LNs, the market introduction of pharmaceutical LN formulations is anticipated in the future.

LIPOSOMAL DRUG DELIVERY CARRIERS

OVERVIEW OF LIPOSOMES

Liposomes were initially reported to serve as a model for cell membranes in biophysical studies (Bangham et al., 1965). Liposomes are phospholipid vesicles (i.e., phospholipid self-assemlies) with varying sizes, ranging from 50 to above 1000 nm, which are formed by unilamellar or multilamellar colloidal self-assemblies with an aqueous phase both inside and between the bilayers (Gershkovich et al., 2008). Recently, liposome formulations are focused on targeting to reduce toxicity and increase accumulation at the target site (Samad et al., 2007). The main advantages of liposomes as a drug delivery system are the following: (1) they have a versatile structure that can be easily tailored in order to possess the properties needed for each specific application; (2) they can accommodate any type of drug molecules either in their aqueous compartments (hydrophilic drugs) or in their bilayers (lipophilic drugs) or both (amphiphilic drugs); and (3) liposomes are non-toxic, non-immunogenic, and biodegradable (Antimisiaris et al., 2008). The structural variance of liposomes is dependent upon phospholipid types (i.e., composition) and process conditions (Barenholz & Crommelin, 1994) which influence size distributions and liposome lamellarities (Table 9-A). Drug release, in vivo stability and biodistribution are determined by the size of the vesicles, their surface charge, surface hydrophobicity, and membrane fluidity (Senior, 1987). Liposomal size is directly correlated with the volume of the aqueous core. The volume of the aqueous core in turn determines the capacity to carry a hydrophilic drug, if the drug is truly entrapped. At identical phospholipid concentrations, dispersions of bigger liposomes carry a bigger aqueous volume than smaller ones, and bigger liposome size is accompanied by a broader size distribution and a higher number of lamellarity (Brandl, 2001). The nature of the polar head functional group determines the net charge. Physical stability of liposomes is influenced by the magnitude and range of electrostatic interactions.

The fundamental interfacial and biophysical property of phospholipids is their amphiphilic nature: a polar head group covalently attached to two hydrophobic hydrocarbon acyl tails (Fig. 9.2). When these phospholipids are exposed to an aqueous environment, interactions between themselves (hydrophilic interactions between polar head groups, hydrophobic interactions between the adjacent acyl chains due to the hydrophobic effect, van der Waals interactions, and hydrogen bonding between the polar headgroups with water molecules) lead to spontaneous formation of closed

Table 9-A. Nomenclature of Liposomes

Name	Description	Size
MLV	Multilamellar large vesicles	>0.5 μm
OLV	Oligolamellar vesicles	0.1–1 μm
UV	Unilamellar vesicles	all sizes
SUV	Small unilamellar vesicles	20–100 nm
LUV	Large unilamellar vesicles	>100 nm
GUV	Giant unilamellar vesicles	>1 μm
MVV	Multivesicular vesicles	usually >1 μm
Name	**Description**	**Preparation Method**
REV	Vesicles prepared by reverse-phase evaporation	Reverse-phase evaporation
MLV-REV	Multilamellar vesicles prepared by reverse-phase evaporation	Reverse-phase evaporation
FATMLV	Multilamellar vesicles prepared by Freeze and thaw method	Frozen and thaw method
VET	Vesicles prepared by extrusion	Extrusion
FPV	Vesicles prepared by French press	French press
DRS	Vesicles prepared by dehydration-rehydration	Dehydration-rehydration

bilayers (Frezard, 1999). Liposomes have been prepared from a variety of synthetic and naturally occurring phospholipids. The choice of phospholipids for liposomal drug delivery carriers is related to the desired physical stability of the liposome formulation both in vitro and in vivo and the drug to be loaded into the liposomes (Brandl, 2001). Cholesterol is sometimes contained in certain types of liposomal formulations to improve the stability of liposomes in serum, to enhance the entrapment of water soluble molecules inside the bilayer membranes, and to increase the fluidity of the bilayer (Symon et al., 1999; Weiner et al., 1989). Nevertheless, liposome formulations containing cholesterol (an oxidant) are known to have stability problems due to cholesterol autoxidation (Crommelin & Storm, 2003). The quality of raw materials of phospholipids depends primarily on the application: parenteral, inhalation, topical, or oral (Uhumwangho & Okor, 2005). Furthermore, it should be noted that regulation of pharmaceutical products differs from those in other industries like the cosmetics and food industries.

The interfacial and spreading properties of phospholipid self-assemblies (Mansour & Zografi, 2007b; Mansour et al., 2008) are important biophysical features that can be exploited for functionality in various applications including nanotechnology, nanomedicine, drug delivery, and cosmetic/nutriceutical applications. In addition, phospholipid bilayer phase behavior and membrane composition are known to be altered in normal vs. disease states, this in turn is reflected in the bilayer interfacial and spreading properties (Mansour et al., 2001). The hydration properties of solid state bilayers at various levels of hydration and the effect on phase behavior/phase transitions (Mansour & Zografi, 2007a) are of considerable recent interest in an effort to stabilize functional cells, biomembranes, and biomedical products in the solid state, such that activity and functionality are retained upon full hydration. This is of notable importance in food, pharmaceuticals, biomedical products, vaccines, and biodefense applications.

CLASSIFICATION OF LIPOSOMAL DRUG DELIVERY CARRIERS

CONVENTIONAL LIPOSOMES. Conventional liposomes (Fig. 9.3) are typically composed of only phospholipids and/or cholesterol, and they can have a variance with their own physicochemical properties such as size, lipid composition, surface charge, and number and fluidity of the phospholipid bilayers (Gregoriadis, 1991; Lasic, 1996). Conventional liposomes are characterized by a relatively short blood circulation time. With parenteral administrations, they are easy to accumulate rapidly in the phagocytic cells of the mononuclear phagocyte system (MPS) and to the reticuloendothelial system (RES) uptake (Gabizon et al., 1994; Martin & Papahadjopoulos, 1982). The major organs of accumulation are the liver and the spleen. With MPS-targeted distribution behavior, conventional liposomes may be attractive candidates for drug delivery to MPS macrophages (Bakker-Woudenberg, 2002; Daemen et al., 1995; Ten Hagen et al., 1995; Killion & Fidler, 1994).

LONG-CIRCULATING LIPOSOMES: STEALTH®. Long-circulating liposomes (Fig. 9.3) were studied by coating the liposome surface with inert, biocompatible polymers, such as poly ethylene glycol(PEG), which form a protective layer over the liposome surface and slow down liposome recognition by opsonins and, therefore, subsequent clearance of liposomes (Blume & Cevc, 1993; Klibanov, 1990; Malmsten, 2002; Torchilin, 2005). Such PEG-coated liposomes are also called "stealth" or "sterically stabilized" liposomes, the former term referring to their MPS-escaping capability and the latter term to the steric stabilization mechanism held responsible for the induction of long circulation times (Storm et al., 1995). The important factor of protective polymers is their flexibility, which allows a relatively small number of surface-grafted polymer molecules to create an impermeable layer over the liposome surface (Torchilin et al., 1994; Torchilin & Trubetskoy, 1995).

Recently, tumor-targeted liposomes were focused for the long-circulating liposomes. After PEG-liposomes accumulate at the target site, through the enhanced permeability and retention (EPR) effect (Maeda et al., 2001), the PEG coating is detached under the action of targeted local pathological conditions, such as the difference of pH between normal and tumor cells. Detachable PEG conjugates have been described as the dithiobenzylurethane linkage between PEG and an amino-

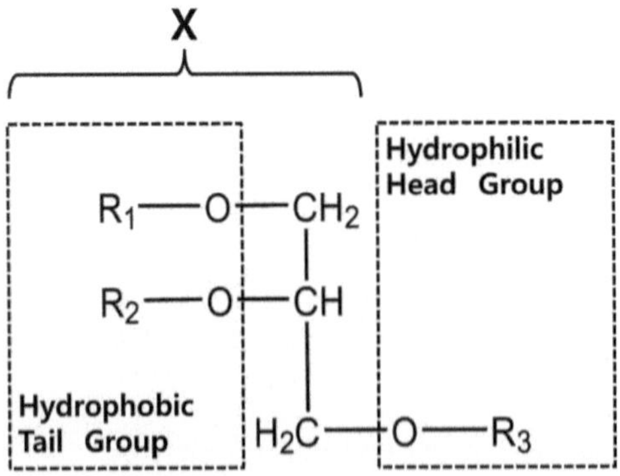

Polar Head Groups

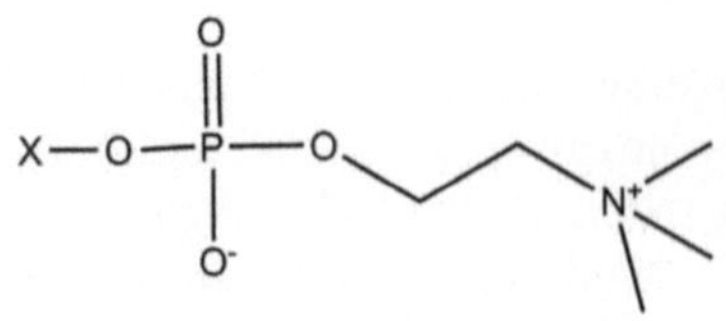

Phosphatidylcholine

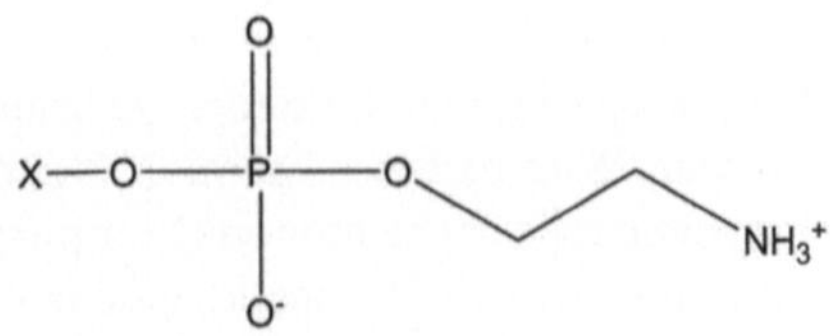

Phosphatidylethanolamine

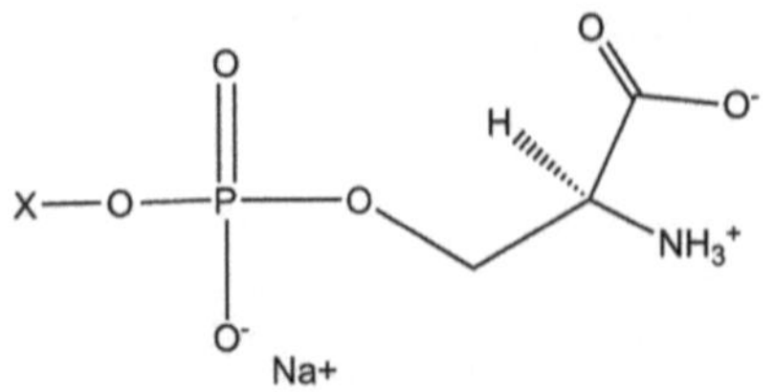

Phosphatidylserine (sodium salt)

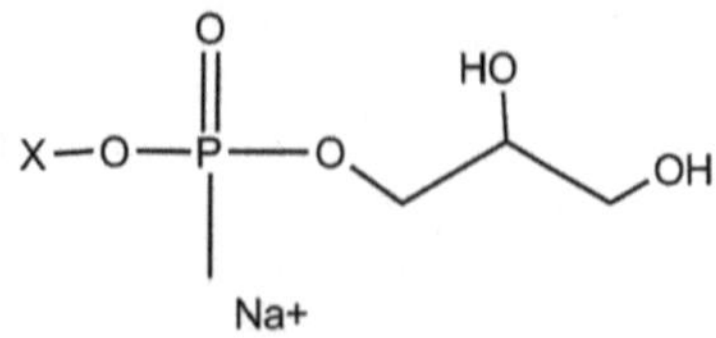

Phosphatidylglycerol (sodium salt)

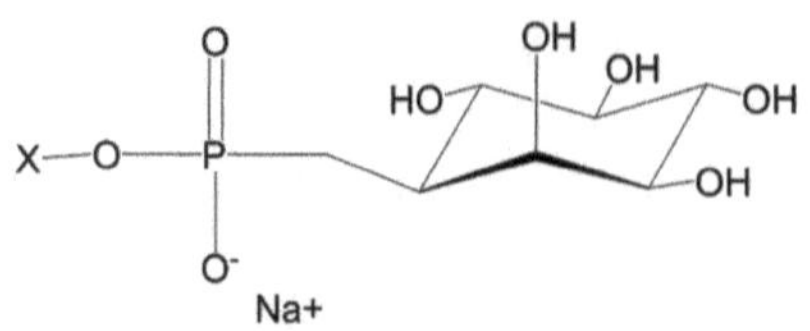

Phosphatidylinositol (sodium salt)

Fig. 9.2. Some examples of chemical structures of phospholipids (ChemDraw Ultra® Ver. 10.0).

Hydrocarbon Acyl Chain Tail Groups

Saturated fatty acids

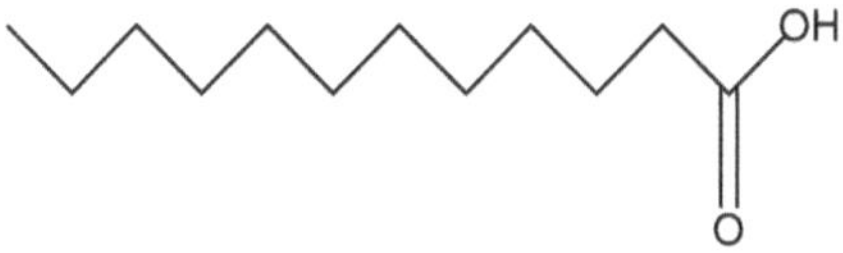

Lauric acid

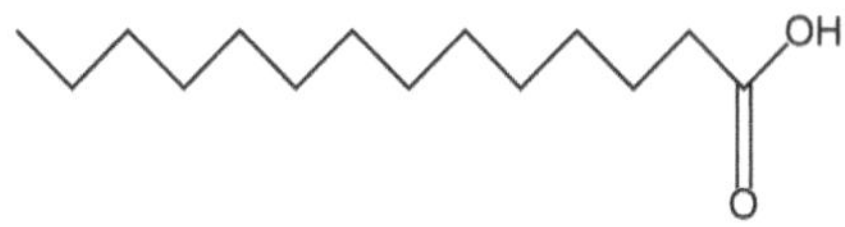

Myristic acid

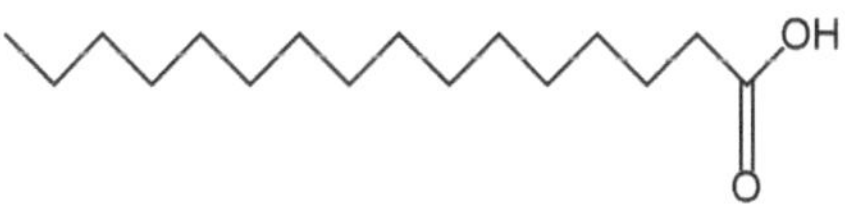

Palmitic acid

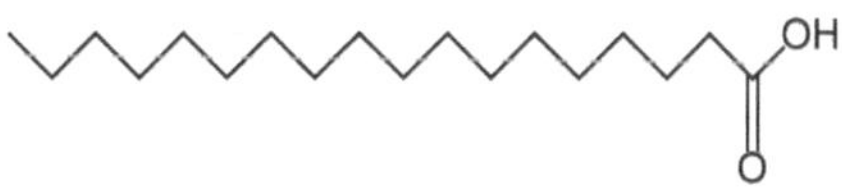

Stearic acid

Unsaturated fatty acids

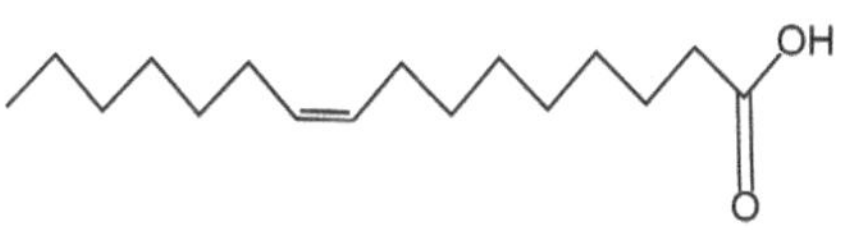

Palmitoleic acid

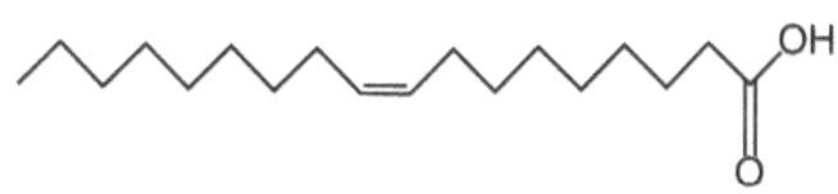

Oleic acid

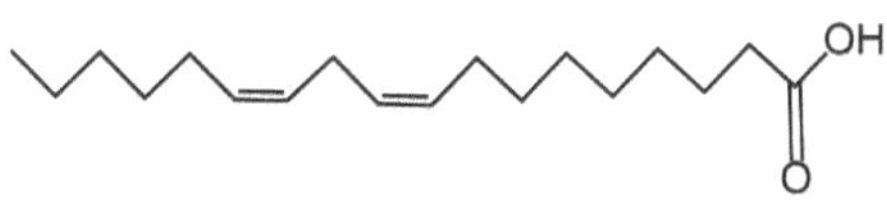

Linoleic acid

Fig. 9.2., cont. Some examples of chemical structures of phospholipids (ChemDraw Ultra® Ver. 10.0).

containing substrate (Zalipsky et al., 1999). Long-circulating liposomes in cancer chemotherapy (Cattel et al., 2003; Gabizon, 2003) were reported for their potential use for other purposes, such as a diagnosis with imaging agents.

IMMUNOLIPOSOMES: TARGETED LIPOSOMES. Immunoliposomes (Fig. 9.3) have specific surface-attached ligands, which are possible to recognize and bind to targeted cells in tissues and organs. Immunoglobulin G (IgG) classes and its fragments are the

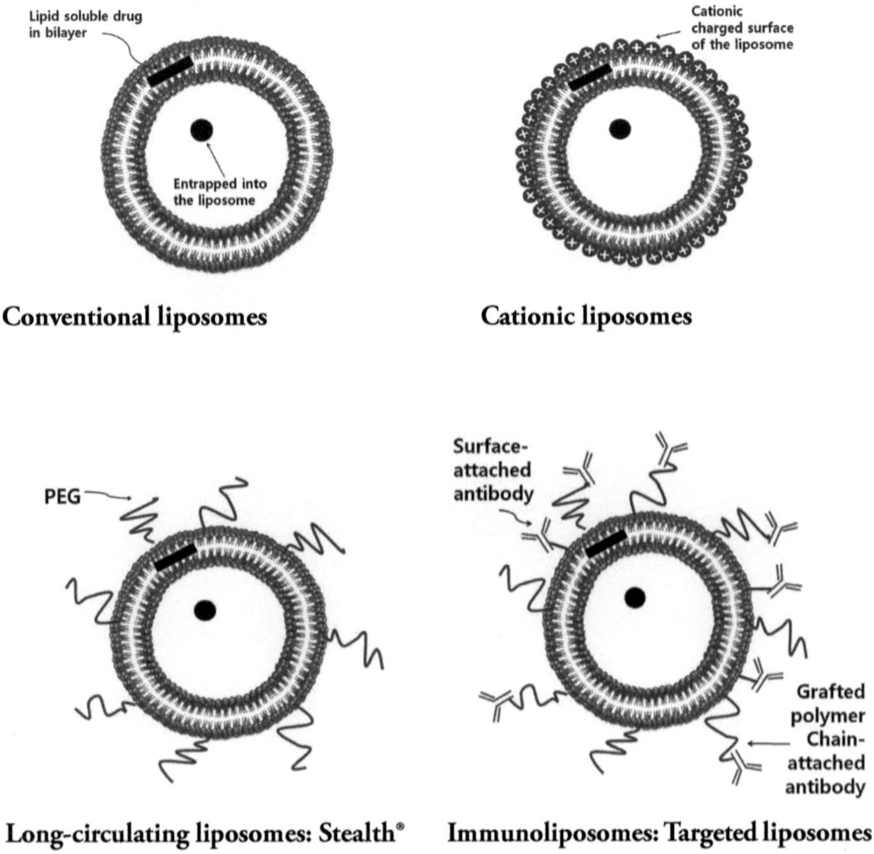

Fig. 9.3. Schematic representations of four liposome categories.

most widely studied for targeting moieties of liposomes, without affecting liposomal integrity or the antibody properties (Torchilin, 1985). Despite the improvements in targeting efficacy of immunoliposomes, there is a limitation with insufficient circulating time in vivo for the interaction between the target cells and targeted liposomes. Long-circulating immunoliposomes may be advanced further by the combination of long-circulating conception with protective polymers and targeted conception with specific surface-attached ligands (Abra et al., 2002; Blume et al., 1993; Torchilin et al., 1992; 2001). Early trial studies have been performed by simple co-coating of an antibody and PEG on the surface of the same liposome, which might allow interferences between the protective polymer and target recognition with the targeting moiety of ligand (Torchilin et al., 1992). To attach the targeting ligand via a PEG spacer arm, the ligand is extended outside of the dense PEG brush, which reduces the steric hindrance of being bound to the target (Blume et al., 1993; Torchilin et al., 2001).

CATIONIC LIPOSOMES. Cationic liposomes (Fig. 9.3) were developed for advanced delivery of genetic material (Storm & Crommelin, 1998) and are described in detail in the previous chapter of this book.

PROLIPOSOMES. Proliposomes are defined as dry free-flowing particles, which have the ability to form a liposomal dispersion when dispersed in an aqueous phase (Rhee & Mansour, 2011). Proliposomes can be used as pulmonary drug delivery carriers as dry powder inhalers (DPIs). Liposomal nanomedicines have the ability to enhance drug residence time in the lungs with controlled release property, to prohibit enzymatic drug degradation and its rapid removal through the clearance mechanism, and to reduce lung irritation with lung-surfactants such as dipalmitoylphosphatidylcholine (DPPC) and dipamytoylphosphatidylglycerol (DPPG). Liposomal nanomedicines can be prepared by various methods such as freeze-drying, spray-drying, spray freeze-drying, and supercritical-fluid technology to achieve long-term stability and overcome problems associated with the suspension form of liposomes (Courrier et al., 2002; Niven, 1995; Rehman et al., 2004; Sweeney et al., 2005; Tom & Debenedetti, 1991).

Pulmonary delivery of drugs encapsulated in conventional liposomal dispersions using a nebulizer or pressurized meter dose inhalers (pMDIs) have been clinically investigated (Patel et al., 2009). Conventional liposomes in suspension form have problems of drug leakage during storage including phospholipid degradation, aggregation, and fusion of liposomes (Hu & Rhodes, 1999). In addition, pMDIs and nebulizers have a tendency for lower drug deposition at the right target site with the required dose, compared to dry powder inhalation (DPI) (Hickey & Mansour, 2009). For oral drug delivery of liposomes, formulation into proliposomes may improve the ability to retain their integrity at the site of absorption (Deshmukh et al., 2008). Free-flowing proliposomal powder formulations can be incorporated into capsules or compressed into tablets (Song et al., 2002). There are some reports (Deshmukh et al., 2008; Payne et al., 1986; Song et al., 2002; Yan-yu et al., 2006) on the usefulness of oral proliposomes but some problems with oral proliposomes still remain in particle size distribution control and low encapsulation efficiency (Junping et al., 2000).

ENGINEERING DESIGN AND PREPARATION TECHNIQUES OF LIPOSOMAL/PROLIPOSOMAL NANOMEDICINES

In the last decade, numerous preparation techniques were successfully established for liposomal nanomedicines, but most of them have not been proven suitable for an industrial scale. Although many preparation methods were investigated in the 1980s and 1990s, less attention has been paid to the transfer of technology to industrial production. Thus, scale-up studies are of great importance for the preparation of clinical material as well as for marketed products that are sterile, well-characterized, and stable (Wagner & Vorauer-Uhl, 2011).

PREPARATION BY FILM METHODS. In this procedure, the phospholipid mixtures are dissolved in an organic solvent, which is then evaporated by rotary evaporation under reduced pressure to form a dried thin film of phospholipids. MLVs form spontaneously when an excess volume of aqueous buffer is added to the dried phospholipids film. Liposomes can be separated by centrifugation or by filtration. There are critical formulation factors: size of the aqueous buffer, time for hydration of the dried film, and agitation condition (Kirby & Gregoriadis, 1984). The film method has several advantages as an easy application of many different kinds of lipid mixtures. In addition, the method can be easily performed with high phospholipid concentrations. Various formulation factors of this method have been reported with modifications of organic solvents, lipid drying, and film rehydration. One major drawback of this method (Wagner & Vorauer-Uhl, 2011) is the difficulty of scaling up to several tens of liters. Furthermore, the process becomes more time and cost intensive since additional processing is needed for a well-defined liposome suspension. Several techniques for size reduction, such as sonication, homogenization, and microfluidization with high-pressure homogenization, have been established in order to make the heterogeneous vesicles more uniform by breaking down the large MLVs.

The French press (Barenholzt & Amselem, 1979) originally was established for breaking up cells under milder and more appropriate conditions compared to the ultrasound technique, since phospholipids as well as protein drugs or other sensitive compounds may be degraded during the sonication procedure (Wagner & Vorauer-Uhl, 2011). This technique was not suitable for large-scale production. However, microfluidization or high pressure homogenization reduced the size of vesicles by continuous and large scale application of the French press technique at high pressure conditions (Brandl et al., 1990; Mayhew et al., 1984). Liposomes are generally formed in the size range of 50 to 100 nm by these processes. These techniques are suitable for large-scale production and sterile liposome preparation. The most prominent scalable downsizing method is the extrusion method (Mayer et al., 1986; Schneider et al, 1994). In this method, preformed vesicles are forced through defined membranes with a much lower pressure than French press method.

REPLACEMENT METHODS OF ORGANIC SOLVENTS BY AQUEOUS MEDIA. Replacement can be performed by injection method, proliposome-liposome method, reverse-phase evaporation method, or double emulsion method.

Ethanol Injection Method is the immediate dilution of the ethanol in the aqueous phase forming liposomal drug delivery carriers with bilayer planar fragments (Lasic, 1995; Maitani et al., 2001; Stano et al., 2004). The formulation parameters (Kremer et al., 1977) influence liposomal size and drug encapsulation efficiency.

They determined the lipid concentration in ethanol as the only liposome formation influencing parameter. Neither stirring rate of the aqueous phase nor injection velocity had a significant influence on liposomes size and size distribution. In a recent report, however, it was noted that liposomal size could be controlled by the lipid concentration at the injection point which was defined by the lipid concentration in ethanol, the injection hole diameter, the injection pressure, and the flow rate of the aqueous phase (Wagner et al., 2006). This method has many advantages, such as ease of scale up and low toxicity via the use of ethanol.

Proliposome-Liposome Method is based on the conversion of the initial proliposome preparation into liposomes dispersion by dilution with an aqueous phase (Maitani et al., 2001). Reproducible liposome preparation is feasible in a controlled manner by exact control of the dilution rate and process temperature. It is claimed that their method is easy to scale up, which makes this method an alternative approach for the production of liposomes for clinical application (Turanek et al., 1997; 2003).

Reverse-Phase Evaporation (REV) is based on oil/water emulsion, which is diluted with further aqueous phase for liposome formation (Turanek et al., 2003). This method has the advantages of high encapsulation rate, compared to other methods. Double emulsion technique had a potential to further improve the encapsulation rates.

DETERGENTS REMOVAL METHODS. Detergents, such as bile salts or alkylglycosides, are used for solubilization of lipids in micellar systems. The size and shape of the resulting vesicles depend on the chemical nature of the detergents, concentration, and phospholipid types. Common detergent removal methods from mixed micelles are dilution, gel chromatography, dialysis through hollow fibers or membrane filters, and by adsorption to hydrophobic resins or cyclodextrins (Wagner & Vorauer-Uhl, 2011). A new technique for detergent removal was reported (Wagner et al., 2002) as the cross-flow injection technique with detergent dilution within one operational step. Lipid vesicles are formed immediately after injection into a micellar protein solution. This multiple injection technique, a one-step detergent dilution/vesicle forming process, was achieved with high yield of encapsulation of water-soluble proteins (Wagner et al., 2007).

PREPARATION TECHNIQUES OF PROLIPOSOMES. Proliposomes have been recently examined as dry powder formulations for pulmonary drug delivery (Bi et al., 2008; Joshi & Misra, 2001; Shah & Misra, 2004a; White et al., 2005). Proliposomes dry powder formulations have been shown to be very promising in pulmonary delivery to control drug release and offer favorable biocompatibility, biodegradability, and non-immunogenicity properties (Mansour et al., 2009; Misra et al., 2009). Various proliposomes have been developed using particle engineering techniques such as supercritical fluid (SCF) technology, solution-enhanced dispersion with supercritical fluids (SEDS), supercritical fluid extraction of emulsions, and spray-drying to achieve

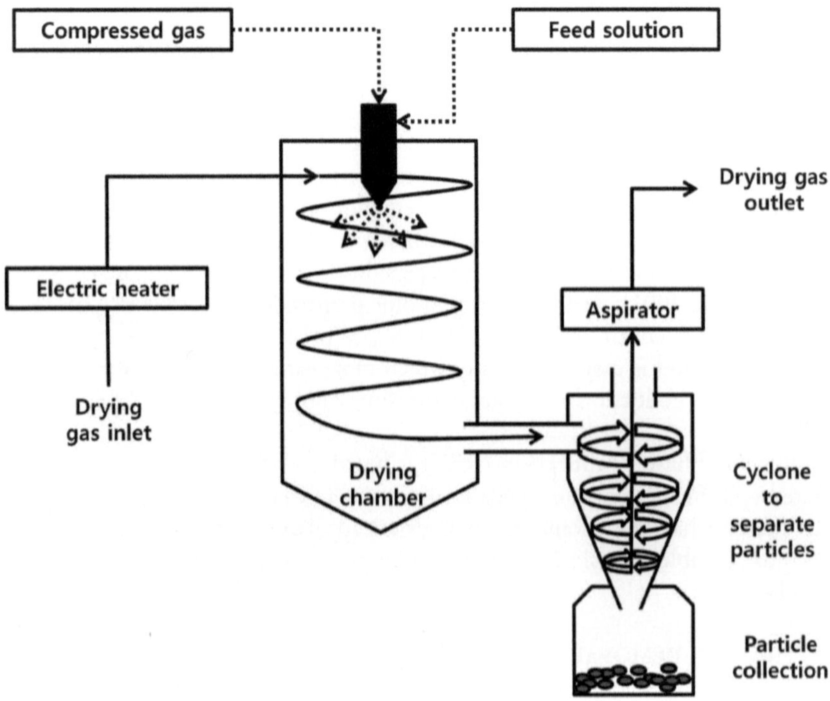

Fig. 9.4. A schematic representation of the spray-drying process.

intelligently designing proliposomes for efficient delivery of a wide variety of therapeutic agents (Hickey & Mansour, 2008).

Spray-Drying Method is an advanced pharmaceutical manufacturing process used to efficiently produce respirable particles in the solid-state (Duddu et al., 2002; Mosen et al., 2004). The feed solution (or feed suspension or feed emulsion) is pumped to the nozzle where it is atomized by gas (Mansour et al., 2009). Many spray drying parameters can be precisely controlled to produce tailored particles. The atomized feed is then dried by preheated drying gas in the larger chamber (first drying process) followed by secondary drying in the smaller vessel/cyclone, thus forming dry particles that are readily collected in the sample collector (Fig. 9.4). This process can be to be used for thermolabile materials such as biological drugs, proteins, and phospholipids (Freitas et al., 2004; Maa et al., 1998; Stahl et al., 2002). Compositional homogeneity in the drug powder can be achieved by spray drying from a multi-component solution (e.g., drug and excipient solution). In addition, spray-drying can result in uniform particle morphology (Gilani et al., 2005; Vidgren et al., 1987; White et al., 2005).

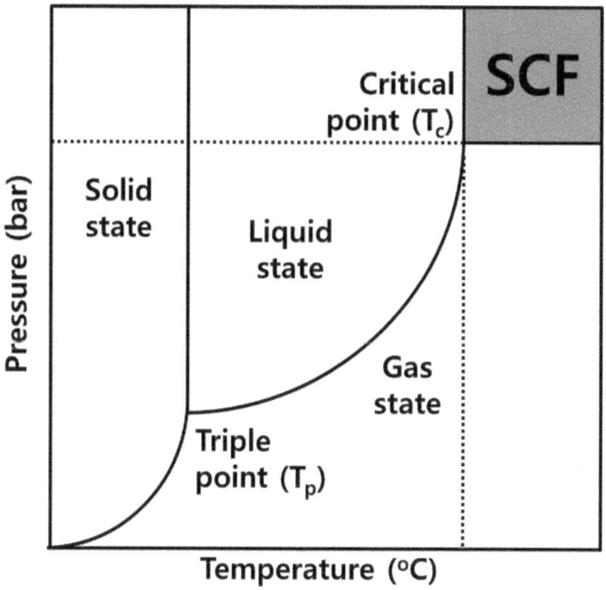

Fig. 9.5. Phase diagram with the SCF region, which is above T_c where a single phase liquefied gas exists. SCF indicates supercritical fluid.

Spray-Freeze-Drying Method (SFD) is an advanced particle engineering method which combines spray-drying and freeze-drying processing steps (Mansour et al., 2009), including the atomization of an aqueous drug solution into a spray chamber filled with a liquid nitrogen or halocarbon refrigerant such as chlorofluorocarbon or fluorocarbon (Rogers et al., 2001). SFD is capable of producing porous particles with high fine particle fraction (FPF) at subambient temperatures (Maa & Prestrelski, 2000). Thermolabile protein and peptide substances such as insulin (Maa et al., 1999; Yu et al., 2004) and plasmid DNA (Kuo & Hwang, 2004) can also be formulated into dry powder inhalation products by this method.

Supercritical Fluid Technology (SCF) is the controlled crystallization of drugs from dispersion in supercritical fluids such as carbon dioxide (Mansour et al., 2009; Steckel et al., 1997). The supercritical antisolvent precipitation (SAS), a type of SCF technology, is based on rapid precipitation when a drug solution is brought into contact with supercritical CO_2, and supercritical fluid extraction of emulsions (SFEE) is based on extraction of the organic phase in oil-in-water or multiple emulsions using supercritical CO_2 (Chattopadhyay et al., 2006; Shekunov et al., 2006). Fig. 9.5 illustrates the general phase diagram of an SCF, which is a liquefied gas existing as a single phase above its critical point, T_c, and possessing desirable properties of both its

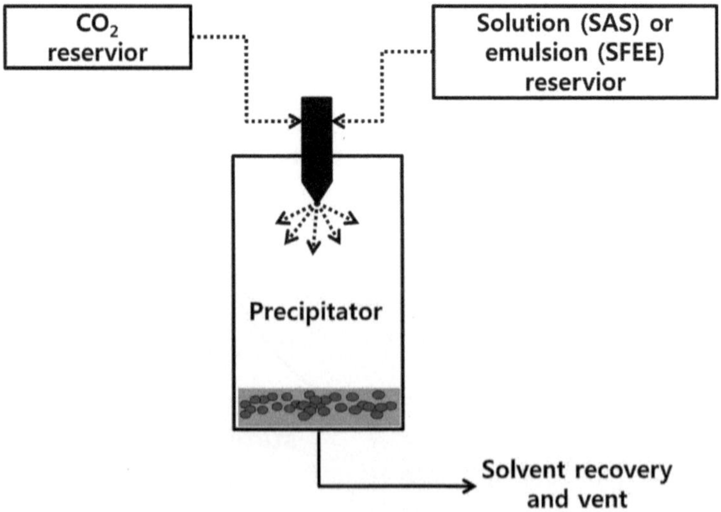

Fig. 9.6. A schematic representation of SAS and SFEE.

liquid state and its gas state (Mansour & Hickey, 2007). The schematic representation of the SAS and SFEE processes is shown in Fig. 9.6.

CHARACTERIZATION AND EVALUATION OF LIPOSOMAL/PROLIPOSOMAL NANOMEDICINES

Physicochemical characterization of nanomedicines is critical for product quality and performance (Burgess et al., 2004; Campos & Shapiro, 1999; Chen, 2008; Gabizon & Barenholz, 1999).

Percent Drug Entrapment (PDE) of the encapsulated drug in liposomes is measured by separating the unentrapped drug from hydrated liposomal suspensions. For water soluble drugs, centrifugation at high speed to form the liposomal pellet or passing through the Sephadex column is possible. In contrast, in the case of water insoluble drugs, low speed centrifugation is needed. After this centrifugation process, liposomes are lysed with the surfactant triton X 100 and the concentration of drug is analyzed with a suitable analytical method (Akwete & Hsu, 1993; Vyas et al., 2005).

Size and Size Distribution of liposomes is examined by particle size analyzers based on laser size diffraction analysis and photon correlation analysis (Amselem et al., 1993; Schneider et al., 1995).

Zeta Potential is determined by the presence of high positive or negative surface charge on liposomes and is significantly correlated with liposomes physical stability and inhibition of aggregation by charge repulsion between similarly charged liposomal particles (Crommelin & Storm, 2003).

Morphology is widely evaluated by various microscopy, such as polarizing microscopy, scanning electron microscopy (SEM), transmission electron microscopy (TEM), and Atomic force microscopy (AFM). SEM is used to visualize the surface morphology of particles (Betagiri et al., 1993; Mansour et al., 2009). The resolution allows identification of specific surface features and asperities that lead to mechanical interlocking (i.e., structural cohesion) and that contain high-energy "active" sites on the surface which influence surface energetic properties and interparticulate interactions and ultimately influence aerosol dispersion performance (Hickey et al., 2007). Atomic force microscopy (AFM) can image surface nanotopography, measure surface energy, and measure interparticulate forces in the mesoscopic scale from10^{-6} to 10^{-9} m, (Bunker et al., 2005).

Entrapped Volume is the aqueous entrapped volume per unit quantity of lipid (μl/μmol or μl/mg) and is a representative value of the lamellarity of liposomes. Higher lamellarity indicates high entrapment value. The entrapped volume can be enhanced using charged lipids so that the interlamellar distance can increase for increased water accumulation (Shah & Misra, 2004b).

Oxidative Index is characterized with percent phospholipid oxidation at various time intervals. Unsaturated phospholipids are more easily oxidized (due to the presence of double bonds) than saturated phospholipids, and their oxidative index is a measure of their oxidized state and hence indicates their stability (Kent et al., 2007; Mina et al., 2002).

X-Ray Powder Diffraction (XRPD) directly measures the degree of molecular long-range vs. short-range order (Mansour et al., 2009) in order to assess the presence/absence of crystallinity and non-crystallinity, respectively. It gives information about the extent and nature of crystallinity and molecular order for solid-state materials (Bates et al., 2006; Newman & Byrn, 2003).

Differential Scanning Calorimetry (DSC) is a thermal analytical method for phase behavior studies at various hydration levels, polymorphs, hydrates, binding interactions, amorphocity, thermotropic, and lyotropic phase transitions of pharmaceutical materials (Mansour et al., 2009). DSC directly measures the gain and loss of enthalpy (as a function of temperature) associated with order-to-disorder (e.g., melting) and disorder-to-order (e.g., crystallization) phase transitions (Saleki-Gerhart et al., 1994).

Karl Fischer Titration (KFT) is used to quantify the water content of proliposomal nanomedicines, which can influence flow properties, solid-state phase behavior and solid-state stability of proliposomes.

PHARMACEUTICAL APPLICATIONS OF LIPOSOMAL NANOMEDICINES

PARENTERAL LIPOSOMAL NANOMEDICINES. Liposomal nanomedicines as parenteral drug carrier systems for both hydrophobic and hydrophilic compounds have advantages, such as controlled drug release, high biocompatibility, low toxicity, lack of immune system activation/suppression, reduction in drug toxicity, and increased drug therapeutic efficacy (Metselaar et al., 2002). In past decades, several parenteral liposomal nanomedicines have been successfully developed as commercial marketed products for lipophilic drugs, like amphothericin B (Albelcet®, AmBisome®) and benzoporphyrin (Visudyne®, Verteporfin for injection) (Banerjee, 2001; Gregoriadis, 1995). Recently there have been reports on an "active targeting" effect of intelligent liposomal drug delivery systems based on liposomal surface modification, such as site-specific ligands, trigger-release systems, and multifunctional liposome systems (Dwivedi et al., 2009; Mikhaylova et al., 2009; Simard & Leroux, 2009; Yamada et al., 2008). Although they have many pharmaceutical and therapeutic advantages, commercial marketed liposomal products are still somewhat limited due to some hurdles in the large-scale manufacturing process, such as sterilization, poor storage stability, low encapsulation efficiency, and relatively high manufacturing cost of liposomal formulations (Joshi & Muller, 2009; Wagner & Vorauer-Uhl, 2011).

ORAL LIPOSOMAL NANOMEDICINES. There have been limitations to oral delivery of liposomes due to unpredictable absorption profiles and insufficient protective effects of conventional liposomes against the rapid enzymatic degradation within the gastrointestinal tract through the interaction with bile salts (Singh et al., 2008). Advanced surface-modified liposomal systems have been alternatively prepared for intestinal absorption enhancement of protein and peptide drugs. There were early attempts to use polymerized liposomes as potential carriers for oral protein and peptide drugs (Arien et al., 1993; Chen et al., 1996; Choudhari et al., 1994). Chitosan-coated liposomes for insulin oral delivery have been shown with hypoglycaemic effects in the mice animal model (Wu et al., 2004). Ganglioside-loaded liposomes were reported to be stable in different biological media and to survive the gastrointestinal tract (Taira et al., 2006). PEG-coated liposomes have been investigated for oral delivery of protein and peptide drugs (which are inherently unstable in the gastrointestinal tract) such as recombinant human epidermal growth factor, salmon calcitonin, glucagon-like peptide-1, and vaccines (Li et al., 2003; Minato et al., 2003; Xing et al., 2003;Yamabe et al., 2003). Proliposomes have been studied for the oral delivery of protein and peptide drugs with the extension of liposome integrity at the intestinal absorption site (Deshmukh et al., 2008; Junping et al., 2000; Song et al., 2002; Yan-yu et al., 2006).

PULMONARY LIPOSOMAL NANOMEDICINES. Liposomal medicines can be comprised of phospholipids, which are endigenous to lung surfactants, and this property makes liposomes attractive candidates for drug delivery vehicles (Justo & Moraes, 2003). Currently marketed in the USA and several countries around the world are lung surfactant replacement pharmaceutical products for the treatment of respiratory distress syndrome (RDS) (Rhee & Mansour, 2011). The marketed pharmaceutical phospholipid product for the treatment of respiratory distress syndrome, Alveofact®, is an efficacious synthetic lung surfactant product (Muller et al., 2000).

The effect of combined liposomal nanomedicine of paclitaxel and cyclosporin A on pulmonary metastases of renal-cell carcinoma in the mice animal model have been reported (Koshkina et al., 2004). As liposomal paclitaxel nanomedicine, pulmonary metastases were also achieved in a murine renal carcinoma model (Koshkina et al., 2000). Proliposomes were investigated as spray-dried powder formulations of liposomal nanomedicines with SOD (Lo et al., 2004), tacrolimus (Chougule et al., 2007), insulin (Bi et al., 2008), and rifampicin (Changsan et al., 2009). Another rifampicin liposomal nanomedicine aerosol was reported in the treatment of tuberculosis (Vyas et al., 2004). Pulmonary gene delivery using cationic liposomes has the advantage of self-assembly with DNA material through favorable cationic–anionic electrostatic interaction, evasion from complement inactivation after in vivo administration, low cost, and relative ease of production of nucleic acid–liposome complexes on a large scale (Densmore, 2006; Morille et al., 2008; Templeton, 2009).

DERMAL AND TRANSDERMAL LIPOSOMAL NANOMEDICINES. Liposomes comprising of dipalmitoylphosphatidylcholine (DPPC) and cholesterol (CH) increased the concentration of triamcinolone acetonide in the epidermis and dermis and reduced percutaneous absorption compared with a standard ointment (Mezei & Gulasekharam, 1980; 1982). Liposomal local anaesthetics were investigated to provide local anaesthesia to intact skin with tetracaine (Foldvari, 1994; Gesztes &Mezei, 1988), lidocaine (Foldvari et al., 1990), lidocaine and tetracaine (Planas et al., 1992), eutectic mixture of local anaesthetics (EMLA, 2.5% lidocaine and 2.5% prilocaine) in humans (Hung et al., 1997), and benzocaine (Mura et al., 2007). Liposome-encapsulated antibodies were reported to have rapid distribution into the deep cutaneous regions of piglet skin with increased percutaneous absorption compared to aqueous solutions (Artman et al., 1990a; 1990b). Localization and penetration of liposomes into human skin has also been of interest (Maghraby et al., 2008). Cevc and Blume (1992) reported that certain types of lipid vesicles (ultradeformable vesicles) can penetrate intact to the deep layers of the skin and may progress far enough to reach systemic circulation. Furthermore, successful systemic delivery of insulin by ultradeformable vesicles has been reported in mice and human studies (Cevc et al., 1995; 1996).

OCULAR LIPOSOMAL NANOMEDICINES. Liposomal nanomedicines for ocular administration present unique advantages for anterior and posterior segment administration (Ebrahim et al., 2005). In the anterior segment, liposomal formulations are investigated for achieving superior corneal penetration. Posterior segment applications have been focused to extend clearance times, to reduce toxicity, and to be delivered at the specific retinal pigment epithelium (RPE). A marketed liposomal spray, reported to combine long activity with minimal visual distortion, is indicated for use in mild-to-moderate dry-eye syndromes with spray onto the closed eyelids of the patients (Angelucci, 2001). Several antibiotics have been investigated in liposomal form such as norfloxacin (Lin et al., 1996), tobramycin (Assil et al., 1991; Frucht-

Perry et al., 1992), and gentamicin (Barza et al., 1984). It is reported that negative charged acyclovir liposomes are superior to positive charged liposomes, and free drugs can penetrate into the cornea (Law et al., 2000). Liposomal tacrolimus has been reported to give higher ocular tissue concentrations than an oil-dissolved preparation of the drug (Pleyer et al., 1993). Intravitreal injections and depot systems of liposomal formulations have been investigated to deliver drug in adequate posterior segment. Intravitreal formulation of liposomal cyclosporine, an immunosuppressive drug, also has been investigated to take a longer elimination half life as compared to free drug in rabbits (Alghadyan et al., 1988a; 1988b).

CYTOSOLIC LIPOSOMAL NANOMEDICINES. Gene delivery using non-viral approaches has been extensively studied as a basic tool for intracellular gene transfer and gene therapy (Gao et al., 2007). As compared with viral gene delivery systems, non-viral systems have also been developed such as cationic lipid-based liposomes with obvious advantages in that they are easy to prepare, reasonably cheap and non-immunogenic (Felgner & Ringold, 1989). Cationic lipids can be mixed with DNA to form a positively charged complex termed a lipoplex (Wasungu & Hoekstra, 2006). There are various reagents for cationic lipid transfection such as N-[1-(2,3-dioleyloxy) propyl]-N,N,N-trimethylammonium chloride(DOTMA) (Behr, 1994; Budker et al., 1996; Farhood et al., 1994; Felgner et al., 1987; Kikuchi et al., 1999; Wheeler et al., 1996), [1,2-bis(oleoyloxy)-3-(trimethylammonio)propane] (DOTAP) (Leventis & Silvius, 1990), 3β[N-(N',N'-dimethylaminoethane)-carbamoyl] cholesterol (DC-Chol) (Gao & Huang, 1991), and dioctadecylamidoglycylspermine (DOGS) (Behr et al., 1989). Dioleoylphosphatidylethanolamine (DOPE), a neutral lipid, is often used for conjunction with cationic lipids because of its membrane destabilizing effects at low pH (Farhood et al., 1995). Each lipid has different structural aspects, such as head group size and hydrocarbon tail length. These aspects confer distinct characteristics to the lipid/DNA complex, which in turn affect association with and uptake into the cell (Balazs & Godbay, 2011). Several have reported (DeCastro et al., 2006; Immordino et al., 2006; Kim et al., 2003; Needham et al., 1992; Shi et al., 2002) that PEGylated lipoplexes increased transfection efficiency in the presence of serum as compared to conventional liposomal transfection methods. Additionally, PEGylated lipoplexes had the potential for improved stability and long circulation times in the blood stream. L-amino acid-based polymers for lipoplex modification could achieve an extended circulation time and reduced clearance by macrophages at levels similar to those seen with lipoplexes modified with PEG (Metselaar et al., 2003).

PRODUCT DEVELOPMENT AND RELATED ISSUES IN LIPOSOMAL NANOMEDICINES

Liposomal nanomedicines have been extensively investigated for several decades. However, they did not become commercially available in U.S. until 1995 when the U.S. FDA approved the first liposome drug product, Doxil®, a doxorubicin HCl

Table 9-B. Examples of Liposomal Nanomedicines Approved for Clinical Application or Undergoing Clinical Evaluation

Route	Drug	Therapeutic area	Brand name	Company	Development Status
Parenteral	Amphotericin B	Antifungal	AmBisome®	Astellas	Launched
			Amphotec®	Alza	Launched
			Abelcet®	Elan	Launched
	Doxorubicin	Anticancer	Doxil®	Alza	Launched
			Myocet®	Elan	Launched
	Daunorubicin	Anticancer	DaunoXome®	Gilead Science	Launched
	Cytarabine	Anticancer	DepoCyt®	Enzon	Launched
	Vincristine	Anticancer	Onco® TCS	Enzon	Phase II
	Lurtotecan	Anticancer	OSI® 211	OSI	Phase II
	Platinum compounds	Anticancer	Aroplatin®	Aronex	Phase I | II
Pulmonary	Lung surfactant	Respiratory Distress Syndrome	Surfaxin®	Drug Discovery Lab	Phase III
			Survanta®	Abbott Labs	Launched
			Exosurf®	GlaxoSmith-Kline	Launched
			Curosurf®	Dey	Launched
			Alveofact®	Boehringer Ingelheim	Launched
	Amikacin	Cystic Fibrosis	Arikace®	Transave Inc.	Phase I | II
Topical	Lidocaine	Local anesthesia	LMX®-4	Ferndale Laboratories	Launched
	Nystatin	Local antifungal	Nyotran®	Aronex Pharmaceuticals	Phase III

liposome injection (Lian & Ho, 2001). The currently marketed commercial products are listed in Table 9-B. For regulatory application and setting, the U.S. FDA defined the nomenclature of liposomes and liposome drug products in FDA Guidance for Industry (U.S. FDA Web site, 2011). These are defined as (1) liposomes are microvesicles composed of one or more bilayers of amphipathic lipid molecules enclosing one or more aqueous compartments, and (2) liposome drug products are those drug products that contain drug substances encapsulated or intercalated in liposomes.

FDA Guidance for Industry (U.S. FDA Web site, 2011) mentioned that it may be necessary to assess the effect of each manufacturing change on the identity, strength, quality, purity, and potency of the liposome drug product. In vivo studies may be warranted to demonstrate that the changed product is equivalent to the original product with respect to safety and efficacy. The long-term in vitro release test can be applied for real-time monitoring of product release over the dosing interval. The short-term in vitro release test (i.e., "accelerated in vitro test") can be used for setting specifications for batch release. The real-time long-term release test should be initially set for simulation of the in vivo environment, while the short-term release test (based on the real-time release test) could be developed for batch-to-batch quality control (Chen, 2008).

CONCLUSION

SLN, NLC, and liposomal carrier systems are attractive carriers for nanomedicines and drug delivery. SLNs themselves represent a solid technology whereas NLCs are comprised of liquid lipid in a solid matrix. Liposomal drug delivery carriers are biomembrane colloidal dispersion systems in an aqueous bulk phase. Proliposomes are solid-state formulation of dry, free-flowing particles, which have the ability to form liposomal colloidal dispersions following hydration. Despite the fact that the use of lipid nanoparticulate drug delivery for nanomedicine is very promising and a highly attractive application platform, further basic research needs to be done in this exciting area.

REFERENCES

Abra, R.M.; et al. The next generation of liposome delivery systems: recent experience with tumor-targeted, sericallystabilized immunoliposomes and active-loading gradients. J. Liposome Res. 2002, 12, 1–3.

Aji Alex, M.R.; Chacko, A.J.; et al. Lopinavir loaded solid lipid nanoparticles (SLN) for intestinal lymphatic targeting. Eur. J. Pharm. Sci. 2010, 42, 11–18.

Akwete L.A.; Hsu L. Leuprolide and Other LH-RH Analogues. In Stability and Characterization of Protein and Peptide Drugs: Case Histories; Plenum Press: New York, 1993; pp. 159–199.

Alghadyan A.A.; Peyman, G.A.; Khoobehi, B.; et al. Liposomebound cyclosporine: retinal toxicity after intravitreal injection. Int. Ophthalmol. 1988a, 12, 105–107.

Alghadyan A.A.; Peyman, G.A.; Khoobehi, B.; et al. Liposomebound cyclosporine: clearance after intravitreal injection. Int. Ophthalmol. 1988b, 12, 109–12.

Amselem, S.; Gabizon, A.; Barenholz, Y. A large-scale method for the preparation of sterile and non pyrogenic liposomal formulations of defined size distributions for clinical use. In Liposome Technology 2nd Edition. Gregoriadis, G., Ed.; CRC Press: Boca Raton, FL, 1993; pp. 501–525.

Angelucci, D. New Solutions for Dry Eye, EyeWorld online, 2001 (Internet) http://eyeworld.org.pastissue.php

Antimisiaris, S.G.; Kallinteri, P.; Fatouros, D.G. Liposome and drug delivery. Pharmaceutical Manufacturing Handbook. John Wiley & Sons: Hobokin, New Jersey, 2008; 10, pp. 443– 534.

Aquilano, D.; Cavalli, R.; et al. Solid lipospheres obtained from hot microemulsions in the presence of different concentrations of cosurfactant: the crystallization of stearic acid polymorphs. Thermochim. Acta 1993, 230, 29–37.

Arien, A.; Goigoux, C.; Baquey, C; Dupuy, B. Study of in vitro and in vivo stability of liposomes loaded with calcitonin or indium in the gastrointestinal tract. Life Sci. 1993, 53, 1279–1290.

Artman, C.; Roding, J.; Ghyczy, M.; Prazel, H.G. Liposomes from soya phospholipids as percutaneous drug carriers: qualitative in vivo investigations with antibody-loaded liposomes. Arzneium. Forsch Drug Res. 1990a, 40, II, 1363–1365.

Artman, C.; Roding, J.; Ghyczy, M.; Prazel, H.G. Liposomes from soya phospholipids as percutaneous drug carriers: quantitative in vivo investigations with radioactivity labelled liposomes. Arzneium. Forsch Drug Res. 1990b, 40, II, 1365–1368.

Assil, K.K.; Frucht-Perry, J.; Ziegler, E.; et al. Tobramycin liposomes. Single subconjunctival therapy of pseudomonal keratitis. Invest. Ophthalmol. Vis. Sci. 1991, 32, 16–20.

Attama, A.A.; Reichl, S.; et al. Diclofenac sodium delivery to the eye: in vitro evaluation of novel solid lipid nanoparticle formulation using human cornea construct. Int. J. Pharm. 2008, 355, 307–13.

Attama, A.A.; Reichl, S.; et al. Sustained release and permeation of timolol from surface-modified solid lipid nanoparticles through bioengineered human cornea. Curr. Eye Res. 2009, 34, 698–705.

Averina, E.S.; Seewald, G.; et al. Nanostructured lipid carriers (NLC) on the basis of Siberian pine (Pinus sibirica) seed oil. Pharmazie 2010, 65, 25–31.

Bakker-Woudenberg, I.A. Long-circulating sterically stabilized liposomes as carriers of agents for treatment of infection or for imaging infectious foci. Int. J. Antimicrob. Agents 2002, 19, 299–311.

Balazs, D.A.; Godbey, W.T. Liposomes for use in gene delivery. J. Drug Del. 2011, 326497, 1–12.

Banerjee, R. Liposomes: applications in medicine. J. Biomater. 2001, 114, 3–21.

Bangham, A.D.; Standish, M.M.; Watkins, J.C. Diffusion of univalent ion across the lamellae of swollen phospholipids, J. Mol. Bio. 1965, 13, 238–252.

Barenholzt, Y.; Amselem, S.; Litchtenberg, D. A new method for preparation of phospholipid vesicles (liposomes)—french press. FEBS Letters 1979, 99, 210–214.

Barenholz,Y.; Crommelin, D.J.A. Encyclopedia of Pharmaceutical Technology. Marcel Dekker: New York, 1994, 1–39.

Barza, M.; Baum, J.; Szoka, F. Jr. Pharmacokinetics of subconjunctival liposome-encapsulated gentamicin in normal rabbit eyes. Invest. Ophthalmol. Vis. Sci. 1984, 25, 486–90.

Basaran, E.; Demirel, M.; et al. Cyclosporine-A incorporated cationic solid lipid nanoparticles for ocular delivery. J. Microencapsul. 2010, 27, 37–47.

Bates, S.; Zografi, G.; Engers, D.; Morris, K.; Crowley, K.; Newman, A. Analysis of amorphous and nanocrystalline solids from their X-ray diffraction patterns. Pharm. Res. 2006, 23, 2333–2349.

Battaglia, L.; Gallarate, M.; et al. Solid lipid nanoparticles produced through a coacervation method. J. Microencapsul. 2010, 27, 78–85.

Behr, J.P. Gene transfer with synthetic cationic amphiphiles: prospects for gene therapy. Bioconjugate Chem. 1994, 5, 382–389.

Behr, J.P.; Demeneix, B.; Loeffler, J.P.; Perez-Mutul, J. Efficient gene transfer into mammalian primary endocrine cells with lipopolyamine-coated DNA. Proceedings of the National Academy of Sciences of the United States of America 1989, 86, 6982–6986.

Betagiri, J.V.; Jenkins, S.A.; Parsons, D.L. Liposomes drug delivery system. Technomic Publishing: Lanchaster, Basel, Switzerland, 1993, pp. 120.

Bhalekar, M.R.; Pokharkar, V.; et al. Preparation and evaluation of miconazole nitrate-loaded solid lipid nanoparticles for topical delivery. AAPS PharmSciTech 2009, 10, 289–96.

Bi, R.; Shao, W.; Wang, Q.; Zhang, N. Spray-freeze-dried dry powder inhalation of insulin-loaded liposomes for enhanced pulmonary delivery. J. Drug Target 2008, 16, 639–648.

Blasi, P.; Giovagnoli, S.; et al. Solid lipid nanoparticles for targeted brain drug delivery. Adv. Drug Deliv. Rev. 2007, 59, 454–477.

Blume, G.; Cevc, G. Molecular mechanism of the lipid vesicle longevity in vivo. Biochim. Biophys. Acta 1993, 1146, 157–168.

Blume, G.; et al. Specific targeting with poly(ethylene glycol)-modified liposomes: coupling of homing devices to the ends of polymeric chains combines effective target binding with long circulation times. Biochim. Biophys. Acta 1993, 1149, 180–184.

Bondi, M.L.; Craparo, E.F. Solid lipid nanoparticles for applications in gene therapy: a review of the state of the art. Expert Opin. Drug Deliv. 2010, 7, 7–18.

Bondi, M.L.; Fontana, G.; et al. Preparation and characterization of solid lipid nanoparticles containing cloricromene. Drug Deliv. 2003, 10, 245–250.

Brandl, M. Liposomes as drug carriers: a technological approach. Biotech. Ann. Rev. 2001, 7, 59–85.

Brandl, M.; Bachmann, D.; Drechsler, D.; Bauer, K.H. Lipopsome preparation by a new high pressure homogenizer Gaulin Micron LAB 40. Drug Dev. Ind. Pharm. 1990, 16, 2167–2191.

Brioschi, A.; Zara, G.P.; et al. Cholesterylbutyrate solid lipid nanoparticles as a butyric acid prodrug. Molecules 2008, 13, 230–254.

Budker, V.; Gurevich, V.; Hagstrom, J.E.; et al. pH-sensitive, cationic liposomes: a new synthetic virus-like vector. Nature Biotech. 1996, 14, 760–764.

Bunker, M.; Davies, M.; Roberts, C. Towards screening of inhalation formulations: measuring interactions with atomic force microscopy. Expert Opin. Drug Deliv. 2005, 2, 613–624.

Burgess, D.J.; Crommelin, D.J.A.; Hussain, A.S.; Chen, M.-L. Assuring quality and performance of sustained and controlled release parenterals: EUFEPS workshop report, Eur. J. Pharm. Sci. 2004, 21, 769–790.

Caliph, S.M.; Charman, W.N.; et al. Effect of short-, medium-, and long-chain fatty acid-based vehicles on the absolute oral bioavailability and intestinal lymphatic transport of halofantrine and assessment of mass balance in lymph-cannulated and non-cannulated rats. J. Pharm. Sci. 2000, 89, 1073–1084.

Campos, S.; Shapiro, C.L. Chapter 15. Clinical evaluation of liposomal anthracyclines, Liposomes—Rational Design. Marcel Dekker: New York, 1999, pp. 363–377.

Castro, G.A.; Orefice, R.L.; et al. Development of a new solid lipid nanoparticle formulation containing retinoic acid for topical treatment of acne. J. Microencapsul. 2007, 24, 395–407.

Cattel, L.; Ceruti, M.; Dosio, F. From conventional to stealth liposomes: a new frontier in cancer chemotherapy. Tumori. 2003, 89, 237–249.

Cavalli, R.; Morel, S.; et al. Preparation and evaluation in vitro of colloidal lipospheres containing pilocarpine as ion pair. Int. J. Pharm. 1995, 117, 243–246.

Cavalli, R.; Caputo, O.; et al. Sterilization and freeze-drying of drug-free and drug-loaded solid lipid nanoparticles. Int. J. Pharm. 1997, 148, 47–54.

Cavalli, R.; Gasco, M.R.; et al. Evaporative drying of aqueous dispersions of solid lipid nanoparticles. Drug Dev. Ind. Pharm. 2001, 27, 919–924.

Cavalli, R.; Gasco, M.R.; et al. Solid lipid nanoparticles (SLN) as ocular delivery system for tobramycin. Int. J. Pharm. 2002, 238, 241–245.

Cevc, G.; Blume, G. Lipid vesicles penetrate into intact skin owing to the transdermal osmotic gradients and hydration force. Biochim. Biophys. Acta 1992, 1104, 226–232.

Cevc, G.; Schatzlein, A.; Blume, G. Transdermal drug carriers: basic properties, optimization and transfer efficiency in the case of epicutaneously applied peptides. J. Control Rel. 1995, 36, 3–16.

Cevc, G.; Blume, G.; Schatzlein, A.; et al. The skin: a pathway for systemic treatment with patches and lipid-based agent carriers. Adv. Drug Deliv. Rev. 1996, 18, 349–378.

Changsan, N.; Chan, H.K.; Separovic, F.; Srichana, T. Physicochemical characterization and stability of rifampicin liposome dry powder formulations for inhalation. J. Pharm. Sci. 2009, 98, 628–639.

Charcosset, C.; El-Harati, A.; et al. Preparation of solid lipid nanoparticles using a membrane contactor. J. Control Rel. 2005, 108, 112–120.

Chattopadhyay, P.; Huff, R.; Shekunov, B.Y. Drug encapsulation using supercritical fluid extraction of emulsions. J. Pharm. Sci. 2006, 95, 667–679.

Chattopadhyay, P.; Shekunov, B.Y.; et al. Production of solid lipid nanoparticle suspensions using supercritical fluid extraction of emulsions (SFEE) for pulmonary delivery using the AERx system. Adv. Drug Deliv. Rev. 2007, 59, 444–453.

Chavanpatil, M.D.; Khdair, A.; et al. Surfactant-polymer nanoparticles overcome P-glycoprotein-mediated drug efflux. Mol. Pharm. 2007, 4, 730–738.

Chen, H.; Torchilin, V.; Langer, R. Lectin-bearing polymerized liposomes as potential oral vaccine carriers. Pharm. Res. 1996, 13, 1378–1383.

Chen, H.; Chang, X.; et al. Podophyllotoxin-loaded solid lipid nanoparticles for epidermal targeting. J. Control Rel. 2006, 110, 296–306.

Chen, M.L. Lipid excipients and delivery systems for pharmaceutical development: a regulatory perspective. Adv. Drug Deliv. Rev. 2008, 60, 768–777.

Choi, S.H.; Jin, S.E.; et al. Novel cationic solid lipid nanoparticles enhanced p53 gene transfer to lung cancer cells. Eur. J. Pharm. Biopharm. 2008, 68, 545–554.

Choudhari, K.B.; Labhasetwar, V.; Dorle, A.K. Liposomes as a carrier for oral administration of insulin: effect of formulation factors. J. Microencapsulation 1994, 11, 319–325.

Chougule, M.; Padhi, B.; Misra, A. Nano-liposomal dry powder inhaler of tacrolimus: preparation, characterization, and pulmonary pharmacokinetics. Int. J. Nanomed. 2007, 2, 675–688.

Clark, M.A.; Jepson, M.A.; et al. Exploiting M cells for drug and vaccine delivery. Adv. Drug Deliv. Rev. 2001, 50, 81–106.

Courrier, H.M.; Butz, N.; Andamme, F. Pulmonary drug delivery systems: developments and prospects. Crit. Rev. Ther. Drug Carrier Syst. 2002, 19, 425–498.

Crommelin, D.J.; Storm, G. Liposomes: from the bench to the bed. J. Liposome Res. 2003, 13, 33–36.

Daemen,T.; et al. Opportunities in targeted drug delivery to Kupffer cells: delivery of immunomodulators to Kupffer cells-activation of tumoricidal properties. Adv. Drug Deliv. Rev. 1995, 17, 21–30.

Deamer, D.; Bangham, A.D. Large volume liposomes by an ether vaporization method. Biochim. Biophys. Acta 1976, 443, 629–634.

DeCastro, M.; Saijoh, Y.; Schoenwolf, G.C. Optimized cationic lipid-based gene delivery reagents for use in developing vertebrate embryos. Dev. Dynam. 2006, 235, 2210–2219.

del Pozo-Rodriguez, A.; Delgado, D.; et al. Solid lipid nanoparticles: formulation factors affecting cell transfection capacity. Int. J. Pharm. 2007, 339, 261–268.

del Pozo-Rodriguez, A.; Delgado, D.; et al. Solid lipid nanoparticles for retinal gene therapy: transfection and intracellular trafficking in RPE cells. Int. J. Pharm. 2008, 360, 177–183.

del Pozo-Rodriguez, A.; Pujals, S.; et al. A proline-rich peptide improves cell transfection of solid lipid nanoparticle-based non-viral vectors. J. Control Rel. 2009, 133, 52–59.

Densmore, C.L. Advances in noninvasive pulmonary gene therapy. Curr. Drug Deliv. 2006, 3, 55–63.

Deshmukh, D.D.; Ravis, W.R.; Betageri, G.V. Improved delivery of cromolyn from oral proliposomal beads. Int. J. Pharm. 2008, 358, 128–136.

Dingler, A.; Blum, R.P.; et al. Solid lipid nanoparticles (SLN/Lipopearls)—a pharmaceutical and cosmetic carrier for the application of vitamin E in dermal products. J. Microencapsul. 1999, 16, 751–767.

Duddu, S.P.; Sisk, S.A.; Walter, Y.H.; et al. Improved lung delivery from a passive dry powder inhaler using an Engineered PulmoSphere powder. Pharm. Res. 2002, 19, 689–695.

Dwivedi, V.; Vasco, A.; Vedi, S.; Dangi, A.; Arif, K.; Bhattacharya, S.M.; Owais, M. Adjuvanticity and protective immunity of Plasmodium yoelii nigeriensis blood-stage soluble antigens encapsulated in fusogenic liposome. Vaccine 2009, 27, 473–482.

Ebrahim, S.; Peyman, G.A.; Lee, P.J. Applications of Liposomes in Ophthalmology. Surv. Ophthalmol 2005, 50, 167–182.

El-Harati, A.A.; Charcosset, C.; et al. Influence of the formulation for solid lipid nanoparticles prepared with a membrane contactor. Pharm. Dev. Technol. 2006, 11, 153–157.

Fang, J.Y.; Fang, C.L.; et al. Lipid nanoparticles as vehicles for topical psoralen delivery: solid lipid nanoparticles (SLN) versus nanostructured lipid carriers (NLC). Eur. J. Pharm. Biopharm. 2008, 70, 633–640.

Farhood, H.; Gao, X.; Son, K. Cationic liposomes for direct gene transfer in therapy of cancer and other diseases. Anals New York Academy Sci. 1994, 716, 23–35.

Farhood, H.; Serbina, N.; Huang, L. The role of dioleoyl phosphatidylethanolamine in cationic liposome mediated gene transfer. Biochim. Biophys. Acta 1995, 1235, 289–295.

Felgner, P.L.; Gadek, T.R.; Holm M.; et al. Lipofection: a highly efficient, lipid-mediated DNA-transfection procedure. Proceedings of the National Academy of Sciences of the United States of America 1987, 84, 7413–7417.

Felgner, P.L.; Ringold, G.M. Cationic liposome-mediated transfection. Nature 1989, 337, 387–388.

Foldvari, M. In vitro cutaneous and percutaneous delivery and in vivo efficacy of tetracaine from liposomal and conventional vehicles. Pharm. Res. 1994, 11, 1593–1598.

Foldvari, M.; Gesztes, A.; Mezei, M. Dermal drug delivery by liposome encapsulation: clinical and electron microscopic studies. J. Microencapsu. 1990, 7, 479–489.

Fontana, G.; Maniscalco, L.; et al. Solid lipid nanoparticles containing tamoxifen characterization and in vitro antitumoral activity. Drug Deliv. 2005, 12, 385–392.

Freitas, C.; Muller, R.H. Correlation between long-term stability of solid lipid nanoparticles (SLN) and crystallinity of the lipid phase. Eur. J. Pharm. Biopharm. 1999, 47, 125–132.

Freitas, S.; Merkle, H.P.; Gander, B. Ultrasonic atomisation into reduced pressure atmosphere-envisaging aseptic spray-drying for microencapsulation. J. Control Rel. 2004, 95, 185–195.

Frezard, F. Liposomes: from biophysics to the design of peptide vaccines. Braz. J. Biol. Res. 1999, 32, 181–189.

Frucht-Perry, J.; Assil, K.K.; Ziegler, E.; et al. Fibrin-enmeshed tobramycin liposomes: single application topical therapy of Pseudomonas keratitis. Cornea 1992, 11, 393–397.

Fundaro, A.; Cavalli, R.; et al. Non-stealth and stealth solid lipid nanoparticles (SLN) carrying doxorubicin: pharmacokinetics and tissue distribution after i.v. administration to rats. Pharmacol. Res. 2000, 42, 337–343.

Gabizon, A. Emerging role of liposomal drug carrier systems in cancer chemotherapy. J. Liposome Res. 2003, 13, 17–20.

Gabizon, A.; Barenholz, Y. Liposomal Anthracyclines: From Basics to Clinical Approval of PEGylated Liposomal Doxorubicin. In Liposomes—Rational Design; Marcel Dekker: New York, 1999; pp. 343–362.

Gabizon, A.; Catane, R.; Uziely. B.; et al. Prolonged circulation tima and enhanced accumulation in malignant exudates of doxorubicin encapsulated in polyethylene-glycol coated liposomes. Cancer Res. 1994, 54, 987–992.

Gallarate, M.; Trotta, M.; et al. Preparation of solid lipid nanoparticles from w/o/w emulsions: preliminary studies on insulin encapsulation. J. Microencapsul. 2009, 26, 394–402.

Gao, X.; Huang, L. A novel cationic liposome reagent for efficient transfection of mammalian cells. Biochem. Biophys. Res. Comm. 1991, 179, 280–285.

Gao, X.; Kim, K.S.; Liu, D. Nonviral gene delivery: what we know and what is next. AAPS J. 2007, 9, 92–104.

Garcia-Fuentes, M.; Torres, D.; et al. Design of lipid nanoparticles for the oral delivery of hydrophilic macromolecules. Colloids Surf. B-Biointer. 2002, 27, 159–168.

Gershkovich, P.; Wasan, K.M.; Barta, C.A. A review of the application of lipid-based systems in systemic, dermal/transdermal, and ocular drug delivery. Crit. Rev. Ther. Drug Carrier Syst. 2008, 25, 545–584.

Gesztes, A.; Mezei, M. Topical anaesthesia of skin by liposome-encapsulated tetracaine. Anesth. Analg. 1988, 67, 1079–1081.

Gilani, K.; Najafabadi, A.R.; Barghi, M.; Rafiee-Tehrani, M. The effect of water to ethanol feed ratio on physical properties and aerosolization behavior of spray dried cromolyn sodium particles. J. Pharm. Sci. 2005, 94, 1048–1059.

Gokce, E.H.; Sandri, G.; et al. Cyclosporine A-loaded solid lipid nanoparticles: ocular tolerance and in vivo drug release in rabbit eyes. Curr. Eye Res. 2009, 34, 996–1003.

Goppert, T.M.; Muller, R.H. Polysorbate-stabilized solid lipid nanoparticles as colloidal carriers for intravenous targeting of drugs to the brain: comparison of plasma protein adsorption patterns. J. Drug Target 2005, 13, 179–187.

Gregoriadis, G. Overview of liposomes. J. Antimicrob. Chemother. 1991, 28(supp. B), 39–48.

Gregoriadis, G. Engineering liposomes for drug delivery: progress and problems. Trends Biotechnol. 1995, 13, 527–537.

Gulbake, A.; Jain, A.; et al. Solid lipid nanoparticles bearing oxybenzone: in-vitro and in-vivo evaluation. J. Microencapsul. 2010, 27, 226–233.

Gupta, Y.; Jain, A.; et al. Transferrin-conjugated solid lipid nanoparticles for enhanced delivery of quinine dihydrochloride to the brain. J. Pharm. Pharmacol. 2007, 59, 935–940.

Hatziantoniou, S.; Deli, G.; et al. Scanning electron microscopy study on nanoemulsions and solid lipid nanoparticles containing high amounts of ceramides. Micron. 2007, 38, 819–823.

Hickey, A.J.; Mansour, H.M. Chapter 44: Formulation challenges of powders for the delivery of small molecular weight molecules as aerosols. Modified-Release Drug Delivery Technology Second Edition. Informa Healthcare: New York, 2008; pp. 573–602.

Hickey, A.J.; Mansour, H.M. Chapter 5: Delivery of drugs by the pulmonary route. Modern Pharmaceutics Volume 2: Applications and Advances Fifth Edition. Taylor & Francis, New York, 2009; pp. 191–219.

Hickey, A.J.; Mansour, H.M.; Telko, M.J.; et al. Physical characterization of component particles included in dry powder inhalers. I. Strategy review and static characteristics. J. Pharm. Sci. 2007, 96, 1282–1301.

Hiroshi, Y.; Shozo, M.; et al. Bifunctional delivery system for selective transfer of bleomycin into lymphatics via enteral route. Int. J. Pharm. 1981, 8, 291–302.

Hong, Y.; Hu, F.Q.; et al. Effect of PEG2000 on drug delivery characterization from solid lipid nanoparticles. Pharmazie 2006, 61, 312–315.

Hsu, S.H.; Wen, C.J.; et al. Physicochemical characterization and in vivo bioluminescence imaging of nanostructured lipid carriers for targeting the brain: apomorphine as a model drug. Nanotechnol. 2010, 21, 405101.
Hu, C.; Rhodes, D.G. Proniosomes: A novel drug carrier preparation. Int. J. Pharm. 1999, 185, 23–35.
Hu, F.Q.; Yuan, H.; et al. Preparation of solid lipid nanoparticles with clobetasol propionate by a novel solvent diffusion method in aqueous system and physicochemical characterization. Int. J. Pharm. 2002, 239, 121–128.
Hu, F. Q.; Hong, Y.; et al. Preparation and characterization of solid lipid nanoparticles containing peptide. Int. J. Pharm. 2004a, 273, 29–35.
Hu, F.Q.; Wu, M.; et al. A novel preparation of solid lipid nanoparticles with cyclosporin A for prolonged drug release. Pharmazie 2004b, 59, 683–685.
Hu, F.Q.; Jiang, S.P.; et al. Preparation and characterization of stearic acid nanostructured lipid carriers by solvent diffusion method in an aqueous system. Colloids Surf B Biointerfaces 2005, 45, 167–173.
Hu, F. Q.; Zhang, Y.; et al. Nimodipine loaded lipid nanospheres prepared by solvent diffusion method in a drug saturated aqueous system. Int. J. Pharm. 2008, 348, 146–152.
Hu, L.; Jia, H.; et al. Improvement of digoxin oral absorption in rabbits by incorporation into solid lipid nanoparticles. Pharmazie 2010a, 65, 110–113.
Hu, L.; Jia, Y.; et al. Preparation and characterization of solid lipid nanoparticles loaded with epirubicin for pulmonary delivery. Pharmazie 2010b, 65, 585–587.
Hu, L.; Xing, Q.; et al. Preparation and enhanced oral bioavailability of cryptotanshinone-loaded solid lipid nanoparticles. AAPS PharmSciTech 2010c, 11, 582–587.
Hung, O.R.; Comeau, L.; Riley, M.R.; Tan, S.; Whynot, S.; Mezei, M. Comparative topical anaesthesia of EMLA and liposome-encapsulated tetracaine. Can. J. Anaesth. 1997, 44, 707–711.
Hussain, N.; Jani, P.U.; et al. Enhanced oral uptake of tomato lectin-conjugated nanoparticles in the rat. Pharm. Res. 1997, 14, 613–618.
Immordino, M.L.; Dosio, F.; Cattel, L. Stealth liposomes: review of the basic science, rationale, and clinical applications, existing and potential. Int. J. Nanomed. 2006, 1, 297–315.
Iscan, Y.; Wissing, S.A.; et al. Solid lipid nanoparticles (SLN) for topical drug delivery: incorporation of the lipophilic drugs N,N-diethyl-m-toluamide and vitamin K. Pharmazie 2005, 60, 905–909.
Jain, S.K.; Chourasia, M.K.; et al. Solid lipid nanoparticles bearing flurbiprofen for transdermal delivery. Drug Deliv. 2005, 12, 207–215.
Jee, J. P.; Lim, S.J.; et al. Stabilization of all-trans retinol by loading lipophilic antioxidants in solid lipid nanoparticles. Eur. J. Pharm. Biopharm. 2006, 63, 134–139.
Jenning, V.; Gohla, S.H. Encapsulation of retinoids in solid lipid nanoparticles (SLN). J. Microencapsul. 2001, 18, 149–158.
Jenning, V.; Gysler, A.; et al. Vitamin A loaded solid lipid nanoparticles for topical use: occlusive properties and drug targeting to the upper skin. Eur. J. Pharm. Biopharm. 2000a, 49, 211– 218.
Jenning, V.; Schafer-Korting, M.; et al. Vitamin A-loaded solid lipid nanoparticles for topical use: drug release properties. J. Control Rel. 2000b, 66, 115–126.
Jensen, L.B.; Magnussson, E.; et al. Corticosteroid solubility and lipid polarity control release from solid lipid nanoparticles. Int. J. Pharm. 2010, 390, 53–60.
Jores, K.; Mehnert, W.; et al. Investigations on the structure of solid lipid nanoparticles (SLN) and oil-loaded solid lipid nanoparticles by photon correlation spectroscopy, field-flow fractionation and transmission electron microscopy. J. Control Rel. 2004, 95, 217–227.

Jores, K.; Haberland, A.; et al. Solid lipid nanoparticles (SLN) and oil-loaded SLN studied by spectrofluorometry and Raman spectroscopy. Pharm. Res. 2005, 22, 1887–1897.

Joshi, M.; Misra, A.N. Pulmonary disposition of budesonide from liposomal dry powder inhaler. Methods Find. Exp. Clin. Pharmacol. 2001, 23, 531–536.

Joshi, M.D.; Muller, R.H. Lipid nanoparticles for parenteral delivery of actives. Eur. J. Pharm. Biopharm. 2009, 71, 161–172.

Junping, W.; Maitani, Y.; Takayama, K.; Nagai, T. In vivo evaluation of doxorubicin carried with long circulating and remote loading proliposome. Int. J. Pharm. 2000, 203, 61–69.

Justo, O.R.; Moraes, A.M. Incorporation of antibiotics in liposomes designed for tuberculosis therapy by inhalation. Drug Deliv. 2003, 10, 201–207.

Kalam, M. A.; Sultana, Y.; et al. Preparation, characterization, and evaluation of gatifloxacin loaded solid lipid nanoparticles as colloidal ocular drug delivery system. J. Drug Target 2010, 18, 191–204.

Kassem, M.A.; Abdel Rahman, A.A.; et al. Nanosuspension as an ophthalmic delivery system for certain glucocorticoid drugs. Int. J. Pharm. 2007, 340, 126–133.

Kaur, I.P.; Kapila, M.; et al. Role of novel delivery systems in developing topical antioxidants as therapeutics to combat photoageing. Ageing Res. Rev. 2007, 6, 271–288.

Kaur, I.P.; Bhandari, R.; et al. Potential of solid lipid nanoparticles in brain targeting. J. Control Rel. 2008, 127, 97–109.

Kent, J.; Huiling, M.; Xuebing, X.; et al. Oxidative stability of liposomes composed of docosahexaenoic acid-containing phospholipids. J. Am. Oil Chem. Soc. 2007, 84, 631–637.

Khoo, S.M.; Edwards, G.A.; et al. A conscious dog model for assessing the absorption, enterocyte-based metabolism, and intestinal lymphatic transport of halofantrine. J. Pharm. Sci. 2001, 90, 1599–607.

Kikuchi, A.; Aoki, Y.; Sugayama, S.; et al. Development of novel cationic liposomes for efficient gene transfer into peritoneal disseminated tumor. Human Gene Ther. 1999, 10, 947–955.

Killion, J.J.; Fidler, I.J. Systemic targeting of liposome-encapsulated immunomodulators to macrophages for treatment of cancer metastasis. Immunomethods 1994, 4, 273–279.

Kim, H.R.; Kim, I.K.; et al. Cationic solid lipid nanoparticles reconstituted from low density lipoprotein components for delivery of siRNA. Mol. Pharm. 2008, 5, 622–631.

Kim, J.-K.; Choi, S.-H.; Kim, C.-O.; Park, J.-S.; Ahn, W.-S.; Kim, C.-K. Enhancement of polyethylene glycol (PEG)-modified cationic liposome-mediated gene deliveries: effects on serum stability and transfection efficiency. J. Pharm. Pharmacol. 2003, 55, 453–460.

Kim, S.T.; Jang, D.J.; et al. Topical administration of cyclosporin A in a solid lipid nanoparticle formulation. Pharmazie 2009, 64, 510–514.

Kirby, C.; Gregoriadis, G. A simple procedure for preparing liposomes capable of high encapsulation efficiency under mild conditions. Liposomes Technology. CRC Press: Boca Raton, FL, 1984; pp. 19–28.

Klibanov, A.L.; Maruyama, K.; Torchilin, V.P.; Huang, L. Amphipatic polyethyleneglycols effectively prolong the circulation time of liposomes. FEBS Lett. 1990, 268, 235–238.

Koshkina, N.V.; et al. 9-Nitrocamptothecin liposome aerosol treatment of melanoma and osteosarcoma lung metastases in mice. Clin. Cancer Res. 2000, 6, 2876–2880.

Koshkina, N.V.; Golunski, E.; Roberts, L.E.; Gilbert, B.E.; Knight, V. Cyclosporin A aerosol improves the anticancer effect of paclitaxel aerosol in mice. J. Aerosol. Med. 2004, 17, 7–14.

Kremer, J.M.H.; Esker, M.W.J.V.D.; Pathmamanoharan, C.; Wiersema, P.H. Vesicles of variable diameter prepared by a modified injection method. Biochem. 1977, 16, 3932–3935.

Kumar, V.V.; Chandrasekar, D.; et al. Development and evaluation of nitrendipine loaded solid lipid nanoparticles: influence of wax and glyceride lipids on plasma pharmacokinetics. Int. J. Pharm. 2007, 335, 167–175.

Kuo, J.H.; Hwang, R. Preparation of DNA dry powder for non-viral gene delivery by spray-freeze drying: effect of protective agents (polyethyleneimine and sugars) on the stability of DNA. J. Pharm. Pharmacol. 2004, 56, 27–33.

Kuo, Y.C.; Lin, S.C. Effect of glutamate on the electrical properties of cationic solid lipid nanoparticles containing stearylamine and dioctadecyldimethyl ammonium bromide. J. Phys. Chem. B 2008, 112, 4454–4460.

Kuo, Y.C.; Chen, H.H. Entrapment and release of saquinavir using novel cationic solid lipid nanoparticles. Int. J. Pharm. 2009, 365, 206–213.

Lasic, D.D. Mechanisms of liposome formation. J. Liposome Res. 1995, 5, 431–441.

Lasic, D.D. Liposomes. Sci. Med. 1996, 3, 34–43.

Law, S.L.; Huang, K.J.; Chiang, C.H. Acyclovir-containing liposomes for potential ocular delivery. Corneal penetration and absorption. J. Control Rel. 2000, 63, 135–140.

Leventis, R.; Silvius, J.R. Interactions of mammalian cells with lipid dispersions containing novel metabolizable cationic amphiphiles. Biochim. Biophys. Acta 1990, 1023, 124–132.

Li, H.; Song, J.H.; Park, J.S.; Han, K. Polyethylene glycol-coated liposomes for oral delivery of recombinant human epidermal growth factor. Int. J. Pharm. 2003, 258, 11–19.

Li, H.; Zhao, X.; et al. Enhancement of gastrointestinal absorption of quercetin by solid lipid nanoparticles. J. Control Rel. 2009, 133, 238–244.

Li, Y. Z.; Sun, X.; et al. Inhalable microparticles as carriers for pulmonary delivery of thymopentin-loaded solid lipid nanoparticles. Pharm. Res. 2010, 27, 1977–1986.

Lian, T.; Ho, R.J.Y. Trends and developments in liposome drug delivery systems. J. Pharm. Sci. 2001, 90, 667–680.

Lin, H.H.; Ko, S.M.; Hsu, L.R.; Tsai, Y.H. The preparation of norfloxacin-loaded liposomes and their in-vitro evaluation in pig's eye. J. Pharm. Pharmacol. 1996, 48, 801–805.

Lin, Y.K.; Huang, Z.R.; et al. Combination of calcipotriol and methotrexate in nanostructured lipid carriers for topical delivery. Int. J. Nanomed. 2010, 5, 117–128.

Liu, J.; Gong, T.; et al. Solid lipid nanoparticles loaded with insulin by sodium cholate-phosphatidylcholine-based mixed micelles: preparation and characterization. Int. J. Pharm. 2007a, 340, 153–162.

Liu, J.; Hu, W.; et al. Isotretinoin-loaded solid lipid nanoparticles with skin targeting for topical delivery. Int. J. Pharm. 2007b, 328, 191–195.

Liu, J.; Gong, T.; et al. Solid lipid nanoparticles for pulmonary delivery of insulin. Int. J. Pharm. 2008, 356, 333–344.

Liu, Z.; Zhong, Z.; et al. Folate receptor mediated intracellular gene delivery using the charge changing solid lipid nanoparticles. Drug Deliv. 2009, 16, 341–347.

Liu, D.; Ge, Y.; et al. Solid lipid nanoparticles for transdermal delivery of diclofenac sodium: preparation, characterization and in vitro studies. J. Microencapsul. 2010, 27, 726–734.

Lo, Y.L.; Tsai, J.C.; Kuo, J.H. Liposomes and disaccharides as carriers in spray-dried powder formulations of superoxide dismutase. J. Control Rel. 2004, 94, 259–272.

Lu, W.; He, L.C.; et al. The use of solid lipid nanoparticles to target a lipophilic molecule to the liver after intravenous administration to mice. Int. J. Biol. Macromol. 2008a, 43, 320–324.

Lu, X.; Howard, M.D.; et al. Nanoparticles containing anti-inflammatory agents as chemotherapy adjuvants: optimization and in vitro characterization. AAPS J. 2008b, 10, 133–140.

Lu, X.; Howard, M.D.; et al. Nanoparticles containing anti-inflammatory agents as chemotherapy adjuvants II: role of plasma esterases in drug release. AAPS J. 2009, 11, 120–122.

Luo, Y.; Chen, D.; et al. Solid lipid nanoparticles for enhancing vinpocetine's oral bioavailability. J. Control Rel. 2006, 114, 53–59.

Luo, Q.; Zhao, J.; et al. Nanostructured lipid carrier (NLC) coated with Chitosan Oligosaccharides and its potential use in ocular drug delivery system. Int. J. Pharm. 2010, 403, 185–191.

Lv, Q.; Yu, A.; et al. Development and evaluation of penciclovir-loaded solid lipid nanoparticles for topical delivery. Int. J. Pharm. 2009, 372, 191–198.

Ma, P.; Dong, X.; et al. Development of idarubicin and doxorubicin solid lipid nanoparticles to overcome Pgp-mediated multiple drug resistance in leukemia. J. Biomed. Nanotechnol. 2009, 5, 151–161.

Maa, Y.F.; Prestrelski, S.J. Biopharmaceutical powders: particle formation and formulation considerations. Curr. Pharm. Biotechnol. 2000, 1, 283–302.

Maa, Y.F.; Nguyen, P.A.; Sit, K.; Hsu, C.C. Spray-drying performance of a bench-top spray dryer for protein aerosol powder preparation. Biotechnol. Bioeng. 1998, 60, 301–309.

Maa, Y.F.; Nguyen, P.A.; Sweeney, T.; Shire, S.J.; Hsu, C.C. Protein inhalation powders: spray drying vs spray freeze drying. Pharm. Res. 1999, 16, 249–254.

Maeda, H.; Sawa, T.; Konno, T. Mechanism of tumortargeted delivery of macromolecular drugs, including the EPR effect in solid tumor and clinical overview of the prototype polymeric drug SMANCS. J. Control Rel. 2001, 74, 47–61.

Maghrabya, G.M. E; Barryc, B.W.; Williamsd, A.C. Liposomes and skin: from drug delivery to model membranes. Eur. J. Pharm. Sci. 2008, 34, 203–222

Maia, C.S.; Mehnert, W.; et al. Solid lipid nanoparticles as drug carriers for topical glucocorticoids. Int. J. Pharm. 2000, 196, 165–167.

Maitani, Y.; Soeda, H.; Junping, W.; Takayama, K. Modified ethanol injection method for liposomes containing β-sitosterol β-d-glucoside. J. Liposome Res. 2001, 11, 115–125.

Malmsten, M. Surfactants and Polymers in Drug Delivery; Marcel Dekker: New York, 2002, pp. 87–131.

Manjunath, K.; Venkateswarlu, V. Pharmacokinetics, tissue distribution and bioavailability of nitrendipine solid lipid nanoparticles after intravenous and intraduodenal administration. J. Drug Target 2006, 14, 632–645.

Mansour, H.M.; Hickey, A.J. Raman characterization and chemical imaging of biocolloidal self-assemblies, drug delivery systems, and pulmonary inhalation aerosols: a review. AAPS PharmSciTech 2007, 8, E99.

Mansour, H.M.; Zografi, G. The relationship between water vapor absorption and desorption by phospholipids and bilayer phase transitions. J. Pharm. Sci. 2007a, 96, 377–396.

Mansour, H.M.; Zografi, G. Relationships between equilibrium spreading pressure and phase equilibria of phospholipid bilayers and monolayers at the air-water interface. Langmuir 2007b, 23, 3809–3819.

Mansour, H.M.; Wang, D.-S.; Chen C.-S.; Zografi, G. Comparison of bilayer and monolayer properties of phospholipid systems containing dipalmitoylphosphatidylglycerol and dipalmitoylphosphatidylinositol. Langmuir 2001, 17, 6622–6632.

Mansour, H.M.; Damodaran, S.; Zografi, G. Characterization of the in situ structural and interfacial properties of the cationic hydrophobic heteropolypeptide, kl4, in lung surfactant bilayer and monolayer models at the air-water interface: implications for pulmonary surfactant delivery. Mol. Pharmaceut. 2008, 5, 681–695.

Mansour, H.M.; Rhee, Y.S.; Wu, X. Nanomedicine in pulmonary delivery. Int. J. Nanomed. 2009, 4, 299–319.
Martin, F.; Papahadjopoulos, D. Irreversible coupling of immunoglobulin fragments to preformed vesicles. J. Biol. Chem. 1982, 257, 286–288.
Martins, S.; Sarmento, B.; et al. Lipid-based colloidal carriers for peptide and protein delivery--liposomes versus lipid nanoparticles. Int. J. Nanomed. 2007, 2, 595–607.
Mayer, L.D.; Hope, M.J.; Cullis, P.R. Vesicles of variable sizes produced by a rapid extrusion procedure. Biochim. Biophys. Acta 1986, 858, 161–168.
Mayhew, E.; Lazo, R.; Vail, W.J.; King, J.; Green, A.M. Characterization of liposomes prepared using a microemulsifier. Biochim. Biophys. Acta 1984, 775, 169–174.
Mehnert, W.; Mader, K. Solid lipid nanoparticles: production, characterization and applications. Adv. Drug Deliv. Rev. 2001, 47, 165–196.
Mei, Z.; Chen, H.; et al. Solid lipid nanoparticle and microemulsion for topical delivery of triptolide. Eur. J. Pharm. Biopharm. 2003, 56, 189–196.
Metselaar, J.M.; Mastrobattista, E.; Storm, G. Liposomes for intravenous drug targeting: design and applications. Mini-Rev. Med. Chem. 2002, 2, 319–329.
Metselaar, J.M.; Bruin, P.; DeBoer, L.W.T.; et al. A novel family of L-amino acid-based biodegradable polymerlipid conjugates for the development of long-circulating liposomes with effective drug-targeting capacity. Bioconjugate Chem. 2003, 14, 1156–1164.
Mezei, M.; Gulasekharam, V. Liposomes—a selective drug delivery system for topical route of administration. 1. Lotion dosage form. Life Sci. 1980, 26, 1473–1477.
Mezei, M.; Gulasekharam, V. Liposomes—a selective drug delivery system for topical route of administration: gel dosage form. J. Pharm. Pharmacol. 1982, 34, 473–474.
Mikhaylova, M.; Stasinopoulos, I.; Kato, Y.; Artemov, D.; Bhujwalla, Z.M. Imaging of cationic multifunctional liposome-mediated delivery of COX-2 siRNA. Cancer Gene Ther. 2009, 16, 217–226.
Mina, A.; Kanako, Y.; Kazuo, M. Oxidative stability of polyunsaturated fatty acid in phaophatidylcholine liposomes. Biosci. Biotechnol. Biochem. 2002, 66, 2573 –2577.
Minato, S.; Iwanaga, K.; Kakemi, M.; Yamashita, S.; Oku, N. Application of polyethyleneglycol (PEG)-modified liposomes for oral vaccine: effect of lipid dose on systemic and mucosal immunity. J. Control Rel. 2003, 89, 189–197.
Misra, A.; Jinturkar, K.; Patel, D.; Lalani, J.; Chougule, M. Recent advances in liposomal dry powder formulations: preparation and evaluation. Expert Opin. Drug Deliv. 2009, 6, 71–89.
Morel, S.; Terreno, E.; et al. NMR relaxometric investigations of solid lipid nanoparticles (SLN) containing gadolinium(III) complexes. Eur. J. Pharm. Biopharm. 1998, 45, 157–163.
Morille, M.; Passirani, C.; Vonarbourg, A.; Clavreul, A.; Benoit, J.P. Progess in developing cationic vectors for non-viral systemic gene therapy against cancer. Biomater. 2008, 29, 3477–3496.
Mosen, K.; Backstrom, K.; Thalberg, K. Particle formation and capture during spray drying of inhalable particles. Pharm. Dev. Technol. 2004, 9, 409–418.
Muchow, M.; Maincent, P.; et al. Lipid nanoparticles with a solid matrix (SLN, NLC, LDC) for oral drug delivery. Drug Dev. Ind. Pharm. 2008, 34, 1394–1405.
Muller, R.H.; Heinemann, S. Fat emulsions for parenteral nutrition. I: Evaluation of microscopic and laser light scattering methods for the determination of the physical stability. Clin. Nutr. 1992, 11, 223–236.
Muller, R.H.; Mader, K.; Gohla, S. Solid lipid nanoparticles (SLN) for controlled drug delivery—a review of the state of the art. Eur. J. Pharm. Biopharm. 2000, 50, 161–177.

Muller, R.H.; Radtke, M.; et al. Solid lipid nanoparticles (SLN) and nanostructured lipid carriers (NLC) in cosmetic and dermatological preparations. Adv. Drug Deliv. Rev. 2002, 54 Suppl 1, S131–S155.

Muller, R.H.; Runge, S.; et al. Oral bioavailability of cyclosporine: solid lipid nanoparticles (SLN) versus drug nanocrystals. Int. J. Pharm. 2006, 317, 82–89.

Muller, R.H.; Petersen, R.D.; et al. Nanostructured lipid carriers (NLC) in cosmetic dermal products. Adv. Drug Deliv. Rev. 2007, 59, 522–530.

Muller, R.H.; Runge, S.A.; et al. Cyclosporine-loaded solid lipid nanoparticles (SLN): drug-lipid physicochemical interactions and characterization of drug incorporation. Eur. J. Pharm. Biopharm. 2008, 68, 535–544.

Mura, P.; Maestrelli, F.; Rodriguez, L.M.G.; Michelacci, I.; Ghelardini, C.; Rabasco, A.M. Development, characterization and in vivo evaluation of benzocaine-loaded liposomes. Eur. J. Pharm. Biopharm. 2007, 67, 86–95.

Muranishi, S.; Fujita, T.; et al. Potential for lymphatic targeting of peptides. J. Control Rel. 1997, 46, 157–164.

Needham, D.; McIntosh, T.J.; Lasic, D.D. Repulsive interactions and mechanical stability of polymer-grafted lipid membranes. Biochim. Biophys. Acta 1992, 1108, 40–48.

Newman, A.W.; Byrn, S.R. Solid-state analysis of the active pharmaceutical ingredient in drug products. Drug Discov. Today 2003, 8, 898–905.

Niven, R.W. Delivery of biotherapeutics by inhalation aerosol. Crit. Rev. Ther. Drug Carrier Syst. 1995, 12, 151–231.

Paliwal, R.; Rai, S.; et al. Effect of lipid core material on characteristics of solid lipid nanoparticles designed for oral lymphatic delivery. Nanomed. 2009, 5, 184–91.

Pandey, R.; Khuller, G.K. Solid lipid particle-based inhalable sustained drug delivery system against experimental tuberculosis. Tuberculosis (Edinburgh, Scotland) 2005, 85, 227–234.

Pandey, R.; Sharma, S.; et al. Oral solid lipid nanoparticle-based antitubercular chemotherapy. Tuberculosis (Edinburgh, Scotland) 2005, 85, 415–420.

Pandita, D.; Ahuja, A.; et al. Characterization and in vitro assessment of paclitaxel loaded lipid nanoparticles formulated using modified solvent injection technique. Pharmazie 2009, 64, 301–310.

Panyam, J.; Labhasetwar, V. Targeting intracellular targets. Curr. Drug Del. 2004, 1, 235–247.

Pardeike, J.; Hommoss, A.; et al. Lipid nanoparticles (SLN, NLC) in cosmetic and pharmaceutical dermal products. Int. J. Pharm. 2009, 366, 170–184.

Patel, G.; Chougule, M.; Singh, M.; Misra, A. Nanoliposomal dry powder formulations. Meth. Enzymol. 2009, 464, 167–191.

Patton, J.S.; Byron, P.R. Inhaling medicines: delivering drugs to the body through the lungs. Nature Rev. Drug Disc. 2007, 6, 67–74.

Payne, N.I.; Timmins, P.; Ambrose, C.V.; Ward, M.D.; Ridgway, F. Proliposomes: a novel solution to an old problem. J. Pharm. Sci. 1986, 75, 325–329.

Pedersen, N.; Hansen, S.; et al. Solid lipid nanoparticles can effectively bind DNA, streptavidin and biotinylated ligands. Eur. J. Pharm. Biopharm. 2006, 62, 155–162.

Pegi, A.; Julijana, K.; et al. The effect of lipophilicity of spin-labeled compounds on their distribution in solid lipid nanoparticle dispersions studied by electron paramagnetic resonance. J. Pharm. Sci. 2003, 92, 58–66.

Pison, U.; Welte, T.; et al. Nanomedicine for respiratory diseases. Eur. J. Pharmacol. 2006, 533, 341–350.

Planas, M.E.; Gonzalez, P.; Rodriguez, L.; Sanchez, S.; Cevc, G. Noninvasive percutaneous induction of topical analgesia by a new type of drug carrier, and prolongation of local pain insensitivity by anesthetic liposomes. Anesth. Analg. 1992, 75, 615–621.

Pleyer, U.; Lutz, S.; Jusko, W.J.; et al. Ocular absorption of topically applied FK506 from liposomal and oil formulations in the rabbit eye. Invest. Ophthalmol. Vis. Sci. 1993, 34, 2737–2742.

Pople, P.V.; Singh, K.K. Development and evaluation of topical formulation containing solid lipid nanoparticles of vitamin A. AAPS PharmSciTech 2006, 7, 91.

Porter, C.J.; Charman, W.N. Intestinal lymphatic drug transport: an update. Adv. Drug Deliv. Rev. 2001, 50, 61–80.

Rahman, Z.; Zidan, A.S.; et al. Non-destructive methods of characterization of risperidone solid lipid nanoparticles. Eur. J. Pharm. Biopharm. 2010, 76, 127–137.

Rehman, M.; Shekunov, B.Y.; York P.; et al. Optimisation of powders for pulmonary delivery using supercritical fluid technology. Eur. J. Pharm. Sci. 2004, 22, 1–17.

Rhee, Y.S.; Mansour, H.M. Nanopharmaceuticals I: nanocarrier systems in drug delivery. Int. J. Nanotechnol. 2011, 8, 84–114.

Rogers, T.; Johnston, K.; Williams, R. Solution-based particle formation of pharmaceutical powders by supercritical or compressed Fluid CO2 and cryogenic spray-freezing technologies. Drug Dev. Ind. Pharm. 2001, 27, 1003–1016.

Rudolph, C.; Schillinger, U.; et al. Application of novel solid lipid nanoparticle (SLN)-gene vector formulations based on a dimeric HIV-1 TAT-peptide in vitro and in vivo. Pharm. Res. 2004, 21, 1662–1669.

Saleki-Gerhardt, A.; Ahlneck, C.; Zografi, G. Assessment of disorder in crystalline solids. Int. J. Pharm. 1994, 101, 237–247.

Samad, A.; Sultana, Y.; Aqil, M. Liposomal drug delivery systems: an update review. Curr.Drug Del. 2007, 4, 297–305.

Sanjula, B.; Shah, F.M.; et al. Effect of poloxamer 188 on lymphatic uptake of carvedilol-loaded solid lipid nanoparticles for bioavailability enhancement. J. Drug Target 2009, 17, 249–256.

Sanna, V.; Gavini, E.; et al. Solid lipid nanoparticles (SLN) as carriers for the topical delivery of econazole nitrate: in-vitro characterization, ex-vivo and in-vivo studies. J. Pharm. Pharmacol. 2007, 59, 1057–1064.

Sarmento, B.; Martins, S.; et al. Oral insulin delivery by means of solid lipid nanoparticles. Int. J. Nanomed. 2007, 2, 743–749.

Saupe, A.; Gordon, K. C.; et al. Structural investigations on nanoemulsions, solid lipid nanoparticles and nanostructured lipid carriers by cryo-field emission scanning electron microscopy and Raman spectroscopy. Int. J. Pharm. 2006, 314, 56–62.

Schafer-Korting, M.; Mehnert, W.G.; et al. Lipid nanoparticles for improved topical application of drugs for skin diseases. Adv. Drug Deliv. Rev. 2007, 59, 427–443.

Schneider, T.; Sachse, A.; Rossling, G.; Brandl, M. Largescale production of liposomes of defined size by a new continuous high pressure extrusion device. Drug Dev. Ind. Pharm. 1994, 20, 2787–2807.

Schneider, T.; Sacfse, A.; Robling, G.; Brandl, M. Generation of contrast-carrying liposomes of defined size with a new continuous high pressure extrusion method. Int. J. Pharm. 1995, 119, 1–12.

Schubert, M.A.; Muller-Goymann, C.C. Solvent injection as a new approach for manufacturing lipid nanoparticles—evaluation of the method and process parameters. Eur. J. Pharm. Biopharm. 2003, 55, 125–131.

Senior J.H. A review of controlling factors. Crit. Rev. Ther. Drug Carrier Syst, 1987, 3, 23–93.

Shah, M.; Pathak, K. Development and statistical optimization of solid lipid nanoparticles of simvastatin by using 2(3) full-factorial design. AAPS PharmSciTech 2010, 11, 489–496.

Shah, S.P.; Misra, A. Development of liposomal Amphotericin B dry powder inhaler formulation. Drug Deliv. 2004a, 11, 247–253.

Shah, S.P.; Misra, A. Liposomal amphotericin B dry powder inhaler: effect of fines on in vitro performance. Pharmazie 2004b, 59, 812–813.

Shah, K.A.; Date, A.A.; et al. Solid lipid nanoparticles (SLN) of tretinoin: potential in topical delivery. Int. J. Pharm. 2007, 345, 163–171.

Shahgaldian, P.; Quattrocchi, L.; et al. AFM imaging of calixarene based solid lipid nanoparticles in gel matrices. Eur. J. Pharm. Biopharm. 2003, 55, 107–113.

Shekunov, B.Y.; Chattopadhyay, P.; Seitzinger, J.; Huff, R. Nanoparticles of poorly water-soluble drugs prepared by supercritical fluid extraction of emulsions. Pharm. Res. 2006, 23, 196–204.

Shi, F.; Wasungu, L.; Nomden A.; et al. Interference of poly(ethylene glycol)-lipid analogues with cationic-lipid-mediated delivery of oligonucleotides; role of lipid exchangeability and non-lamellar transitions. Biochem. J. 2002, 366, 333–341.

Siddiqui, A.; Patwardhan, G.A.; et al. Mixed backbone antisense glucosylceramide synthase oligonucleotide (MBO-asGCS) loaded solid lipid nanoparticles: in vitro characterization and reversal of multidrug resistance in NCI/ADR-RES cells. Int. J. Pharm. 2010, 400, 251–259.

Siekmann, B.; Westesen, K. Sub-micron sized parenteral carrier systems based on solid lipids. Pharm. Pharmacol. Lett. 1992, 1, 123–126.

Siekmann, B.; Westesen, K. Investigations on solid lipid nanoparticles prepared by precipitation in o/w emulsions. Eur. J. Pharm. Biopharm. 1996, 42, 104–109.

Silva, A.C.; Santos, D.; et al. Minoxidil-loaded nanostructured lipid carriers (NLC): characterization and rheological behaviour of topical formulations. Pharmazie 2009, 64, 177–182.

Simard, P.; Leroux, J.C. pH-sensitive immunoliposomes specific to the CD33 cell surface antigen of leukemic cells. Int. J. Pharm. 2009, 381, 86–96.

Singh, R.; Singh, S.; Lillard Jr., J.W. Past, present, and future technologies for oral delivery of therapeutic proteins. J. Pharm. Sci. 2008, 97, 2497–2523.

Singh, S.; Dobhal, A.K.; et al. Formulation and evaluation of solid lipid nanoparticles of a water soluble drug: Zidovudine. Chem. Pharm. Bull. (Tokyo) 2010, 58, 650–655.

Sjöström, B.; Bergenstahl, B.; et al. A method for the preparation of submicron particles of sparingly water-soluble drugs by precipitation in oil-in-water emulsions. II: Influence of the emulsifier, the solvent, and the drug substance. J. Pharm. Sci. 1993a, 82, 584–589.

Sjöström, B.; Kronberg, B.; et al. A method for the preparation of submicron particles of sparingly water-soluble drugs by precipitation in oil-in-water emulsions. I: Influence of emulsification and surfactant concentration. J. Pharm. Sci. 1993b, 82, 579–583.

Sjöström, B.; Westesen, K.; et al. Preparation of submicron drug particles in lecithin-stabilized o/w emulsions: II. Characterization of cholesteryl acetate particles. Int. J. Pharm. 1993c, 94, 89–101.

Sjöström, B.; Kaplun, A.; et al. Structures of nanoparticles prepared from oil-in-water emulsions. Pharm. Res. 1995, 12, 39–48.

Song, K.H.; Chung, S.J.; Shim, C.K. Preparation and evaluation of proliposomes containing salmon calcitonin. J. Control Rel. 2002, 84, 27–37.

Souto, E.B.; Anselmi, C.; et al. Preparation and characterization of n-dodecyl-ferulate-loaded solid lipid nanoparticles (SLN). Int. J. Pharm. 2005, 295, 261–268.

Stahl, K.; Claesson, M.; Lilliehorn, P.; Linden, H.; Backstrom, K. The effect of process variables on the degradation and physical properties of spray dried insulin intended for inhalation. Int. J. Pharm. 2002, 233, 227–237.

Stano, P.; Bufali, S.; Pisano, C.; et al. Novel camptothecin analogue (Gimatecan)-containing liposomes prepared by the ethanol injection method. J. Liposome Res. 2004, 14, 87–109.

Steckel, H.; Thies, J.; Muller, B.W. Micronizing of steroids for pulmonary delivery by supercritical carbon dioxide. Int. J. Pharm. 1997, 152, 99–110.

Stevens, P.J.; Sekido, M.; et al. Synthesis and evaluation of a hematoporphyrin derivative in a folate receptor-targeted solid-lipid nanoparticle formulation. Anticancer Res. 2004, 24, 161–165.

Storm, G.; Crommelin, D.J.A. Liposomes: quo vadis? PSTT 1998, 1, 19–31.

Storm, G.; et al. Surface modification of nanoparticles to oppose uptake by the mononuclear phagocyte system. Adv. Drug Deliv. Rev. 1995, 17, 31–48.

Sung, J.C.; Pulliam, B.L.; et al. Nanoparticles for drug delivery to the lungs. Trends in Biotechnol. 2007, 25, 563–570.

Suresh, G.; Manjunath, K.; et al. Preparation, characterization, and in vitro and in vivo evaluation of lovastatin solid lipid nanoparticles. AAPS PharmSciTech 2007, 8, 24.

Sweeney, L.; Wang, Z.; Loebenberg, R.; Wong, J.P.; Lange, C.F.; Finlay, W. Spray-freeze-dried liposomal ciprofloxacin powder for inhaled aerosol drug delivery. Int. J. Pharm. 2005, 305, 180–185.

Symon, Z.; et al. Selective delivery of doxorubicin to patients with breast carcinoma metastases by stealth liposomes. Cancer 1999, 86, 72–78.

Tabatt, K.; Sameti, M.; et al. Effect of cationic lipid and matrix lipid composition on solid lipid nanoparticle-mediated gene transfer. Eur. J. Pharm. Biopharm. 2004, 57, 155–162.

Taira, M.C.; Chiaramoni, N.S.; Pecuch, K.M.; Alonso-Romanowski, S. Stability of liposomal formulations in physiological conditions for oral drug delivery. Drug Deliv. 2004, 11, 123–128.

Templeton, N.S. Nonviral delivery for genomic therapy of cancer. World J. Surg. 2009, 33, 685–697.

Ten Hagen, T.L.M.; Van Vianen, W.; Bakker-Woudenberg, I.A.J.M. Modulation of nonspecific antimicrobial resistance of mice to Klebsiella Pneumoniae septicemia by liposome-encapsulated muramyl tripeptide phosphatidylethanolamine and interferon-ã alone or combined J. Infect. Dis. 1995, 171, 385–392.

Tiyaboonchai, W.; Tungpradit, W.; et al. Formulation and characterization of curcuminoids loaded solid lipid nanoparticles. Int. J. Pharm. 2007, 337, 299–306.

Tom, J.W.; Debenedetti, P.G. Particle formation with supercritical fluids-a review. J. Aerosol. Sci. 1991, 22, 555–584.

Torchilin, V.P. Liposomes as targetable drug carriers. Crit. Rev. Ther. Drug Carrier Syst. 1985, 1, 65–115.

Torchilin, V.P. Recent advances with liposomes as pharmaceutical carriers. Nature Reviews Drug Discovery 2005, 4, 145–160.

Torchilin, V.P.; Trubetskoy, V.S. Which polymers can make nanoparticulate drug carriers long-circulating? Adv. Drug Deliv. Rev. 1995, 16, 141–155.

Torchilin, V.P.; et al. Targeted accumulation of polyethylene glycol-coated immunoliposomes in infarcted rabbit myocardium. FASEB J. 1992, 6, 2716–2719.

Torchilin, V.P.; et al. Poly(ethylene glycol) on the liposome surface: on the mechanism of polymer-coated liposome longevity. Biochim. Biophys. Acta 1994, 1195, 11–20.

Torchilin, V.P.; et al. p-Nitrophenylcarbonyl-PEG-PEliposomes: fast and simple attachment of specific ligands, including monoclonal antibodies, to distal ends of PEG chains via p-nitrophenylcarbonyl groups. Biochim. Biophys. Acta 2001, 1511, 397–411.

Trevaskis, N.L.; Charman, W.N.; et al. Lipid-based delivery systems and intestinal lymphatic drug transport: a mechanistic update. Adv. Drug Deliv. Rev. 2008, 60, 702–716.

Trotta, M.; Gallarate, M.; et al. Emulsions containing partially water-miscible solvents for the preparation of drug nanosuspensions. J. Control Rel. 2001, 76, 119–128.
Trotta, M.; Debernardi, F.; et al. Preparation of solid lipid nanoparticles by a solvent emulsification-diffusion technique. Int. J. Pharm. 2003, 257, 153–160.
Tsai, M.J.; Huang, Y.B.; et al. Oral apomorphine delivery from solid lipid nanoparticles with different monostearate emulsifiers: Pharmacokinetic and behavioral evaluations. J. Pharm. Sci. 2010, 100, 547-557.
Turanek, J.; Zaluska, D.; Ne˘ca, J. Linkup of a fast protein liquid chromatography system with a stirred thermostated cell for sterile preparation of liposomes by the proliposome liposome method: application to encapsulation of antibiotics, synthetic peptide immunomodulators, and a photosensitizer. Analytical Biochem. 1997, 249, 131–139.
Turanek, J.; Kasna, A.; Zaluska, D.; Neca, J. Preparation of sterile liposomes by proliposome-liposome method. Methods Enzymol. 2003, 367, 111–125.
U.S. Department of Health and Human Services, Food and Drug Administration, Center for Drug Evaluation and Research. Draft Guidance for Industry: Liposome Drug Products — Chemistry, Manufacturing and Controls; Human Pharmacokinetics and Bioavailability; and Labeling Documentation. Office of Training and Communications, Division of Drug Information, HFD-240, Rockville, MD. (Internet) http://www.fda.gov/cder/guidance/2191dft.pdf.
Ugazio, E.; Cavalli, R.; et al. Incorporation of cyclosporin A in solid lipid nanoparticles (SLN). Int. J. Pharm. 2002, 241, 341–344.
Uhumwangho, M.U.; Okor, R.S. Current trends in the production and biomedical applications of liposomes: a review. J. Biomed. Sci. 2005, 4, 9–21.
Uner, M. Preparation, characterization and physico-chemical properties of solid lipid nanoparticles (SLN) and nanostructured lipid carriers (NLC): their benefits as colloidal drug carrier systems. Pharmazie 2006, 61, 375–386.
Varia, J. K.; Dodiya, S. S.; et al. Cyclosporine A loaded solid lipid nanoparticles: optimization of formulation, process variable and characterization. Curr. Drug Deliv. 2008, 5, 64–69.
Vasir, J.K.; Labhasetwar, V. Biodegradable nanoparticles for cytosolic delivery of therapeutics. Adv. Drug Deliv. Rev. 2007, 59, 718–728.
Venkateswarlu, V.; Manjunath, K. Preparation, characterization and in vitro release kinetics of clozapine solid lipid nanoparticles. J. Control Rel. 2004, 95, 627–638.
Vidgren, M.T.; Vidgren, P.A.; Paronen, T.P. Comparison of physical and inhalation properties of spray-dried and mechanically micronized disodium cromoglycate. Int. J. Pharm. 1987, 35, 139–144.
Vighi, E.; Ruozi, B.; et al. Re-dispersible cationic solid lipid nanoparticles (SLNs) freeze-dried without cryoprotectors: characterization and ability to bind the pEGFP-plasmid. Eur. J. Pharm. Biopharm. 2007, 67, 320–328.
Vighi, E.; Montanari, M.; et al. Nuclear localization of cationic solid lipid nanoparticles containing Protamine as transfection promoter. Eur. J. Pharm. Biopharm. 2010a, 76, 384–393.
Vighi, E.; Ruozi, B.; et al. pDNA condensation capacity and in vitro gene delivery properties of cationic solid lipid nanoparticles. Int. J. Pharm. 2010b, 389, 254–261.
Vyas, S.P.; Kannan, M.E.; Jain, S.; Mishra, V.; Singh, P. Design of liposomal aerosols for improved delivery of rifampicin to alveolar macrophages. Int. J. Pharm. 2004, 269, 37–49.
Vyas S.P.; Quraishi, S.; Gupta, S.; et al. Aerosolized liposome-based delivery of amphotericin B to alveolar macrophages. Int. J. Pharm. 2005, 296, 12–25.
Wagner, A.; Vorauer-Uhl, K. Liposome technology for industrial purposes. J. Drug Deliv. 2011, 591325, 1–9.

Wagner, A.; Vorauer-Uhl, K.; Kreismayr, G.; Katinger, H. Enhanced protein loading into liposomes by the multiple crossflow injection technique. J. Liposome Res. 2002, 12, 271–283.
Wagner, A.; Platzgummer, M.; Kreismayr, G.; et al. GMP production of liposomes—a new industrial approach. J. Liposome Res. 2006, 16, 311–319.
Wagner, A.; Stiegler, G.; Vorauer-Uhl, K. et al. One step membrane incorporation of viral antigens as a vaccine candidate against HIV. J. Liposome Res. 2007, 17, 3–4.
Wang, Y.; Wu, W. In situ evading of phagocytic uptake of stealth solid lipid nanoparticles by mouse peritoneal macrophages. Drug Deliv. 2006, 13, 189–192.
Wang, Y.J.; Pearlman, R. Formation, Characterization, and Stability of Protein Drugs. Plenum Press: New York, 1993, 159–199.
Wang, J.X.; Sun, X.; et al. Enhanced brain targeting by synthesis of 3',5'-dioctanoyl-5-fluoro-2'-deoxyuridine and incorporation into solid lipid nanoparticles. Eur. J. Pharm. Biopharm. 2002, 54, 285–290.
Wang, J.J.; Liu, K.S.; et al. Skin permeation of buprenorphine and its ester prodrugs from lipid nanoparticles: lipid emulsion, nanostructured lipid carriers and solid lipid nanoparticles. J. Microencapsul. 2009, 26, 734–747.
Wang, T.; Wang, N.; et al. Solvent injection-lyophilization of tert-butyl alcohol/water cosolvent systems for the preparation of drug-loaded solid lipid nanoparticles. Colloids Surf. B Biointer. 2010, 79, 254–261.
Wartewig, S.; Neubert, R.H. Pharmaceutical applications of Mid-IR and Raman spectroscopy. Adv. Drug Deliv. Rev. 2005, 57, 1144–1170.
Wasungu, L.; Hoekstra, D. Cationic lipids, lipoplexes and intracellular delivery of genes. J. Control Rel. 2006, 116, 255–264.
Weiner, N.; Martin, F.; Riox, M. Liposomes as drug delivery system. Drug Dev. Ind. Pharm. 1989, 15, 1523–1554.
Wheeler, C.J.; Felgner, P.L.; Tsai, Y. J. et al. A novel cationic lipid greatly enhances plasmid DNA delivery and expression in mouse lung. Proceedings of the National Academy of Sciences of the United States of America 1996, 93, 11454–11459.
White, S.; Bennett, D.B.; Cheu, S.; et al. EXUBERA: pharmaceutical development of a novel product for pulmonary delivery of insulin. Diabetes Technol. Ther. 2005, 7, 896–906.
Wissing, S.A.; Muller, R.H. Solid lipid nanoparticles as carrier for sunscreens: in vitro release and in vivo skin penetration. J. Control Rel. 2002, 81, 225–233.
Wissing, S.A.; Muller, R.H. Cosmetic applications for solid lipid nanoparticles (SLN). Int. J. Pharm. 2003a, 254, 65–68.
Wissing, S.A.; Muller, R.H. The influence of solid lipid nanoparticles on skin hydration and viscoelasticity--in vivo study. Eur. J. Pharm. Biopharm. 2003b, 56, 67–72.
Wissing, S.A.; Kayser, O.; et al. Solid lipid nanoparticles for parenteral drug delivery. Adv. Drug Deliv. Rev. 2004a, 56, 1257–1272.
Wissing, S.A.; Muller, R.H.; et al. Structural characterization of Q10-loaded solid lipid nanoparticles by NMR spectroscopy. Pharm. Res. 2004b, 21, 400–405.
Wong, H.L.; Bendayan, R.; et al. Development of solid lipid nanoparticles containing ionically complexed chemotherapeutic drugs and chemosensitizers. J. Pharm. Sci. 2004, 93, 1993–2008.
Wu, L.; Tang, C.; et al. Preparation and characterization of paclitaxel delivery system based on semi-solid lipid nanoparticles coated with poly (ethylene glycol). Pharmazie 2010, 65, 493–499.
Wu, Z.H.; Ping, Q.N.; Wei, Y.; Lai, J. M. Hypoglycemic efficacy of chitosan-coated insulin liposomes after oral administration in mice. Acta Pharmacol. Sin. 2004, 25, 966–972.

Xiang, Q.Y.; Wang, M.T.; et al. Lung-targeting delivery of dexamethasone acetate loaded solid lipid nanoparticles. Arch. Pharm. Res. 2007, 30, 519–525.

Xing, L.; Dawei, C.; Liping, X.; Rongqing, Z. Oral colonspecific drug delivery for bee venom peptide: development of a coated calcium alginate gel beads-entrapped liposome. J. Control Rel. 2003, 93, 293–300.

Yamabe, K.; Kato, Y.; Onishi, H.; Machida, Y. Potentiality of double liposomes containing salmon calcitonin as an oral dosage form. J. Control Rel. 2003, 89, 429–436.

Yamada, A.; Taniguchi, Y.; Kawano, K.; Honda, T.; Hattori, Y.; Maitani, Y. Design of folate-linked liposomal doxorubicin to its antitumor effect in mice. Clin. Cancer Res. 2008, 14, 8161–8168.

Yang, S.; Zhu, J.; et al. Body distribution of camptothecin solid lipid nanoparticles after oral administration. Pharm. Res. 1999a, 16, 751–757.

Yang, S.C.; Lu, L.F.; et al. Body distribution in mice of intravenously injected camptothecin solid lipid nanoparticles and targeting effect on brain. J. Control Rel. 1999b, 59, 299–307.

Yang, W.; Peters, J.I.; et al. Inhaled nanoparticles—A current review. Inter. J. Pharm. 2008a, 356, 239–247.

Yang, W.; Tam, J.; et al. High bioavailability from nebulized itraconazole nanoparticle dispersions with biocompatible stabilizers. Inter. J. Pharm. 2008b, 361, 177–188.

Yang, L.; Geng, Y.; et al. Enhancement the oral bioavailability of praziquantel by incorporation into solid lipid nanoparticles. Pharmazie 2009, 64, 86–89.

Yan-yu, X.; Yun-mei, S.; Zhi-peng, C.; Qi-neng, P. Preparation of silymarin proliposome: a new way to increase oral bioavailability of silymarin in beagle dogs. Int. J. Pharm. 2006, 319, 162–168.

Ye, J.; Wang, Q.; et al. Injectable actarit-loaded solid lipid nanoparticles as passive targeting therapeutic agents for rheumatoid arthritis. Int. J. Pharm. 2008, 352, 273–279.

Yu, B.T.; Sun, X.; et al. Enhanced liver targeting by synthesis of N1-stearyl-5-Fu and incorporation into solid lipid nanoparticles. Arch. Pharm. Res. 2003, 26, 1096–1101.

Yu, Z.; Garcia, A.S; Johnston, K.P.; Williams, R.O. 3rd. Spray freezing into liquid nitrogen for highly stable protein nanostructured microparticles. Eur. J. Pharm. Biopharm. 2004, 58, 529–537.

Zalipsky, S.; Qazen, M.; Walker, J.A.; Mullah, N.; Quinn, Y.P.; Huang, S.K. New detachable poly(ethylene glycol) conjugates: cysteine-cleavable lipopolymers regenerating natural phospholipid, diacyl phosphatidylethanolamine. Bioconjug. Chem. 1999, 10, 703–707.

Zara, G.P.; Cavalli, R.; et al. Intravenous administration to rabbits of non-stealth and stealth doxorubicin-loaded solid lipid nanoparticles at increasing concentrations of stealth agent: pharmacokinetics and distribution of doxorubicin in brain and other tissues. J. Drug Target 2002, 10, 327–335.

Zhang, X.; Pan, W.; et al. Preparation of a dispersible PEGylate nanostructured lipid carriers (NLC) loaded with 10-hydroxycamptothecin by spray-drying. Chem. Pharm. Bull. (Tokyo) 2008, 56, 1645–1650.

Zhao, X.L.; Yang, C.R.; et al. Preparation and characterization of nanostructured lipid carriers loaded traditional Chinese medicine, zedoary turmeric oil. Drug Dev. Ind. Pharm. 2010, 36, 773–780.

Zhuang, C.Y.; Li, N.; et al. Preparation and characterization of vinpocetine loaded nanostructured lipid carriers (NLC) for improved oral bioavailability. Int. J. Pharm. 2010, 394, 179–185.

Zimmermann, E.; Souto, E.B.; et al. Physicochemical investigations on the structure of drug-free and drug-loaded solid lipid nanoparticles (SLN) by means of DSC and 1H NMR. Pharmazie 2005, 60, 508–513.

zur Muhlen, A.; zur Muhlen, E.; et al. Atomic force microscopy studies of solid lipid nanoparticles. Pharm. Res. 1996, 13, 1411–1416.

CONTRIBUTORS

EDGAR J. ACOSTA

Edgar J. Acosta received his B.Sc. in Chemical Engineering (Summa Cum Laude) from the Universidad del Zulia (Venezuela) in 1996, and his M.Sc. and Ph.D. in Chemical Engineering from the University of Oklahoma in 2000 and 2004, respectively. He is currently an associate professor at the Department of Chemical Engineering and Applied Chemistry of the University of Toronto. Dr. Acosta has received the Provost Dissertation Award from the University of Oklahoma (2005), the Akzo-Nobel "Ralph Potts" Award (2002), the AOCS S&D Best Paper Award (2004 and 2008), and the AOCS Young Scientist Award (2010). Dr. Acosta has published over 60 research articles, 6 book chapters, 3 patent applications, and has been author or coauthor of over 100 presentations at international conferences. His research encompasses the area of colloids, complex fluids, and formulation engineering.

IMRAN AHMAD

Dr. Imran Ahmad is president and chief executive officer of Jina Pharmaceuticals Inc., Libertyville, IL and serves as a board member of several pharmaceutical companies. He is also the head of the Scientific Advisory Board of Intas Pharmaceuticals and holds an adhoc position of professor at Lombardi Cancer Center, Georgetown University, Washington, DC. Prior to joining Jina Pharmaceuticals, Dr. Ahmad was chief scientific officer and executive vice president at NeoPharm Inc., Waukegan, IL. At NeoPharm, he was responsible for building the research and development infrastructure and led the drug development program. Before joining NeoPharm, he held leadership positions at The Liposome Company, New Jersey. He has over 25 years of experience and played a pioneer role in designing and developing several lipid-based drug delivery systems resulting in commercialization of drugs. Dr. Ahmad did post-doctoral research at the University of Alberta, Department of Pharmacology, Edmonton, Canada. He is the author of over 70 scientific publications and more than 50 patents and patent applications.

MOGHIS U. AHMAD

Moghis U. Ahmad obtained his B.Sc. (Honors), M.Sc., M.Phil., and Ph.D. from Department of Chemistry, AMU, Aligarh, India. Dr. Ahmad did his post-doctoral research at Department of Biochemistry & Biophysics, Texas A&M University, College Station, Texas; and Department of Food Science & Technology, Oregon State University, Corvallis, Oregon. Currently, Dr. Ahmad is vice president of Chemical Technology & Manufacturing at Jina Pharmaceuticals Inc., Libertyville, Illinois. Earlier Dr. Ahmad held the position of vice president, Lipids Division at NeoPharm, Inc., Waukegan, Illinois. He also held research positions at Lipids Department of Sigma-Aldrich, St. Louis, Missouri, where he led the development and production of several new lipid products. Dr. Ahmad has made decisive accomplishments in research for the improvement and development of products including lipid-based products over the last 30 years. He has published more than 50 peer reviewed articles, book chapters, and 15 patent and patent applications. He has presented his work in national and international meetings. He has held leadership positions in various scientific societies and their committees. Dr. Ahmad is a Fellow of the Royal Society of Chemistry (FRSC).

SHOUKATH ALI

Dr. Shoukath Ali received his Ph.D. in Chemistry from the University of Memphis, Tennessee in 1996. Since then he has been working in the areas of medicinal chemistry, lipid chemistry, drug development, and drug delivery. He is currently director of Product Development at Jina Pharmaceuticals, Inc. He has published more than 20 papers in peer-reviewed journals, has about 10 patent applications pending, and has presented several papers in national and international meetings. He is also a reviewer of several journal articles. He is an active member of the American Oil Chemists' Society (AOCS), the American Chemical Society (ACS), and the American Association of Pharmaceutical Scientists (AAPS).

ABRAHAM ASERIN

Dr. Aserin is a researcher at the Casali Institute of Applied Chemistry at The Hebrew University of Jerusalem. His research is focused on many practical facets of surfactant chemistry, such as emulsions, microemulsions, lyotropic liquid crystals, dispersions, and the improvement of drug bioavailability. He received a B.Sc., a M.Sc., and a Ph.D. from The Hebrew University of Jerusalem, Israel, in 1978, 1981, and 1987, respectively. Dr. Aserin is the author of more than 150 peer-reviewed scientific articles, 40 reviews, 2 edited books and holds 12 patents. He has educated and tutored 45 M.Sc. and 15 Ph.D. students. He is the co-founder of NutraLease Ltd.

LING-ZHI CHEONG

Ling-Zhi Cheong is a post-doctoral research fellow in Department of Molecular Biology, Aarhus University, Denmark. She obtained her Ph.D. in enzyme and food technology from Universiti Putra Malaysia, Malaysia in 2008. Her research areas of interests include enzyme technology, food and lipid technology, food functionality, food colloids, and encapsulation. Currently, she is actively researching surface active lipids with unique self assembling property for use in encapsulation of sensitive compounds. She has actively published in the related areas.

MILENA CORREDIG

Milena Corredig received her Ph.D. from the University of Guelph in 1998. She is now a professor and Canada Research Chair in Food Nanostructures at the University of Guelph. Her research focuses on food colloids, delivery systems, and value-added ingredients.

PATRICK P. DELUCA

Dr. DeLuca is Emeritus Professor at the University of Kentucky, College of Pharmacy. His teaching and research interests are in novel drug delivery, formulation and process development of proteins and peptides, and freeze-drying. He is the author of 220 publications. Professor DeLuca is the recipient of several AAPS awards, including the Research Achievement Award in Pharmaceutical Technology, the Outstanding Educator Award and the Dale Wurster Research Achievement Award in Pharmaceutics. He is a Fellow and founding member of AAPS and recently served as the President of AAPS.

LEVENTE L. DIOSADY

Levente L. Diosady is the professor of Food Engineering in the Department of Chemical Engineering and Applied Chemistry at the University of Toronto. His research interests include oilseed processing, membrane processes, extrusion, advanced separation processes, and micronutrient fortification of food. He is the author of over 120 publications in refereed journals, including 17 patents. Dr. Diosady is the recipient of the Canada Award for Business Excellence, the Professional Engineers of Ontario Engineering Medal the Eva Award of the Canadian Institute of Food Science and Technology, the KY Lo Medal of the Engineering Institute of Canada and the Babcock-Hart Award of IFT. Levente is a Fellow of the Chemical Institute of Canada, the Canadian Institute of Food Science and Technology, the American Oil Chemists Society, the Hungarian Academy of Engineering, and the Canadian Academy of Engineering. As well, he is a member of the Order of Ontario, the province's highest civilian honor.

BITA FARHANG

Bita Farhang is currently undertaking studies toward a Ph.D. in the Department of Food Science at the University of Guelph, Canada. Her research focuses on milkfat phospholipids and their application in liposomal delivery system for bioactive compounds.

NISSIM GARTI

Nissim Garti obtained his B.Sc., M.Sc., and Ph.D. from the Hebrew University of Jerusalem. He has been a full professor since 1990 and holds the Ratner Chair of Chemistry in the Department of Chemistry and Applied Chemistry. He also has served as a Board Member Elect and Director of the Hebrew University Governors Executive Board since January 2011. Nissim is the recipient of numerous prestigious awards including the Rockefeller Award, the Israel President Award for one of the most innovative inventions in 60 years of the existence of the country, Life-Time Achievement Award of the Food Society, the Chang Award of the AOCS, the Corporate Research Achievement Award of the AOCS for 2011, and many others. He has published over 380 original (research) papers in peer reviewed journals, over 60 book chapters, edited 7 books, and granted over 80 patents. Nissim's expertise, competence, and active research are in colloid chemistry, emulsion technology, dispersed systems, delivery new vehicles, microemulsions and lyotropic liquid crystals, crystallization phenomena, interfacial reactions and reactivity, amphiphilic proteins, hydrocolloids, dendrimers, nutraceuticals, and food science.

ZHENG GUO

Zheng Guo is Associate Professor at the Department of Molecular Biology, Aarhus University, Denmark. He received his Ph.D. in Chemical Engineering in 2004 at Tianjin University, China. From 2004 to 2008, he worked as a research assistant professor at the Technical University of Denmark. In 2011, he was sponsored by Novartis as a visiting scholar at Massachusetts Institute of Technology. His research area involves lipids processing technology, lipid enzymology and reaction, lipid materials, and lipid microbiology.

SOSAKU ICHIKAWA

Sosaku Ichikawa received his Ph.D. in Chemical Engineering from the University of Tokyo (Japan) in 1995. He is now an associate professor of the Graduate School of Life and Environmental Sciences, University of Tsukuba (Japan). His interests focus on the development of nano/micro-dispersion systems for improving food and biochemical technologies, including micelles, emulsions, vesicles, and gel beads.

ISAO KOBAYASHI

Isao Kobayashi received his Ph.D. in Agricultural Sciences from the University of Tsukuba (Japan) in 2003. After working at the University of Tsukuba as a JSPS post-doctoral research fellow, he is now a senior researcher at the National Food Research Institute, National Agriculture and Food Research Organization (Japan). His current interests include emulsification, food micro/nanotechnology, micro/nanofluidics, and in vitro gastrointestinal digestion.

TAKASHI KUROIWA

Takashi Kuroiwa received his Ph.D. in Biotechnology from the University of Tsukuba (Japan) in 2005. After working at the University of Tsukuba and the National Food Research Institute, National Agriculture and Food Research Organization as a postdoctoral fellow, he is now an associate professor of the Faculty of Engineering, Tokyo City University. His interests are biochemical engineering, including fabrication of functional materials via the self-assembly of biomolecules, and application of multiphase enzymatic reactions.

DIMA LIBSTER

Dr. Libster is a leading expert in the field of formation and characterization of new liquid crystalline carriers for delivery and controlled release of peptide and protein drugs. Dr. Libster has interdisciplinary education both in chemical engineering (from the Technion) and chemistry (Hebrew University). Dr. Libster is the author of more than 40 peer-reviewed publications and holds two patents.

BENA-MARIE LUE

Bena-Marie Lue is a research scientist with Detergent Applications at Novozymes A/S, Denmark. She obtained her B.Sc. (2000) and M.Sc. (2004) from the Department of Food Science & Agricultural Chemistry, McGill University, Canada, and her Ph.D. (2009) from Aarhus University, Denmark. To date, her research has focused on enzyme and lipid technology, particularly with respect to the use of lipases in non-conventional media and biocatalysis of antioxidant compounds.

HEIDI M. MANSOUR

Dr. Heidi M. Mansour, Ph.D., R.Ph., is assistant professor of pharmaceutics and drug delivery in the Drug Development Division at the University of Kentucky, College of Pharmacy and faculty associate at the University of Kentucky, Center for Membrane Sciences. She currently holds faculty appointments and graduate faculty appointments at the College of Pharmacy at the University of Kentucky, University of North Carolina-Chapel Hill, and in the NSF IGERT and NSF REU research training programs joint with the University of Kentucky, College of Engineering. She is an active long-time member of several professional organizations.

RIKKE MIKLOS

Rikke Miklos is a Ph.D. student in the Department of Food Science, Faculty of Life Science, University of Copenhagen, Denmark. She made her master study at the Technical University of Denmark. Her Ph.D. study is on the modification of fat using innovative enzyme technology approaches for meat production.

MITSUTOSHI NAKAJIMA

Mitsutoshi Nakajima received his Ph.D. in Chemical Engineering from the University of Tokyo, Japan in 1983. After working at Kyushu University and the National Food Research Institute, National Agriculture and Food Research Organization, he is now a professor of the Graduate School of Life and Environmental Sciences, University of Tsukuba, as well as director of Alliance for Research on North Africa, University of Tsukuba. His interest is food engineering, including micro/nano food technology.

MARCOS NEVES

Marcos Neves obtained his Ph.D. in Agricultural Sciences from the University of Tsukuba (Japan) in 2006. He is now an assistant professor at the University of Tsukuba and has also been working as a visiting researcher at the National Food Research Institute (Tsukuba). His major research topics are development of micro/nanotechnology for advanced food processing, functional foods, microchannel emulsification, and biomass conversion technology.

TOMOKO OKADA

Tomoko Okada received her Ph.D. in Food Science and Technology from Oregon State University in 2006. Her experience includes post-doctoral fellow at Hokkaido University, 2006-2009 and assistant professor at Obihiro University of Agriculture and Veterinary Medicine, 2009-2011. She is now a senior researcher of Product Development Division of Cosmo Foods Co., Ltd. Her research interest focuses mainly on development of novel functional components from natural resources.

WORAWAN PANPIPAT

Worawan Panpipat is a Ph.D. student in the Department of Molecular Biology, Aarhus University, Denmark. She obtained her M.S. in Food Technology from Suranaree University of Technology, Thailand. Her current research interest is in synthesizing novel sterol-based compounds with unique self-assembling properties.

CHUN-WOONG PARK

Dr. Chun-Woong Park, Ph.D., received his M.S. in pharmaceutics in 2005 and Ph.D. in Physical Pharmacy in 2008 from Sungkyunkwan University, Republic of Korea. Currently, he is a post-doctoral visiting scholar at the University of Kentucky, College of Pharmacy, Department of Pharmaceutical Sciences, Drug Development Division. His research focus is on advanced particle engineering design for dry powder inhalation aerosols.

A. VENKETESHWER RAO

Dr. Rao completed his M.Sc. and Ph.D. in Food Science from Oregon State University, Oregon. After being a full professor in the Department of Nutritional Sciences, Faculty of Medicine, he currently holds the position of professor emeritus at the University of Toronto. He has established a major focus in the area of diet and health with particular emphasis on the role of dietary phytochemicals and intestinal microflora. More recently his research has focused on the role of oxidative stress and antioxidants in the causation and prevention of chronic diseases, with particular emphasis on the role of carotenoids and polyphenols. He is one of the pioneering researchers to study the bioavailability, metabolism, mechanisms of action and biological role of lycopene, a carotenoid antioxidant present in tomatoes and other fruits and vegetables. Dr. Rao served as a member of the Provincial and National Expert Committees in Canada in the areas of nutrition, health, food safety, and agriculture. He serves as a senior scientific consultant to provincial and federal government agencies and food and pharmaceutical industries globally.

YUN-SEOK RHEE

Dr. Yun-Seok Rhee, Ph.D., is an assistant professor in the College of Pharmacy at Gyeongsang National University (Republic of Korea). He received his Ph.D. in Pharmacy from Sungkyunkwan University School of Pharmacy (Republic of Korea) in 2001. He completed post-doctoral research at the University of Kentucky, College of Pharmacy, Department of Pharmaceutical Sciences, Drug Development Division. His research interests include advanced pulmonary drug delivery using colloidal drug carrier systems and the design of new oral drug delivery system for BCS Class III drugs.

DÉRICK ROUSSEAU

Dérick Rousseau is a professor in the Department of Chemistry and Biology at Ryerson University in Toronto, Canada. His research efforts focus on understanding the physical and chemical factors that control the formation and stability of dispersed systems. His group has made important contributions in a number of areas, including emulsion stabilization in food and non-food applications, chocolate functionality, dairy product fortification, lipid crystallization, and the development of controlled release matrices.

SHUANG SONG

Shuang Song is a Ph.D. student in the College of Life Science and Technology, Beijing University of Chemical Technology, China. She is a guest Ph.D. student in the Department of Molecular Biology, Aarhus University, Denmark. She is now a senior Ph.D. student and will graduate in 2012. Her research activities are directed towards designing structured surfactants with unique self-assembling properties and protection of antioxidants.

KORETARO TAKAHASHI

Koretaro Takahashi received his Ph.D. in Fisheries Science from Hokkaido University (Japan) in 1984. He is now professor of the Faculty of Fisheries Sciences, Hokkaido University (Japan). His interests include the utilization of marine complex lipids contained in fisheries by-products and the novel approach to food quality analysis by applying electromagnetic waves in the tera hertz range.

MASATOSHI TANOUCHI

Masatoshi Tanouchi received his M.Eng. in Synthetic Organic Chemistry from Kyoto University (Japan) in 1964. He is now a representative of Health Food & AM Laboratory (Tokyo, Japan). His interests include the development of new business concerned with polyunsaturated fatty acid containing lipids. He is trying to launch polyunsaturated fatty acid containing phospholipids into the market as a new functional food.

XUEBING XU

Xuebing Xu is a professor of Agro-Biotechnology Science, Aarhus School of Engineering, Department of Molecular Biology, Aarhus University, Denmark. He obtained his Ph.D. in Chemical Engineering from the Technical University of Denmark in 2000. He has more than 180 publications in lipid/enzyme technology, food functionality, industrial biocatalysis, etc.

CHRISTOPHER YIP

Christopher Yip obtained his Ph.D. in Chemical Engineering from the University of Minnesota in 1996. Currently, he is a professor in the Departments of Chemical Engineering and Applied Chemistry, Biochemistry, the Institute of Biomaterials and Biomedical Engineering, and the Donnelly Centre for Cellular and

Biomolecular Research at the University of Toronto. His research interests are single molecule biophysics and the design and application of combinatorial microscopies and spectroscopies to assess molecular conformation, function, dynamics, and self-assembly.

INDEX

M

N

Q

R

Kolibri 360 GmbH
Ferdinand-Jühlke-Straße 7
99095 Erfurt, Deutschland
produktsicherheit@kolibri360.de